Transforming Tertiary Agricultural Education in Africa

Transforming Tertiary Agricultural Education in Africa

Edited by

David Kraybill, John Lynam and Adipala Ekwamu

RUFORUM
Capacity Building in Agriculture

CABI

CABI is a trading name of CAB International

CABI
Nosworthy Way
Wallingford
Oxfordshire OX10 8DE
UK

Tel: +44 (0)1491 832111
Fax: +44 (0)1491 833508
E-mail: info@cabi.org
Website: www.cabi.org

CABI
WeWork
One Lincoln St
24th Floor
Boston, MA 02111
USA

Tel: +1 (617)682-9015
E-mail: cabi-nao@cabi.org

A catalogue record for this book is available from the British Library, London, UK.

References to Internet websites (URLs) were accurate at the time of writing.

ISBN-13: 978 1 78924 654 4 (hardback)
978 1 78924 655 1 (ePDF)
978 1 78924 656 8 (ePub)

DOI: 10.1079/9781789246544.0000

Commissioning Editor: David Hemming
Editorial Assistant: Emma McCann
Production Editor: James Bishop

Typeset by SPi, Pondicherry, India
Printed and bound in the UK by Severn, Gloucester

Contents

List of Contributors

Serge Abihona, University of Abomey Calavi Foundation, BP. 36 Abomey-Calavi, Benin
Might Kojo Abreh, Institute for Educational Planning and Administration, University of Cape Coast, Cape Coast, Ghana
Festus Annor-Frempong, Department of Agricultural Economics and Extension, School of Agriculture, College of Agriculture and Natural Sciences, University of Cape Coast, PMB University Post Office Cape Coast, Ghana
Nienke Beintema, independent consultant and former head of ASTI, based in The Netherlands
Salome Bukachi, Institute of Anthropology, Gender and African Studies, University of Nairobi, PO Box 30197,00100, Nairobi, Kenya
Anthony Egeru, Regional Universities Forum for Capacity Building in Agriculture (RUFORUM), PO Box 16811 Wandegeya, Kampala, Uganda
Adipala Ekwamu, Regional Universities Forum for Capacity Building in Agriculture (RUFORUM), PO Box 16811 Wandegeya, Kampala, Uganda
Young Kafui Abel Etsey, University of Cape Coast, Cape Coast, Ghana
Keba Hulela, Botswana University of Agriculture and Natural Resources, Gaborone, Botswana
Mabel Imbuga, Jomo Kenyatta University of Agriculture and Technology, PO Box 62000–00200 Nairobi, Kenya
Joseph Amooti Kasozi, School of Education, Botswana Open University, Gaborone, Botswana
Farai Kapfudzaruwa, Centre for the Advancement of Scholarship, University of Pretoria, Private Bag X20, Hatfield 0028, Pretoria, South Africa
Peter Koehn, Department of Political Science, University of Montana, 32 Campus Way, Missoula, MT, USA 59812
David Kraybill, Department of Agricultural, Environmental, and Development Economics, Ohio State University, 250 Agricultural Administration Building, 2120 Fyffe Road, Columbus, OH USA 43210
Kay Muir Leresche, retired professor of agricultural and natural resource economics, South Africa
John Lynam, independent scholar, PO Box 16526, Nairobi 00620, Kenya
Violet Makuku, The Association of African Universities, PO Box AN 5744, Accra-North, Ghana
Sylvanus Mensah, College of Agricultural and Environmental Sciences, Makerere University, PO Box 7062, Kampala, Uganda
Melody Mentz-Coetzee, Centre for the Advancement of Scholarship, University of Pretoria, Private Bag X20, Hatfield 0028, Pretoria, South Africa

Isaac Minde, Department of Agricultural, Food, and Resource Economics, College of Agriculture and Natural Resources, Michigan State University, 426 Auditorium Rd, East Lansing, Michigan, USA 48824

Eusebius Mukhwana, Kenya National Qualifications Authority, PO Box 72635-00200, Nairobi, Kenya

Joseph Mukuni, School of Education, Virginia Tech, Blacksburg, VA, USA 24061

Nancy W. Mungai, Egerton University, PO Box 536, Egerton-Njoro, 20115, Kenya

Caroline Mutepfa, African Union Development Agency (AUDA-NEPAD), 230 15th Road, Midrand, South Africa

Milcah Mulu Mutuku, Egerton University, PO Box 536, Egerton-Njoro, 20115, Kenya

Florence Nakayiwa, Regional Universities Forum for Capacity Building in Agriculture (RUFORUM), PO Box 16811 Wandegeya, Kampala, Uganda

Paul Nampala, Department of Agriculture and Biological Sciences, Faculty of Science and Technology, Uganda Christian University, PO Box 4, Mukono, Uganda

David Nielson, North America Agricultural Advisory Network and Colorado State University, Denver, CO, USA

Jemimah Njuki, International Food Policy Research Institute (IFPRI), Africa Region, PO Box 30709, 00100, Nairobi, Kenya

Patrick Okori, International Crops Research Institute for the Semi-Arid Tropics (ICRISAT), PO Box 1096 Lilongwe, Malawi

Duncan Ongeng, Gulu University, PO Box 166, Laroo-Pece City Division, Gulu City, Uganda

Moses Osiru, International Centre of Insect Physiology and Ecology, Nairobi, Kenya

Abraham Sarfo, independent consultant, ABSL Consulting, 21 Regents Hills, 5th Road, Midrand, South Africa

Daniel Sherrard, EARTH University, PO Box 4442-1000, San José, Costa Rica

Daniel N. Sila, Jomo Kenyatta University of Agriculture and Technology, PO Box 62000–00200 Nairobi, Kenya

Adnane Alaoui Soulimani, Mohammed VI Polytechnic University (UM6P), Lot 660, Hay Moulay Rachid, Ben Guerir 43150, Morocco

Aldo Stroebel, National Research Foundation, Box 2600, Pretoria 0001, South Africa

Frans Swanepoel, Centre for the Advancement of Scholarship, University of Pretoria, Private Bag 20, Hatfield 0028, Pretoria, South Africa

Clesensio Tizikara, Agrotiz Limited, PO Box 16409, Kampala, Uganda

John Wesonga, Jomo Kenyatta University of Agriculture and Technology, PO Box 62000–00200 Nairobi, Kenya

Preface

Background

Projected growth in demand for food production has increased the urgency of African leaders' response to current and emerging agricultural and development challenges. Both Africa-wide and globally, the Sustainable Development Goals and African Union Agenda 2063 provide overarching frameworks to anchor national and regional planning. These framewords are founded on the recognition that increased use of science, technology and innovation - particularly for African agriculture, as is highlighted in the Science, Technology and Innovation Strategy for Africa - are required to accelerate social and economic development. Within the food sector, the Comprehensive Africa Agricultural Development Programme and, more recently, the Malabo Declaration seek to increase agricultural output by at least 6% per year and strengthen commodity and other value chains. In sub-Saharan Africa (SSA), agribusiness is growing fast, and including agriculture, is projected to reach a value of more than US$1 trillion by 2030 - up from US$313 billion in 2010. Agriculture in SSA already employs more than 60% of Africa's population and contributes nearly 30% of GDP (closer to 50% when combined with agribusiness). As a result, agriculture-led growth is the most efficient strategy for reducing national poverty, with rates of decline shown to be more than four times higher than those of other strategies.

Some African governments have embarked on ambitious programmes and interventions to deliver the much-needed transformation of the agricultural sector to further spur economic growth and, especially, to facilitate an industrialization agenda. These ambitious plans depend on strong institutions employing the critical human-resource capacity needed to utilize and generate science, technology and innovation. Low levels of literacy and education at the farm level necessitate enhanced community education coupled with science-based innovations to drive production and value-chain development. With few exceptions, Africa's national agricultural research systems are characterized by inadequate funding and infrastructure, and a low proportion of PhD-qualified scientists. In addition, a weak technical and vocational education and training (TVET) sector is further challenged by the trend towards transforming these institutions into universities.

The Genesis of this Book

The expansion in basic education without targeting the high-level skills which drive innovation limits Africa's competitiveness globally. Over time, this has prompted greater focus on the need - and ways - to deepen human capital in the agricultural sector, both at the vocational and higher levels of tertiary agricultural education (TAE). One of the mechanisms for achieving this is South–South and North–South collaboration and partnership. This is particularly being advocated because of its ability to: (i) bring greater coherence to investments in African TAE, (ii) increase financial and technical support, as well as attention to TAE and (iii) help facilitate and guide the reform and transformation of Africa's TAE institutions through shared experiences and expertise. Strategy formulation, however, has generally suffered from the lack of quantitative information, evidence and experiences of *what* has worked and *how* it has worked. Where such evidence exists, it is too scattered and not readily available.

The World Bank's 2007 publication, Cultivating Knowledge and Skills to Grow African Agriculture: A Synthesis of an Institutional, Regional, and International Review, has been a key source of information, but rapid and extensive changes across the TAE and agricultural sectors demanded comprehensive up-to-date analyses. With this in mind, in 2016, the Regional Universities Forum for Capacity Building in Agriculture (RUFORUM) entered into a partnership with the Mastercard Foundation through the programme, Transforming African Agricultural Universities to Meaningfully Contribute to Africa's Growth and Development (TAGDev). TAGDev works with African agricultural universities and their graduates to enhance the application of science, technology, entrepreneurship and innovation for rural agricultural transformation. TAGdev also aims to strengthen collaboration among TAE actors in Africa and beyond. Accordingly, a collaborative effort was initiated to bring together key TAE stakeholders to develop an updated volume providing (i) sound analyses of current trends and developments in the TAE sector and (ii) direction and focus for future initiatives to strengthen the sector.

As a network of 130 universities from 38 African countries, RUFORUM relishes facilitating collaborative engagement among its members and network partners. RUFORUM is committed to realizing 'vibrant transformative universities to catalyse sustainable inclusive agricultural development to feed and create prosperity for Africa' as is envisaged in its Vision 2030, *African Universities' Agenda for Agricultural Higher Education, Science, Technology and Innovation*. In developing this volume over the past three years, RUFORUM has organized consultative dialogues and collaborative working sessions in Gaborone (Botswana), Nairobi (Kenya), Benguerir (Morocco) and Cape Coast (Ghana), as well as numerous virtual sessions to ensure the book's content captures insights from across Africa and the rest of the world. Thus, this book is the culmination of the enormous collaborative effort and commitment of highly-experienced and knowledgeable researchers.

The Focus and Structure of the Book

Part I of the book begins with an introduction on agriculture and education within the context of global and continental development goals. Chapter 1 lays the foundation for the issues presented in the book by providing global and regional context in terms of development goals and their implications. Part II presents the sectoral and institutional context underlying TAE. Chapter 2 addresses agricultural and rural transformation of Africa's food system, and TAE's role in that transformation. Chapter 3 discusses TAE's position within a changing policy and institutional context, pointing out the need for TAE institutions to be responsive to the rapid transformations occurring across Africa. Chapter 4 discusses global and African trends in tertiary education, while Chapter 5 discusses African trends in higher agricultural education.

Part III focuses on the pathways of transforming TAE in Africa. Chapter 6 focuses on leadership and change management to transform TAE institutions. The discussion builds on the experiences of university leaders from within Africa who have raised institutions to a competitive level in response to the demand for targeted and relevant education to meet the needs of graduates, communities and employers. Chapter 7 tackles the issue of transformative curricula, teaching methods and learning environments to ensure that TAE meets the demands of the evolving African labour market. Chapter 8 discusses transformative research and innovation capacity in African TAE, while Chapter 9 focuses on transformative outreach in African TAE. Chapter 10 focuses on issues and methods of quality assurance for TAE in Africa, building on experiences and processes from the African Union Commission's Quality Assurance Framework. Chapter 11 discusses the issue of gender and inclusive practices for transforming African TAE. Chapter 12 focuses on entrepreneurship and the role of universities in generating employment opportunities in Africa, especially for youth. This chapter provides experiences and lessons to demonstrate how universities can contribute to employment creation by helping to create a functional entrepreneurship ecosystem. Chapter 13 deals with the role of technical and vocation training in African TAE, pointing out the vital contribution of agricultural TVET institutions in preparing youth both for employment and for further studies. Chapter 14 discusses network approaches for transforming TAE in Africa by exploring the role of networks and consortiums as catalytic actors. Chapter 15 discusses lessons from university transformation in Africa, demonstrating African institutions which have already achieved success in their own transformations and can light the path for others to follow. Chapter 16 discusses valuable lessons from North–South and South–South partnerships in TAE in Africa.

Part IV discusses the way forward for strengthening tertiary agricultural education in Africa. Chapter 17 synthesizes the discussion on tertiary agricultural education in Africa, identifying the implications of the material presented in the book and offering recommendations for transforming TAE systems and institutions. The key message is that Africa's TAE sector is rising to the challenge, but long-term investment is required to sustain the seeds of change. African governments and development partners need to collaborate to leverage synergies in financing TAE to address the surge in student numbers and the acute demand for skilled and entrepreneurial graduates.

It is my sincere intention and hope that this book finds a broad audience, offering them sound, relevant and comprehensive knowledge and approaches to guide their diverse contributions to Africa's TAE sector and, in turn, its agricultural and agribusiness sectors, and overall development agenda.

Adipala Ekwamu
Executive Secretary, RUFORUM
10 May 2021

Acknowledgements

The editors wish to acknowledge the RUFORUM Secretariat for developing the book concept; organizing the book planning workshops in Gaberone, Botswana, in July 2017 and in Lilongwe, Malawi, in October 2017; and managing the overall book project. The editors are also grateful for valuable input from the participants of the aforementioned workshops on the scope and nature of the book. Carl Erik Larsen provided early input into the structure of the book and helped with the technical editing of some of the chapters. Deep appreciation and thanks are expressed to the authors who researched and wrote the commissioned chapters and revised them in response to editorial comments and suggestions. Special mention is due to Mary Jane Banks for substantive copyediting of the draft manuscript, to Anthony Egeru for overall management of the book project, and to David Hemming, commissioning editor at CABI. Funding for the editorial functions and publication of the book was provided by MasterCard Foundation.

List of Acronyms

4IR	Fourth Industrial Revolution
AAU	Association of African Universities
ACE	African Higher Education Centers of Excellence Project
AESIF	Agricultural Education and Skills Improvement Framework
AET	agricultural education and training
AfriQAN	African Quality Assurance Network
Agenda 2030	2030 Agenda for Sustainable Development
AGRA	Alliance for a Green Revolution in Africa
AI	appreciate inquiry
AIS	agricultural innovation system(s)
AQRM	African Quality Rating Mechanism
ARUA	African Research Universities Alliance
ARWU	Academic Ranking of World Universities
ASG-QA	African Standards and Guidelines for Quality Assurance in Higher Education
ATCs	agricultural training centres
ATVET	agricultural technical and vocational education and training
AUC	African Union Commission
BUAN	Botswana University of Agriculture and Natural Resources
C10	Committee of Ten Heads of State to Champion Education, Science and Technology
CAADP	Comprehensive Africa Agriculture Development Programme
CAP	Common African Position on Post-2015 Development Agenda
CBT	competence-based training
CESA	Continental Education Strategy for Africa, 2016–2025
CURAD	Consortium for Enhancing University Responsiveness to Agribusiness Development Ltd
DACUM	Developing a Curriculum
EAAFRO	East African Agricultural and Forestry Organization
ECOWAS	Economic Community of West African States
FAAP	Framework for African Agricultural Productivity
FARA	Forum for Agricultural Research in Africa
FTCs	farmer training centres
FTEs	full-time equivalents
GDP	gross domestic product

GER	gross enrolment ratio
GIZ	German Development Cooperation
GU	Gulu University [Uganda]
HAQAA	Harmonisation of African Higher Education Quality Assurance and Accreditation
HEIs	higher-education institutions
HERANA	Higher Education Research and Advocacy Network in Africa
iAGRI	Innovative Agricultural Research Initiative
ICT	information and communications technology
IMF	International Monetary Fund
IRDPs	integrated rural development programmes
IUCEA	Inter-University Council of East Africa
JKUAT	Jomo Kenyatta University of Agriculture and Technology, Kenya
LAMS	Lycée Agricole Mèdji de Sékou [Benin]
LSMS	living standards measurement study
MALF	Ministry of Agriculture, Livestock, and Fisheries [Tanzania]
MDGs	Millennium Development Goals
MOA	ministry of agriculture
MOOCs	Massive open online courses
MSU	Michigan State University [United States]
NAIPs	national agricultural investment plans
NARI(s)	national agricultural research institute(s)
NFRE	non-farm rural economy
NGOs	Nongovernmental organizations
NICHE	Netherlands Initiative for Capacity Development in Higher Education
NQFs	national qualification frameworks
OECD	Organisation for Economic Co-operation and Development
OSU	Ohio State University [United States]
PI(s)	principal investigator(s)
PRSC	Poverty Reduction Strategy Credits
PRSP	Poverty Reduction Strategy Paper
QAAs	quality-assurance agencies
R&D	research and development
RAIPs	regional agricultural investment plans
RECAP	RUFORUM Entrepreneurship Challenge Programme
RECs	regional economic communities
RMC	Resource Mobilization Committee
RUFORUM	Regional Universities Forum for Capacity Building in Agriculture
S3A	Science Agenda for Agriculture in Africa
S&T	science and technology
SADC	Southern African Development Community
SADCQF	SADC Qualifications Framework
SARUA	Southern Africa Regional Universities Association
SDG(s)	Sustainable Development Goal(s)
SES	student enterprise scheme
SRO(s)	sub-regional organization(s)
SSA	sub-Saharan Africa
STEM	science, technology, engineering and mathematics
STI	science, technology and innovation
STISA	Science, Technology and Innovation Strategy for Africa 2024
SUA	Sokoine University of Agriculture [Tanzania]
T&V	training and visit

TAE	tertiary agricultural education
TAGDev	Transforming African Agricultural Universities to Meaningfully Contribute to Africa's Growth and Development
TVET	technical and vocational education and training
UAC	University of Abomey Calavi [Benin]
UCC	University of Cape Coast [Ghana]
UCR	University of Costa Rica
UM6P	Mohamed VI Polytechnic University [Morocco]
UNESCO	United Nations Educational, Scientific and Cultural Organization
UNIVEN	University of Venda [South Africa]
USAID	United States Agency for International Development

Cover Photograph Credits

Top

RUFORUM-supported students listening to lecture in Taifa Conference Hall at University of Nairobi in Kenya. GoTrolley Films, Cape Town, South Africa.

Bottom Left

A potato-farmer participating in RUFORUM-supported Potato Community Action Research Project in Molo, Kenya. GoTrolley Films, Cape Town, South Africa.

Bottom Center

A USAID-sponsored doctoral student and a laboratory technician documenting a soil science experiment at Sokoine University of Agriculture in Tanzania. Photograph by Molly Kraybill.

Bottom Right

RUFORUM-supported students conducting laboratory research at Gulu University in Uganda. GoTrolley Films, Cape Town, South Africa.

Part I

Introduction

1 Agriculture and Tertiary Education within the Context of Global and African Development Goals

Adipala Ekwamu[1]*, Frans Swanepoel[2], Melody Mentz-Coetzee[2], Farai Kapfudzaruwa[2] and Kay Muir Leresche[3]
[1]Regional Universities Forum for Capacity Building in Agriculture (RUFORUM), Kampala, Uganda; [2]Centre for the Advancement of Scholarship, University of Pretoria, Pretoria, South Africa; [3]Retired professor of agricultural and natural resource economics, based in South Africa

1.1 Introduction

Extreme poverty remains a significant challenge in Africa. Although the global rate of extreme poverty fell from 28% in 1999 to 10% in 2015, as of 2017, an estimated 731 million people still lived in poverty - 443 million in sub-Saharan Africa (SSA) (World Bank, 2020). Forecasts also indicate that by 2030 nearly 90% of the extremely-poor population will live in SSA (World Bank, 2018). Most African youth, who constitute over 60% of the continent's 1.2 billion people, have been exposed to poverty. In SSA, one-third of Africa's 450 million youth, aged 15–35 years, are unemployed, another third are vulnerably employed and only one in six have wage-based employment (African Development Bank Group, 2016). These vulnerable young people predominantly live in rural areas (63%), and two-thirds of them rely on agriculture for their incomes (UNECA, 2017a). Africa is also disproportionately affected by malnutrition. Although the global proportion of under-nourished people fell from 15% during 2000–2002 to 11% during 2014–2016, about 793 million people were still under-nourished as of 2014–2016. Southern Asia and SSA accounted for almost two-thirds of these people (FAO, 2018a).

Apart from providing food for nutrition, agriculture remains one of the most effective pathways out of poverty in Africa, accounting for almost two-thirds of household livelihoods as of 2017 (World Bank, 2018). Yet, Africa's full agricultural potential remains untapped. In the 1970s, Africa provided 17% of the world's total agricultural exports, whereas by 2013 this share had fallen to a negligible 2% (Anderson, 2010; FAO, 2018a). Africa's yearly food import bill of US$35 billion - estimated to rise to US$110 billion by 2025 - reflects the troubling trajectory of the continent's agricultural sector (FAO, 2018a). This is in stark contrast to the fact that 60% of the world's uncultivated arable land is located in Africa (McKinsey Global Institute, 2010).

This current state of Africa's agriculture is mostly linked to the low productivity of the over 33 million smallholder farmers, who own 80% of all farms and contribute up to 90% of food production in some countries (World Bank, 2018). These smallholder farmers face numerous

*Corresponding author: e.adipala@ruform.org

 Transforming Tertiary Agricultural Education in Africa (eds D. Kraybill, J. Lynam and Adipala Ekwamu)
DOI: 10.1079/9781789246544.0001

risks to their agricultural production, including pest and disease outbreaks, extreme weather events and market shocks, which often undermine their household food and income security (O'Brien *et al.*, 2004; Morton, 2007). Climate change is also expected to disproportionately affect smallholder farmers by exacerbating the risks they face (Harvey *et al.*, 2014; Gbegbelegbe *et al.*, 2017). Furthermore, smallholder farming practices are often characterized by intensive and continuous cultivation, resulting in soil erosion and the eventual decline of soil fertility, especially under arable cultivation where soil-conservation measures are absent (Lalani *et al.*, 2016; Roland *et al.*, 1997).

Another major obstacle is the fact that agricultural education and training (AET) needs overhauling (see for example, ASSAf, 2017, for the case of South Africa). In particular, tertiary agricultural education (TAE)[1] systems in Africa are highly fragmented, with limited coordinated systems at the national level to support the development of curricula and infrastructure (Swanepoel *et al.*, 2014; NEPAD, 2015). The approaches and curricula being offered are outdated and preclude the effective development of the necessary human-resource capacity to enable Africa to transform its agricultural sector. The outdated curricula do not adequately integrate science, technology and innovation (STI), which are critical to the co-creation of knowledge to support sustainable agriculture and food systems for inclusive growth (Chakeredza *et al.*, 2008; NEPAD, 2015). Despite the projected opportunities for agribusiness in Africa (WEF, 2017), youth remain largely uninterested in agricultural careers or see agriculture as a low-status option for employment (Leavy and Hossain, 2014).

This chapter explores African and global policy frameworks and strategies in order to highlight the role that the transformation of Africa's agricultural sector and TAE is envisaged to play in addressing developmental challenges in Africa. The main developmental agendas discussed, which anchor the transformation of African agriculture and education, are the global Agenda 2030 Sustainable Development Goals (SDGs) and the Africa-focused Agenda 2063. Agriculture or education strongly feature in all 17 SDGs but most prominently in SDG 2, which aims to eradicate hunger, and in SDG 4, which focuses on high-quality education. At the continental level, elements of Agenda 2063 strongly link to agriculture and focus on achieving a 'prosperous Africa based on inclusive growth and sustainable development' (AUC, 2015a, p. 2). The Agenda specifically identifies modernizing the agricultural sector to increase production, productivity and value addition as key to achieving development outcomes. Agenda 2063 is supported by the Comprehensive Africa Agriculture Development Programme (CAADP) and the Continental Education Strategy for Africa, 2016–2025 (CESA), which provide clear strategies for transforming Africa's agricultural and education sectors (NEPAD, 2003; AUC, 2015b). Supporting these policy frameworks are strategies which focus on enhancing human capital development in agriculture, as well as promoting STI in agriculture. These strategies and frameworks - including the Science, Technology and Innovation Strategy for Africa 2024 (STISA) and the Agricultural Education and Skills Improvement Framework (AESIF) - are explored in greater depth below.

1.2 Agenda 2030 and the Agricultural Transformation of Africa

The adoption of the 2030 Agenda for Sustainable Development (Agenda 2030) by the United Nations General Assembly in 2015 shifted global attention from the Millennium Development Goals (MDGs), which focused on eradicating poverty, to a more inclusive, long-term and sustainable development process. The 17 SDGs and accompanying 169 targets outline a global commitment to transforming the world and creating sustainable development pathways, while also addressing the numerous complex problems facing society. The interconnectedness of the SDGs and their associated targets clearly illustrates the complexity of the problems needing to be addressed and their cross-sectoral nature. It is, thus, not surprising that agriculture permeates the 2030 Agenda in several ways (Tosun and Leininger, 2017). SDG 2 most directly relates to agriculture, calling for global action to end hunger, achieve food security, improve nutrition and promote sustainability. A further 13 of the 17 SDGs either directly or indirectly relate to agriculture and food systems (FAO, 2018a).

Agriculture is the nexus linking the elimination of poverty (SDG 1), food security (SDG 2), human health and well-being (SDG 3), gender equality (SDG 5), clean energy (SDG 7), viable ecosystems (SDGs 6, 14, and 15), climate change (SDG 13) and strong institutional arrangements (SDGs 16 and 17) (Table 1.1). Agriculture's complex interconnectedness with multiple SDGs makes a compelling case for revitalizing strong, multistakeholder partnerships to facilitate engagement and the sharing of knowledge and expertise (SDG 17). Effectively transforming Africa's agricultural landscape will require a wide range of collaborative actors contributing to five crucial components linked to SDG 17 on partnerships: (i) finance; (ii) capacity building,

Table 1.1. Summary of the Sustainable Development Goals' linkages to agriculture

Goal and focus	Linkages to agriculture
SDG 1: End poverty	Agriculture remains the main source of livelihood in Africa, especially for 63% of the continent's rural population. Therefore, growth in the agricultural sector, particularly in the low-income and agrarian economies, is at least twice as effective in reducing poverty as any other sector.
SDG 2: Zero hunger	One in four people in sub-Saharan Africa still suffer from hunger due to food insecurity and lack of sufficient vitamins and minerals. Promoting sustainable agriculture and transforming food systems are important to ensure that enough food is produced for the growing population through sustainable use of natural resources.
SDG 3: Good health and well-being	Malnutrition in all its forms - undernutrition, micronutrient deficiencies, obesity and diet-related non-communicable diseases - impose a high cost on society's health and well-being. Therefore, creating nutrition-sensitive agriculture and food systems at all stages of the food chain is critical in providing nutritious and healthy diets which uplift both the human and environmental well-being of society.
SDG 5: Gender equality	Women represent around half of the total agricultural labour force in developing countries, which means they occupy important roles in the development of agricultural economies. Despite that, land tenure systems in many African countries exclude women from owning agricultural land, which often limits their access to capital, markets and economic opportunities. Therefore, gender equality and women's empowerment are central to transforming Africa's agriculture.
SDG 6: Water	Increasing food production while using less water is one of the greatest challenges to sustainable agriculture. Crops and livestock use 70% of all water withdrawals and up to 95% in some African countries.
SDG 7: Affordable and clean energy	Energy has a key enabling role in achieving food security and better nutrition. Food systems, which currently consume 30% of the world's energy, will gradually need to decouple from fossil fuel dependence in order to deliver more food with less and cleaner energy.
SDG 11: Sustainable cities and communities	As the share of Africans living in urban areas is projected to grow from 36% in 2010 to 50% by 2030, enormous demand will be exerted on food systems. Fluctuating food prices affect urban consumers the most, because they are almost exclusively dependent on food purchases. Sustainable and innovative urban agriculture needs to be promoted to ensure that urban dwellers can access nutritious food at affordable prices.
SDG 12: Sustainable consumption and production	Every year, the world loses or wastes about one-third of the food it produces. Feeding the world more sustainably will require that food be grown in ways that reduce negative environmental impacts and drastically reduce post-harvest losses. At the same time, consumers must be encouraged to shift to nutritious and safe diets with a lower environmental impact and to decrease consumer food waste.
SDG 13: Combat climate change	Agriculture has a major role to play in responding to climate change. While temperature rises pose a real threat to global food production, investment in all sectors of agriculture can simultaneously support climate change adaptation and mitigation.

Continued

Table 1.1. Continued.

Goal and focus	Linkages to agriculture
SDG 14: Oceans, seas and marine resources	Seafood is a critical source of animal protein, minerals and micronutrients for more than 200 million Africans. However, about 29% of the commercially important marine fish stocks are overfished, and 61% are fully fished. Therefore, sustainable management of the ocean ecosystem is imperative to ensure the sustainability of fisheries.
SDG 15: Life on land	The sustainable use and management of terrestrial ecosystems, mountains, forests, land and soils are critical to global agricultural and food systems. Destruction of the biosphere - e.g. deforestation in the tropical forests at an annual rate of 0.8% per year - presents a serious threat to sources of food, medicine and energy.
SDG 16: Strong institutions	Successfully transforming Africa's agriculture hinges on strong and accountable institutions, inclusive and effective decision-making processes and societal equality and equity. Ensuring resilient agriculture and agrobiodiversity in Africa both builds and depends on stronger communities that can participate more actively in sustainable development.
SDG 17: Partnerships	The complexity and interconnectedness of agriculture to various sectors and actors calls for a new transformative process that requires all development actors to engage and share complementary strengths, knowledge and resources.

Source: Constructed by authors from FAO (2018a).
Note: SDG(s) = Agenda 2030 Sustainable Development Goal(s).

(iii) systemic issues, (iv) technology and (v) trade. The need for these key partnerships and components manifests itself across the entire agricultural value chain (Lee *et al.*, 2010; Trienekens, 2011).

1.3 Converging Agendas and the Role of Agriculture in Africa's Sustainable Development

The African Union Commission's (AUC) Agenda 2063 - a Pan-African vision for an inclusive, prosperous, politically-united, culturally-vibrant, peaceful, and sustainable Africa - predates, but is nonetheless complementary to the global Agenda 2030 (AUC, 2015a). It is important to point out that the global Agenda 2030 and its SDGs were influenced by the African Union's Common African Position on Post-2015 Development Agenda (CAP), with Africa being the only region to submit a position in writing (AUC, 2015a). As a result, Agenda 2030 and Agenda 2063 converge on all four primary dimensions of the global Agenda through social development (people), inclusive economic development (prosperity), peaceful and inclusive societies and responsive institutions (peace) and sustainability (the planet) (UNECA, 2017b). The next two sections discuss the high-level alignment between the SDGs and Agenda 2063, and then the specific SDG 2 targets that directly relate to agriculture, as well as their alignment with specific goals and targets under Agenda 2063.

High-Level Alignment between Agenda 2063 and the Sustainable Development Goals

Compared with Agenda 2030 (which is a 15-year development framework), AUC's Agenda 2063 is a 50-year framework anchored by seven aspirations, each supported by corresponding goals, priority areas, targets and strategies. Twenty goals currently apply under the First Ten-Year Implementation Plan (AUC, 2015a). A number of key aspects under Agenda 2063 align with the SDGs (Table 1.2).

Clearly demonstrable synergies arose from the CAP inputs into the development of the SDGs, with all of the SDGs being linked to one or more key priority in Agenda 2063 (Table 1.2). An important element featured in both agendas - and of particular relevance to the content of this

Table 1.2. Alignment of the Agenda 2063 goals and the 2030 Sustainable Development Goals

Agenda 2063 goals	Agenda 2030 Sustainable Development Goals
1. A high standard of living, quality of life and well-being for all citizens	1. End poverty in all its forms everywhere in the world 2. End hunger, achieve food security and improved nutrition and promote sustainable agriculture 8. Promote sustained, inclusive and sustainable economic growth, full and productive and decent work for all 10. Reduce inequality within and among countries
2. Well-educated citizens and skills revolution underpinned by science, technology and innovation	4. Ensure inclusive and equitable quality education and promote lifelong learning for all 9. Build resilient infrastructure, promote inclusive and sustainable industrialization and foster innovation
2. Healthy and well-nourished citizens	2. End hunger, achieve food security and improved nutrition and promote sustainable agriculture 3. Ensure healthy lives and promote well-being for all at all ages
5. Modern agriculture for increased productivity and production	2. End hunger, achieve food security and improved nutrition and promote sustainable agriculture
6. Blue/ocean economy for accelerated economic growth	14. Conserve and sustainably use the oceans, seas and marine resources for sustainable development
7. Environmentally sustainable and climate resilient economies and communities	6. Ensure availability and sustainable management of water and sanitation for all 7. Ensure access to affordable, reliable, sustainable and modern energy for all 11. Make cities and human settlements inclusive, safe, resilient and sustainable 12. Ensure sustainable consumption and production patterns 13. Take urgent action to combat climate change and its impacts 15. Protect, restore and promote sustainable use of terrestrial ecosystems, sustainably manage forests, combat desertification, halt and reverse land degradation and halt biodiversity loss
12. Capable institutions and transformative leadership in place	16. Promote peaceful and inclusive societies for sustainable development, provide access to justice for all and build effective, accountable and inclusive institutions at all levels
17. Full gender equality in all spheres of life	5. Achieve gender equality and empower all women and girls
18. Engaged and empowered youth and children	4. Ensure inclusive and equitable quality education and promote lifelong learning for all 5. Achieve gender equality and empower all women and girls

Source: Adapted by authors from AUC (2014b).

book - is the focus on youth. Africa has the largest population of young people in the world, but their rate of unemployment is twice as high as the rate for adults (African Development Bank Group, 2016). It is thus highly relevant that the vision of Africa expressed in Agenda 2063 is one of an Africa whose development is people driven, especially relying on the potential offered by women and youth (AUC 2015a). In alignment to this vision, SDGs 4 and 8 highlight the need to increase relevant skills in order to increase opportunities for youth employment. (Education as a driver of youth empowerment - a key theme throughout the book - is discussed in Section 1.4 of this chapter).

Despite the symmetry between the two agendas, key implementation challenges for African Member States are (i) the integration of the agendas within national planning frameworks and (ii) appropriate mechanisms for monitoring progress. In a 2017 report, AUC's Commissioner for Economic Affairs asserted that progress had been made on all 13 of Agenda 2063's fast-track programmes and projects (Maruping, 2017). The details of this progress are laid out for each programme and project, but the measurable extent of attainment of the implementation of the goals seems limited. In fact, Commissioner Maruping was only able to make a rather conservative conclusion on the progress made to that point: 'Implementation of Agenda 2063 has definitely commenced' (Maruping, 2017, p. 7).

A key development in determining progress towards implementing Agenda 2063 was AUC's release of an indicator framework to monitor and report on the the Agenda's First Ten-Year Implementation Plan. This framework, finalized in 2017, includes 138 regional monitoring and reporting indicators (Sissoko, 2017). This framework for monitoring, evaluating and reporting guidelines only came to fruition recently, which highlights what has to date been a key obstacle in determining the extent and success of policy implementation in Africa - namely, a lack of data quality, data analysis and monitoring and evaluation systems (Makombe and Benin, 2018).

Data gaps are also key issues in monitoring progress towards international development goals, with only 37.8% of the SDG indicators having data available to measure African progress. Furthermore, coverage of each goal is uneven: SDGs 1 and 2 (i.e. no poverty and zero hunger), which are of particular significance to agriculture, both have less than 60% data coverage. Goal 1 has 12 indicators, of which only 25% have the required data to measure progress. Goal 2 has 14 indicators, of which a more encouraging 57.1% have data coverage (Sissoke, 2017; UNECA, 2017a). These figures highlight why participants in the Sixth Statistical Commission for Africa, held in October 2018, urged governments to allocate more resources to the production, dissemination and use of statistics to support implementation of Agenda 2030, Agenda 2063 and national development plans (Benson-Wahlén, 2018).

Agriculture-Specific Linkages in Global and African Development Agendas

As previously discussed, to varying degrees, both Agenda 2030 and Agenda 2063 recognize the importance of sustainable agriculture and food security in driving prosperity and inclusive growth in Africa (Table 1.3).

The alignment between these two high-level agendas and, ultimately, the role of agriculture in African development, can be summarized around seven key themes, each of which is briefly explored below.

1. *Eliminating hunger and ensuring food security for all.* Even though progress has been made in the past two decades, one-fifth of all Africans are still undernourished, representing over 257 million people (FAO and UNECA, 2018). Africa's projected population growth of 1.3 billion between 2015 and 2050 (more than half of projected global growth) will result in a four-fold increase in food demand (Hilderink *et al.*, 2012). Agriculture clearly has - and will continue to have - a vital role to play in Africa's future food security.

2. *Radically increasing productivity and incomes to eliminate poverty.* Despite possessing a significant proportion of the world's available arable land, Africa only generates 10% of global agricultural output (McKinsey Global Institute, 2010). The fact that 63% of the African population rely on agriculture - not only as a source of food, but also as a source of income - means that transforming the agricultural sector to radically increase productivity, in addition to being directly related to SDG 2, will also be instrumental in responding to SDG 1 (eradicating extreme poverty for all people everywhere by 2030). Africa under-performed in tackling poverty under the MDGs. At the onset of the MDGs in 1990, 47% of the world's extreme poor lived in East Asia and the Pacific, while SSA accounted for 18%. By 2013, however, Africa accounted for 51% of the world's extreme poor, and the share in East Asia had fallen to 9% (World Bank, 2018). Many authors have argued that productivity growth in smallholder farming and rural non-farm employment linked to agriculture offer the greatest opportunities for reducing African poverty (World Bank, 2007; Beegle *et al.*, 2016; De Janvry and Sadoulet, 2017). As evidence, research in Uganda by Christiaensen and Kaminski (2016) showed that 70% of the observed decline in the poverty rate during 2005–2009 was achieved within agriculture, and that only 30% stemmed from employment shifts to

Table 1.3. The relationship between SDG 2 and Agenda 2063 targets focusing on agriculture

Themes	SDG 2 targets	Agenda 2063 targets and priority areas
Eliminate hunger, ensure food security for all	2.1. By 2030, end hunger and ensure access by all people, in particular, the poor and people in vulnerable situations, including infants, to safe, nutritious and sufficient food all year round	72e. Consolidate the modernization of African agriculture and agrobusinesses to completely eliminate hunger and food insecurity
End all forms of malnutrition	2.2. By 2030, end all forms of malnutrition, including achieving, by 2025, the internationally-agreed targets on stunting and wasting in children under 5 years of age, and address the nutritional needs of adolescent girls, pregnant and lactating women and older persons	
Radically increase productivity and incomes Significantly increase the productivity and incomes of small-scale female farmers	2.3. By 2030, double the agricultural productivity and incomes of small-scale food producers, in particular women, indigenous peoples, family farmers, pastoralists and fishers, including through secure and equal access to land, other productive resources and inputs, knowledge, financial services, markets and opportunities for value addition and non-farm employment	72e. Consolidate the modernization of African agriculture and agrobusinesses 72e. Develop and implement affirmative policies and advocacy to ensure women's increased access to land and inputs, and ensure that at least 30% of agricultural financing is accessed by women
Ensure sustainable production systems, and implement climate-smart agriculture	2.4. By 2030, ensure sustainable food production systems and implement resilient agricultural practices that (i) increase productivity and production, (ii) help maintain ecosystems, (iii) strengthen capacity for adaptation to climate change, extreme weather, drought, flooding and other disasters and (iv) progressively improve land and soil quality	72f. Act with a sense of urgency on climate change and the environment through implementation of the Programme on Climate Action in Africa, including: a climate resilient agricultural development programme; national adaptation plans, systems and structures
Maintain biodiversity	2.5. By 2020, maintain the genetic diversity of seeds, cultivated plants and farmed and domesticated animals and their related wild species, including through soundly-managed and diversified seed and plant banks at the national, regional and international levels, and promote access to and fair and equitable sharing of benefits arising from the use of genetic resources and associated traditional knowledge, as internationally agreed	72f. Ensure sustainable exploitation and management of Africa's diversity for the benefit of its people
Increase investment in research, innovation and applied technology for agriculture	2A. Increase investment in rural infrastructure, agricultural research and extension services, technology development and plant and livestock genebanks	72e. Expand the introduction of modern agricultural systems, technology, practices and training, including abolishing the hand-hoe

Continued

Table 1.3. Continued.

Themes	SDG 2 targets	Agenda 2063 targets and priority areas
Address market related challenges to ensure properly functioning markets	2B. Correct and prevent trade restrictions and distortions in world agricultural markets, including through the parallel elimination of all forms of agricultural export subsidies and all export measures with equivalent effect, in accordance with the mandate of the Doha Development Round	72e. Reduce the imports of food and raise intra-Africa trade in agriculture and food to 50% of total formal food and agricultural trade
	2C. Adopt measures to ensure the proper functioning of food-commodity markets and their derivatives and facilitate timely access to market information in order to help limit extreme food price volatility	72d. Establish commodity exchanges for strategic African products

Source: Adapted by authors from AUC *et al.* (2017, p. 28).

non-agricultural activities. The juxtaposition of the failure to address extreme poverty under the MDGs against agriculture's high potential to address extreme poverty underscores the necessity - and urgency - of agricultural transformation in Africa. Agenda 2063's aspiration for a prosperous Africa 'determined to eradicate poverty in one generation' recognizes the importance of eliminating hunger as a priority contribution to national farmer and prosperity (AUC, 2015a, p. 2).

3. *Significantly increasing the productivity and incomes of small-scale female farmers.* Women engaged in agricultural activities are fundamental to sustainable African development. They contribute a significant share of the agricultural labour but have the least access to economic, political and social opportunities (FAO, 2018c). Studies posit that if female farmers had the same access to resources as male farmers, the number of hungry people in the world could be reduced by up to 150 million (Bunch and Mehra, 2008; Brody, 2015). Furthermore, research has shown that when women are economically empowered, significant health benefits accrue at the household level (FAO, 2011). Thus, significantly increasing the productivity and incomes of smallholder female farmers, a target in SDG 2, drives positive poverty reduction (SDG 1), health (SDG 3) and gender equality (SGD 5) outcomes. The equal participation of women in all spheres of life, envisaged in Agenda 2063, is of direct relevance here. Strategies in Agenda 2063, such as implementing 'affirmative policies and advocacy to ensure women's increased access to land and inputs, and ensure that at least 30% of agricultural financing [is] accessed by women' will enable the achievement of this gender equality (AUC 2015a, p. 16).

4. *Ensuring sustainable production systems and maintaining biodiversity.* Tackling SDG 1 on extreme poverty through agricultural transformation (SDG 2) must be achieved in tandem with sustainable use of natural resources (SDGs 6, 7, 12–15). While Green Revolution–driven agricultural intensification increased crop yields in other regions of the world, it also had unintended consequences for water use, soil degradation, and chemical runoff, resulting in serious environmental impacts (Burney *et al.*, 2010). The policy environment around the Green Revolution promoted 'injudicious and overuse of inputs and expansion of cultivation into areas that could not sustain high levels of intensification, such as sloping lands' (Pingali, 2012, p. 12304). In Africa, the ever-increasing demand for farm land, fuel wood and charcoal production coupled with population growth has accelerated the rate of deforestation, with the net loss of forests exceeding 4 million hectares per year between 2000 and 2005 (FAO, 2018b). These ecological challenges point to the need to integrate water conservation (SDG 6), marine resource conservation (SDG 14) and forest and terrestrial ecosystem protection (SDG 15) into sustainable agricultural intensification practices. Such practices are key to achieving SDG 12 on sustainable production and consumption, particularly through target 12.2 - sustainable management and efficient use of natural resources - and target 12.4 - environmentally-sound management of chemicals.

5. *Implementing climate-smart agriculture.* Agriculture is both affected by, and contributes to, climate change. Yet, as is the case with several other sectors, the agricultural sector has only recently begun to prioritize climate change issues (SDG 13), particularly its dramatic impact on millions of smallholder farmers and food processors who produce most of the world's food. Many agriculture-dependent households are located in ecosystems which are the most vulnerable to the effects of climate change (Campbell *et al.*, 2016). The Intergovernmental Panel on Climate Change (2007) reports that, by 2080, SSA's food productivity will fall by 9–21% as a result of climate change. Liliana (2005) also indicates that about two-thirds of Africa's arable land is expected to be lost by 2025 due to low rainfall and drought (see also Thornton *et al.*, 2011; Niang *et al.*, 2014; and Connolly-Boutin and Smit, 2016). On the other hand, agriculture contributes 14% of the world's greenhouse gas emissions and 24% of related land use changes (Niang *et al.*, 2014). Agenda 2030 and Agenda 2063 both recognize the correlation between climate variability and hunger and poverty in Africa. As a result, both agendas promote the principles of climate-smart agriculture, which reliably increases food production while adapting and building resilience to climate change (Lipper *et al.*, 2014).

6. *Addressing market-related challenges.* Agenda 2030 and Agenda 2063 highlight the trade distortions in global agricultural markets. Given US$ 35 billion in African food imports annually, these distortions significantly undermine food security in the region (African Development Bank Group, 2018). SDG 2b specifically highlights the need to prevent and restrict these distortions through the elimination of all forms of agricultural export subsidies and all export measures with equivalent effect, while Agenda 2063 emphasizes reducing food imports and raising intra-African agricultural trade, as well as establishing an African Commodity Exchange (AUC, 2015a).

7. *Increasing investment in research, innovation and applied technology for agriculture.* In recognition of need to invest in and leveraging STI, Agenda 2063 underscores the importance of a knowledge society which promotes 'modern agriculture'. Agenda 2063 sets a target to abolish the hand hoe by 2025 and, through investment in STI, ensure that the agricultural sector is 'modern, profitable, and attractive' to African women and youth (AUC, 2015a, p. 3). Complementing Agenda 2063, Agenda 2030 positions research, innovation and technology as key factors in the implementation of the SDGs to eliminate hunger and malnutrition, and increase food production using sustainable agriculture practices. SDG 9.5 specifically focuses on enhancing scientific research, upgrading technological capacity, encouraging innovation and substantially increasing the number of research and development workers.

Despite the synergies discussed above related to agriculture and food security, Agenda 2063 is not explicit about specific goals and targets to eradicate malnutrition on the continent. In contrast, SDG target 2.2 explicitly focuses on ending all forms of malnutrition - undernutrition, micronutrient deficiencies, obesity, and diet-related non-communicable diseases - as well as stunting and wasting in children. Malnutrition in all its forms imposes an unacceptably high economic and social cost on society (FAO, 2018a). The 2018 Global Nutrition Report reveals that 30 out of 41 African countries are struggling with challenges related to malnutrition. The report notes that only five of the 54 countries are making sufficient progress towards meeting their nutritional targets to reduce stunting for children under five years old. Furthermore, 20% of the total African population is undernourished. African countries also fare badly on various nutrition scores, with 41% of adult women being overweight, and 8% of all adults being diabetic (Development Initiatives, 2018). These challenges negatively affect the health and well-being of Africans. In the context of Agenda 2030, policy-makers have been promoting nutrient-sensitive agriculture - i.e. agricultural development that puts nutritionally-rich foods, dietary diversity, and food fortification at the heart of overcoming malnutrition and micronutrient deficiencies (Balz *et al.*, 2015;

Ruel *et al.*, 2018). As Africa's urban populations increase and diets change, dealing with all forms of malnutrition will gain even greater importance.

Putting the Vision of Agriculture's Contribution to Development into Action

With the highest-level frameworks in place to guide global and regional thinking about agriculture's role in development, the pressing question becomes how these aspirations can be acted upon to achieve the desired impact in Africa. The primary mechanism for this at the pan-African level is CAADP - the policy framework signed in 2003 in Maputo, Mozambique, to promote agricultural transformation, wealth creation, food security and nutrition, economic growth and prosperity (NEPAD, 2003). More than 40 African governments adopted the policy framework. Since its inception, CAADP set broad targets for signatory countries to allocate at least 10% of their public expenditures to the agricultural sector and attain at least 6% annual growth in agricultural GDP.

In its first decade (2003–2013) CAADP's was implemented through round-table processes that established clear subregional and national steps, including signing a CAADP Compact, developing national and regional agricultural investment plans (NAIPs or RAIPs) and holding a CAADP business meeting (Table 1.4). In 2014, the African Union's heads of state and government reaffirmed their commitment to CAADP by adopting the 2014 Malabo Declaration on Accelerated Agricultural Growth and Transformation for Shared Prosperity and Improved Livelihoods, which included seven broad commitments: (i) upholding the CAADP principles and values, (ii) enhancing investment in agriculture, (iii) ending hunger, (iv) halving poverty by 2025, (v) boosting intra-Africa agricultural trade, (vi) enhancing resilience to climate variability and (vii) strengthening mutual accountability for actions by conducting Biennial Reviews (AUC, 2014a). With CAADP now in its second decade, countries and regions are updating or developing their NAIPs/RAIPs to ensure that they comply with these commitments. At the country level, this process starts with the Malabo domestication event - led by AUC (formerly NEPAD), and the regional economic communities (RECs) - at which CAADP constituents convene to discuss and agree on country roadmaps.

Progress has been uneven across the subregions. As of 2018, domestication events had been held in 16 countries, and by August 2018, Malabo Status Assessments and Profiles had been completed for 21 countries, Malabo Goals and Milestone Reports had been completed for 16 countries, and a total of 19 countries had either drafted or validated their Malabo-compliant NAIPs (Table 1.4). Some analysts view these results, documented in the Inaugural CAADP Biennial Review Report (AUC, 2018), as evidence that countries are making good progress towards achieving their Malabo Declaration commitments. This view is supported by research indicating that early adopters of the CAADP goals out-perform non-adopters in various critical areas, including, annual increases in agricultural production and land productivity, annual increases in GDP per capita, and annual declines in the prevalence of malnutrition (African Green Revolution Forum, 2018). Additional encouraging feedback indicates that countries with higher levels of CAADP implementation have higher average annual growth in agricultural expenditures and higher annual agricultural growth compared with those that have not adopted the process (Benin, 2016; Makombe and Benin, 2018).

In contrast, other analysts view the same data more critically, noting that the average score for agricultural transformation for the whole of Africa is only 3.6 (AUC, 2018). Government agricultural expenditure, at an average of 3% of total public expenditure in the period 2008–2017, remained well below CAADP's 10% target (Makombe *et al.*, 2018), and by 2018 only 13 Member States had met or surpassed the 10% target (African Green Revolution Forum, 2018). These indicators are interpreted by some as indicating that the African Union is not on track to meet its Malabo commitments (AUC, 2018).

Explanations identified include failure to secure greater understanding and ownership of CAADP at the country level, and insufficient investment by governments and development partners (Signé, 2017). Furthermore, low per capita GDP and total spending in most African

Table 1.4. Regional progress in implementing CAADP based on number of countries as of 2018

	First generation investment plan					Second generation investment plan				Inaugural biennial review process		
Region	Round table held and compact signed	Plan drafted, reviewed and validated	Business meeting held	GAFSP funding approved (million US$)	JSR assessment conducted or Initiated	Malabo domestication event held	Malabo status assessment and profile finalized	Malabo goals and milestones report finalized	NAIP drafted or reviewed or validated	Report drafted, validated and submitted to REC	Countries on track to meet Malabo commitments	Countries not on track to meet Malabo commitments
Africa	42	34	29	17	30	16	21	16	19	47	20	27
Central Africa	9	6	4	1	3	1			1	9	1	8
East Africa	10	9	7	5	8	5	5	1	4	10	6	4
Northern Africa	1	1	1							4	2	2
Southern Africa	7	3	3	2	7	1	1		1	10	6	4
West Africa	15	15	14	9	12	9	15	14	13	14	5	9

Source: Constructed by authors from Wouterse and Taffesse (2018).

Notes: CAADP = Comprehensive Africa Agriculture Development Programme; GAFSP = Global Agriculture and Food Security Program; JSR = joint sector review; NAIP = national agricultural invest plan and REC = regional economic community.

countries indicate a lack of sufficient resources to make the necessary investments to accelerate growth. The situation is also exacerbated by corruption and misuse of financial resources, in some cases representing amounts in excess of those received through development aid and foreign investments (Benin and Yu, 2013). Two factors that stand out as critical to the effective implementation of development agendas for both agriculture and education are increased agricultural R&D investment and extension services (Benin and Yu, 2013; Laborde *et al.*, 2018) and the role played by RECs (Kolavalli *et al.*, 2010; Signé, 2017).

1.4 Education in the Context of Global and African Development Goals

Transforming Africa's agriculture through Agenda 2030, Agenda 2063 and CAADP requires an adequately-skilled workforce and appropriate investment in education. Indeed, capacity development through education is deeply embedded in both the global and Africa-focused agendas. In an effort to respond to global education challenges, SDG 4 seeks to 'ensure inclusive and equitable quality education and promote lifelong learning opportunities for all' (United Nations, 2015, p. 19). Agenda 2063 sets specific targets on education which complement SDG 4 targets (Table 1.5). It is also worth noting that education was a priority in African development agendas prior to Agenda 2063 when AUC launched the First Decade of Education (1997–2006) and Second Decade of Education (2006–2015), which prioritized gender and culture, education management, teacher development, tertiary education, technical and vocational education and training, and curriculum and teaching materials.

The synergies between Agenda 2030 and Agenda 2063 can be summarized in the five themes briefly discussed below.

1. *Access to affordable, high-quality tertiary and vocational education*. The tertiary-education landscape in Africa has undergone significant expansion in the past three decades. In 1970, fewer than 400,000 tertiary students were enrolled in SSA, but by 2013 this number had increased to 7.2 million representing an annual increase of 4.3% (World Bank, 2013). This increase has driven the expansion of public universities, as well as the diversification of suppliers, including the emergence of private providers. Yet the total enrolment rate in Africa is well below rates in comparable developing states and in the global north. Challenges related to quality, affordability and access also exist and are addressed and inextricably interlinked in the global and African agendas. SDG 4.3 sets targets to ensure equal access to affordable and high-quality technical, vocational and tertiary education, including university education, by 2030. Agenda 2063 emphasizes the need to scale up investments to establish and strengthen technical and vocational education and training (TVET) centres across the continent. It also prioritizes greater access to tertiary education, including graduate education to 'ensure world-class infrastructure for learning and research and support scientific reforms that underpin the transformation of the continent' (AUC, 2015a, p. 3).

2. *Focus on skills to increase employability*. Youth unemployment is one of the biggest challenges threatening Africa's drive towards an inclusive and prosperous society. While 10–12 million youth (as previously mentioned, those aged 15–35 years) enter the workforce each year, only 3.1 million jobs are created every year (African Development Bank Group, 2016). The problem for Africa is not just unemployment, but under-employment, which affects over half the youth demographic in the labour force. As Africa's youth population increases (see Ch. 4, this volume), the need to address unemployment will only intensify. Several studies have identified the lack of key competencies and skills in graduates as a major constraint in employability (African Development Bank Group, 2016; WEF, 2017). Education for skills development must therefore receive priority. While enrolments in secondary and tertiary education have improved across Africa, vocational training - which is skills focused - has lagged (UNESCO, 2018). Moreover, the few existing programmes are often lacking in relevance and misaligned to the needs of industry and employers. Furthermore, most university programmes are also misaligned with graduate employability and the need for a focus on self-employment (African Capacity Building Foundation, 2017). The African education system, including tertiary agricultural education, therefore needs to address this lack of alignment, especially in relation to transferrable skills, such as communication, information technology, decision-making, innovation, adaptability and critical thinking, all of which can increase employability (see Ch. 9, this volume, for a detailed discussion of training for employability). These vital issues are addressed in both Agenda 2030 and Agenda 2063. SDG 4.4 explicitly focuses on skills development to increase the number of youth and adults with the relevant skills. SDG 8.5 focuses on achieving full and productive employment and decent work for all. To complement the SDGs, Aspiration 6 of Agenda 2063 specifically focuses on 'an Africa where development is people-driven, unleashing the potential of its women and youth' and commits to eliminating youth unemployment by guaranteeing full access to education, training, skills and technology for Africa's youth (AUC, 2015a, p. 2).

3. *Knowledge and skills to enable sustainable development.* In addition to enhancing skills development to increase employability, there is also an increased need to focus education at all levels and in all contexts on issues related to sustainable development. The reorientation of education to support development of the skills, knowledge and behaviours needed for sustainable development is critical given that Africa will account for more than half of global population growth by 2050 (1.3 billion people), exerting pressure on the limited natural resources (UNECA, 2017a). Moreover, evidence of a causal relationship between human activities and increased greenhouse gas emissions is overwhelming (Niang *et al.*, 2014), necessitating large-scale changes in societal behaviour. Education for sustainable development enables this by providing students with the knowledge and skills to promote sustainability in their personal and professional lives, empowering them to become agents of change in their communities. SDG 4.7 highlights the importance of education for sustainable development by establishing a target that 'all learners acquire the knowledge and skills needed to promote sustainable development' (United Nations, 2015, p. 19). Similarly, Agenda 2063 seeks to support skills development among young people to encourage their ability to be the drivers of Africa's renaissance. This will be achieved through equitable and sustainable natural resources use, the promotion of sustainable livelihoods, the appreciation of Africa's cultural diversity, and the promotion of human rights and peace.

4. *Gender equality in access to high-quality technical, vocational and tertiary education.* As previously discussed, increased access to tertiary education by women yields significant benefits. To date, however, African women's participation in tertiary education has lagged. As of 2018, the total rate of enrolment in tertiary education in SSA was 8.1% for women and 10.6% for men (UNESCO, 2020a). Moreover, compared with men, fewer women establish academic careers and occupy senior positions at African universities. Ensuring equality in education for women is highlighted explicitly in the global and African agendas. SDG 5 on gender equality seeks to 'end all forms of discrimination against women and girls everywhere' (United Nations, 2015, p. 20). Though there is no explicit reference to education in SDG 5, SDG 4.5 specifically aims to 'eliminate gender disparities in education and ensure access to all levels of education and vocational training' (United Nations, 2015, pp. 19–20). Similarly, Agenda 2063 notes that Africa's development is people-driven, especially 'relying on the potential offered by its women and youth' and explicitly focuses on the 'elimination of gender disparities at all levels of education' in strengthening Africa's human capital (AUC, 2015a, p. 3).

5. *Increased investment in infrastructure for education and availability of financing for education.* Africa has significant physical infrastructure deficits, including deficits in infrastructure for education (World Bank, 2008). SDG 4.A aims to build and upgrade education facilities to provide inclusive and effective learning environments for all, whilst Agenda 2063 explicitly aims to build and expand an African knowledge society through investments in universities, science, technology, research and innovation. Agenda 2063 further seeks to expand access to graduate education to ensure world-class infrastructure for learning and research. Achieving this will require significant investments in educational and science infrastructure. Official development assistance scholarships amounted to US$1 billion in 2015, a decline from US$1.2 billion in 2014 (OECD, 2018). In

Table 1.5. SDG 4 targets alignment with Agenda 2063 priority areas

Sustainable Development Goal 4 targets	Agenda 2063 goals and priority areas on education
4.1 By 2030, ensure that all girls and boys complete free, equitable and quality primary and secondary education leading to relevant and effective learning outcomes	Expand universal access to quality early childhood, primary and secondary education
4.2 By 2030, ensure that all girls and boys have access to quality early childhood development, care and pre-primary education so they are ready for primary education	Expand and consolidate gender parity in education
4.3 By 2030, ensure equal access for all women and men to affordable and quality technical, vocational and tertiary education, including university	Strengthen technical and vocational education and training through scaled up investments, establishment of a pool of high-quality technical and vocational education and training (TVET) centres across Africa Ensure that at least 70% of the public perceive quality improvements in education at all levels
4.4 By 2030, substantially increase the number of youth and adults who have relevant skills, including technical and vocational skills, for employment, decent jobs and entrepreneurship	Ensure that at least 70% of secondary school students not entering the tertiary sector are provided with a range of options for further skills development
4.5 By 2030, eliminate gender disparities in education and ensure equal access to all levels of education and vocational training for the vulnerable, including persons with disabilities, indigenous peoples and children in vulnerable situations	Expand and consolidate gender parity in education Ensure that at least 30% of secondary school leavers go into tertiary education with at least 40% being female
4.6 By 2030, ensure that all youth and a substantial proportion of adults, both men and women, achieve literacy and numeracy	Ensure that Africa will have 100% literacy and numeracy
4.7 By 2030, ensure that all learners acquire the knowledge and skills needed to promote sustainable development, including, among others, through education for sustainable development and sustainable lifestyles, human rights, gender equality, promotion of a culture of peace and non-violence, global citizenship and appreciation of cultural diversity and of culture's contribution to sustainable development	Ensure that Africa will have a universal culture of good governance, democratic values, gender equality, respect for human rights, justice and the rule of law Build and expand an African knowledge society through transformation and investments in universities, science, technology, research and innovation
4.A Build and upgrade education facilities that are child, disability and gender sensitive and provide safe, non-violent, inclusive and effective learning environments for all	Build and expand an African knowledge society through transformation and investments in universities, science, technology, research and innovation
4.B By 2020, substantially expand the global number of scholarships available to developing countries - in particular, least developed countries, small island developing states and African countries - for enrolment in higher education, including vocational training and information and communications technology, technical, engineering and scientific programmes, in developed countries and other developing countries	Build and expand an African knowledge society through transformation and investments in universities, science, technology, research and innovation
4.C By 2030, substantially increase the supply of qualified teachers, including through international cooperation for teacher training in developing countries, especially least developed countries and small island developing states	Harmonization of education standards and mutual recognition of academic and professional qualifications

Source: Constructed by authors from UNESCO (2021).

contrast, SDG 4.C identifies the need to expand the number of scholarships to least developed countries, including those in Africa. Agenda 2063 also stresses the importance of domestic resources to build Africa's human resources, along with the principle of self-reliance in mobilizing African financial resources to strengthen the region's educational infrastructure and support the educational aspirations of the African people.

Some African stakeholders have had success in advocating for tertiary/higher education in African development. A notable example is the Regional Universities Forum for Capacity Building in Agriculture (RUFORUM), which initiated and led the Ministerial Conference on Higher Education in Agriculture co-hosted by the Government of Uganda and other regional and international partners in 2010. Ongoing engagement and Ministerial meetings with the Chair of AUC, working with other African networks, supported the establishment of an AUC committee of ten presidents tasked with championing education, innovation, science and technology. These Ministerial conferences with academia have broadened support and also laid the foundation and momentum for several investments in higher agricultural education in Africa including the African Higher Education Centres of Excellence programme that has already invested over US$350 million in universities. These examples illustrate the potential of partnerships across the continent for raising investment in higher education. Additional partnerships will be needed to achieve the Agenda 2030 and 2063 ideals and aspirations.

The Continental Education Strategy for Africa, 2016–2025

To respond to the challenges in African education, AUC initiated an overarching education strategy, the aforementioned CESA, in alignment with Agenda 2030 and CAP (AUC, 2015b). CESA has 12 strategic objectives, two of which are of direct relevance to tertiary education. The first objective is to expand TVET opportunities at both secondary and tertiary levels and strengthen linkages between the employment world and systems of education and training. The second objective is to revitalize and expand tertiary education, research and innovation to address Africa-relevant challenges and promote global competitiveness. CESA's indicators manual will facilitate the monitoring and evaluation of the implementation of the Strategy (AUC, 2018). The specific strategies for revitalizing tertiary education include: (i) encouraging national governments to honour their commitment to allocate 1% of GDP to research and innovation, (ii) improving the provision of infrastructure and resources and (iii) consolidating and expanding Centres of Excellence (AUC, 2015b). CESA further calls for the expansion of graduate education, but as a high-level pan-African framework, the Strategy provides little nuanced insight on how to go about doing so. This gap was later addressed by the aforementioned AESIF, which is discussed in the next section.

Technical and vocational education and training. CESA's strategy aims to create a paradigm shift in the traditional perception of TVET institutions as refuges for those who failed in general education (Psascharopoulos, 1991; Oketch, 2007). Another dominant perception, that TVETs are 'neither an efficient nor effective policy response to Africa's educational developmental challenges' (McGrath, 2011, p. 35; Palmer, 2007), also needs to be countered. Renewed policy and donor support for TVETs in Africa over the past few decades is intended to prepare students to provide practical services for a productive economy (Finch and Crunkilton, 1999; McGrath, 2012). CESA aims to increase access to TVETs by women and youth, mostly in rural areas. Given their focus on practical and in-field training, TVETs are critical for transforming the agricultural sector because they provide lifelong learning opportunities to lower- and middle-level agricultural professionals, such as agricultural extension officers. This form of training needs to be strengthened, updated using new information technologies and enhanced so it promotes agricultural entrepreneurship and innovation through incubation and research and development (R&D). TVETs, including agricultural colleges, are often administered by the relevant government/ministerial department rather than falling under the umbrella of higher education systems (ASSAf, 2017). As a result, limited linkages have existed between research, teaching and extension (ASSAf, 2017). To address this challenge, CESA aims to establish a coalition of education stakeholders, including state and private actors in the agriculture sector, to jointly identify and develop strategic initiatives. In an effort to strengthen the institutional capacity of TVETs in particular, CESA also seeks to build and rehabilitate education infrastructure and update learning and training facilities, especially in rural and other underserved areas.

University education. The overall strategy is to reorient enrolments, graduate education, research and innovation to interlink them with key economic and industrial sectors, such as agriculture. If the region is to develop competitive agricultural value chains and innovative products, Africa's tertiary and research institutions need to generate high-quality research outputs and innovations beyond the mere 1% of the world's research it was responsible for as of 2018 (Ngongalah *et al.*, 2018).

The Agricultural Education and Skills Improvement Framework

In Africa, agriculture is often perceived as a 'last resort subject' for under-achievers (MIJARC *et al.*, 2012). Not surprisingly, the total number of post-secondary agricultural enrolments, and the quality of students, have declined across all levels over time (ILO, 2018). Traditional training systems are outdated and unsuited to end users' needs, which are gradually shifting towards value addition, a private-sector market-driven orientation and resilience to climate change.

During the first decade of CAADP's implementation, AET actors and institutions were left without concrete guidance and support to adapt their training programmes to provide skills and competences to meet evolving agricultural demands (NEPAD, 2015). To respond to this and other AET challenges, the 2014 Malabo Declaration in support of CAADP took steps to 'enhance agricultural education, skills development and knowledge support' (NEPAD, 2015, p. 4). A key strategic area focusing on skills, knowledge and agricultural education was dedicated to support the CAADP's implementation. To put this into action, AESIF (2015–2025) was developed to stimulate and empower TAE institutions. AESIF was positioned as an integrated policy framework encompassing (i) TAE acquired through institutions conferring certificates, diplomas or (undergraduate and graduate) degrees, (ii) agricultural technical and vocational education and training (ATVET) and (iii) informal learning acquired through active practice, on-the-job training or other forms of in-field learning/training. Through these three levels, AESIF's main aim is to transform AET in order to leverage the quality and quantity of skilled workers needed to generate agricultural growth and achieve the Malabo commitments. In particular, AESIF is intended to address key challenges in AET including (i) misalignment between the type of training provided and the skills required in the workforce, (ii) fragmentation of knowledge actors and isolation of AET from research and extension services, (iii) duplication and inefficient allocation of resources and (iv) poor perceptions of, and hence lack of interest in, AET among women and youth (NEPAD, 2015). The recognition that informal training, education and lifelong learning also occur in everyday life in families and the workplace is significant. This supports informal learning, including through traditional knowledge systems, as an important contributing element to educational development in Africa. Continuous learning through informal education channels is critical to the successful adoption of agriculture innovations through extension services.

Building on the lessons learned from the first decade of CAADP, AESIF identified three strategic areas of action to improve AET efficiency: (i) governance and management policies, (ii) teaching and learning systems and (iii) mobilization of public and private partnerships. The lack of coordination and strategic alignment with national development priorities - in particular, those related to national agricultural priorities - resulted in AESIF recommending the establishment of national AET councils which would be responsible for coordinating and driving the entire national AET system. To match the training and skills requirements in the job market, AESIF seeks to reform and update curricula and teaching methods. The last priority, related to public–private partnerships, is driven by the need to mobilize innovative and sustainable financing from the private sector.

1.5 Science, Technology and Innovation in the Context of African Agriculture

Efforts to transform African agriculture also need to consider the importance of STI in TAE and Africa's broader development. Agenda 2063 and CAP in Agenda 2030 highlight STI as a key enabler of African countries' ability to achieve economic transformation and development. Despite growing emphasis on the STI for development, significant resource and infrastructure constraints, as well as human resource capacity barriers, hinder African countries from leveraging STI to promote national development (African Capacity Building Foundation, 2017). To some extent, the STI lag is linked to African countries' investment priorities, whereby political commitments have yet to be converted into practical programmes. As of 2018, average spending on R&D in Africa represented 0.5% of GDP - less than the expected 1%. Gross domestic expenditure on R&D has risen in several countries, however (UNESCO, 2020b).

Many African countries have not integrated STI in their national development plans (Chaminuka *et al.*, 2019), and although AUC has made strides in improving STI policy and investment in R&D, more needs to be done (Juma and Serageldin, 2016). Encouragingly, Africa's main education strategy, CESA 16–25, prioritizes STI to strengthen 'science and math curricula and dissemination of scientific knowledge and the culture of science in the African society' as its seventh strategic objective (AUC, 2015b, p. 8). African governments have attempted to drive development through efforts to mainstream STI in the region's development policies and actions, including the Monrovia Strategy (1979), the Lagos Plan of Action (1980), the Abuja Treaty (1991) and the African Manifesto for STI. The Manifesto was released in 2010 by the African Technology Policy Studies Network in collaboration with African, Indian and European partners and stakeholders under the auspices of a European VII Framework programme, Science Ethics and Technological Responsibilities in Developing and Emerging Countries.

In 2014, the African heads of state established the aforementioned STISA to 'accelerate Africa's transition to an innovation-led, knowledge-based economy' and build on the Consolidated Plan of Action for Science and Technology in Africa adopted in 2007 (AUC, no date, p. 11). The strategy is part of the long-term people-centred African Union Agenda which emphasizes the benefits of STI as tools and enablers of the region's development goals in critical sectors, including agriculture. STISA is the first phase of a ten-year strategy that aims to enhance the implementation of STI programmes across the continent by increasing technical skills, institutional capacity and economic competitiveness through the promotion of entrepreneurship, innovation and improvements in research and innovation infrastructure. STISA identifies eradicating hunger and achieving food security among its six priority areas. The strategy seeks to use STI to build the region's capacity to tackle emerging agricultural challenges, such as low commodity yields, climate change and variability and water and land management. This indicates that many of the challenges associated with ensuring sustainable agriculture, as well as food and nutrition security, cannot be adequately addressed without the participation of the research community and the application of science. The centrepiece of STISA is R&D flagship programmes and projects that will be developed to effectively address each of the priority areas and the interlinkages among them. In addition, transformative programmes and projects will be developed periodically by the scientific community through the Scientific Research and Innovation Council platform, an organization approved by AUC in 2014 to help facilitate the advancement of science and technology in Africa.

Various initiatives have also been launched to support and complement STISA. These include the African Science, Technology, and Innovation Fund, aimed at sustaining STI programmes and promoting technology-based entrepreneurship, as well as the African Institutes of Science and Technology, which promote linkages among universities and research institutions to improve the quality of tertiary education in science and engineering. Evidence shows that sound and sustained investment in agricultural science and technology is key to increasing agricultural productivity, reducing poverty and fostering economic growth (Glover *et al.*, 2016; Juma and Serageldin, 2016). Acknowledging this link, the Science Agenda for

Agriculture in Africa (S3A) was developed to support the implementation of strategies linked to the agriculture and food security component of STISA. The S3A represents decisive efforts to build and strengthen the capacity required to put STI to work for agriculture in Africa. S3A is aligned with CAADP's targets under the Sustaining the CAADP Momentum strategy, and its implementation is intended to advance the attainment of the CAADP targets. The Agenda has three core elements: (i) integrating science more effectively, (ii) connecting science to end users with greater impact, especially for CAADP at national and regional levels and (iii) strengthening the sciences and their application to agriculture to ensure that the agricultural total factor productivity can be doubled during the period 2015–2025.

Given Africa's inherent heterogeneity, including the complex nature of the African climate systems and crop and livestock production, S3A's implementation will need to be varied across nations so as to develop a series of differentiated 'agricultural revolutions' to suit the region's varied ecological niches (Gengenbach *et al.*, 2018). Nevertheless, critical aspects of S3A can be broadly applied across the continent. These aspects relate to the strengthening institutions, increasing the availability and affordability of improved inputs, expanding high-quality rural infrastructure, creating incentives for the uptake of technologies and ensuring adequate and timely supply of information to support decision-making (Juma, 2015).

STISA and S3A highlight the political will and recognition by African leaders that STI is indispensable as a long-term component of Africa's transformation process. This commitment has most recently been affirmed by AUC in the establishment of the Committee of Ten Heads of State to Champion Education, Science and Technology (C10) - Assembly Decision AU/Dec.671 and Assembly/AU/Dec 572 - to support greater use of STI for development and to accelerate the implementation STISA. C10 comprises two voluntary heads of state from each of Africa's geographical subregions. C10's first meeting was held in November 2018 (in Malawi) and resulted in the development of a declaration and action plan for education, science and technology to facilitate the implementation of both CESA and STISA.

The role of STI in agriculture is even greater, given the new and more complex challenges associated with fragile ecosystems and complex agricultural production systems. For agriculture in Africa to be more productive, competitive, sustainable and inclusive, scientific solutions need to be pursued through an integrated approach as outlined in Agenda 2030 (particularly, SDG 17 on partnerships) and Agenda 2063's inclusive vision.

1.6 Implications for Tertiary Agricultural Education

The discussion presented in this chapter has highlighted recognition by both the global Agenda 2030 and the pan-African Agenda 2063 that sustainable development across Africa is strongly linked with the growth and promotion of a vibrant and sustainable agricultural sector. This has significant implications for TAE, which are summarized below.

Adequate Numbers of Appropriately-Trained Graduates

Central to the desired agricultural transformation in Africa is the need to prepare professionals capable of leading change, with tertiary education playing a critical role (Chakeredza *et al.*, 2008). CAADP's ten-year review showed that Africa's capacity to generate knowledge, foster learning and enable skills development in its workforce is critical to transforming the agricultural sector. Yet, the sector is still struggling with deficits in human resource capacity (NEPAD, 2015). A systematic and coordinated approach is required to popularize agriculture through TAE and STI at both secondary and tertiary education levels to ensure adequate numbers of graduates with relevant skills.

A Diverse, Multilayered and Vibrant Tertiary Agricultural Education System

The tertiary system of agricultural education in many African countries is fragmented and requires significant institutional reform and stimulation of innovation (ASSAf, 2017). Africa-wide, CESA 16–25 aims to instigate a paradigm shift in both universities and TVETs to address capacity challenges. Importantly, CESA 16–25 seeks to increase access to TVETs, as well as revitalize tertiary education, research and innovation. This two-pronged approach is highly relevant to TAE. Specific to agriculture, AESIF is a direct response to critical gaps in human resource capacity. It provides a clear vision and agenda to empower AET, particularly through institutional reform of higher learning institutions.

Updated and Relevant Curricula

Tertiary agricultural training systems used to train the agricultural workforce for decades are seriously outdated. Agriculture is now a complex, multifaceted industry requiring knowledge of sustainable farming practices, STI and the globalized commodity market. The current system of TAE training creates a mismatch between what students are taught and what the industry requires. In response to these systemic challenges in AET, AESIF emphasizes reform and updates of agricultural training approaches and curricula, taking into consideration the dynamic shift the sector has undergone in recent decades. The development of adequately skilled graduates of agricultural and related disciplines could effect a fundamental change to the future of agriculture across Africa. New graduates must possess the knowledge and experience to develop and manage policies and work with multiple stakeholders in agribusiness, while contributing to sustainable development. Emphasis is also needed on imparting practical, entrepreneurial skills and strengthening creativity and flexibility. This enables graduates of agriculture not only to be employable, but also to become job creators able to harness new opportunities and adapt to change.

STI is a key driver of agricultural transformation, but more capacity is required. Both of the global agendas discussed in this chapter recognize the need to increase investment in research, innovation and applied technology for agriculture. STISA and S3A recognize that agricultural transformation relies on the catalytic power of STI to increase agricultural productivity, competitiveness, sustainability and inclusivity. Nevertheless, increasing the contribution of STI to agriculture requires a significant shift in the capacities and skills of those engaged in the sector. The situation is dire at tertiary agricultural institutions, including TVETs and universities, which have too few highly-qualified faculty (Spielman *et al.*, 2008; ASSAf, 2017; Kirui and Kozicka, 2018). The ability of African educators and faculty to teach secondary- and tertiary-level agriculture, and to carry out high-quality research, is of serious concern (Walker and Hofstetter, 2016; ASSAf, 2017). To address these capacity limitations, STISA emphasizes expanding the availability of high-quality graduate education and, in particular, programmes leading to doctoral qualifications. Educating highly-qualified researchers also requires upgrades to science laboratories and the establishment of world-class STI infrastructure. STISA also urges tertiary institutions, including tertiary agricultural institutions, to adopt multidisciplinary and multisector approaches to collaboration, 'open-innovation[2]' and entrepreneurship with public and private stakeholders. The co-creation, adaptation and commercialization of research and innovation is critical in the context of a complex, multifaceted sector, such as agriculture.

Increased Focus on Systems and Multidisciplinarity

Given the interconnectedness of agricultural value chains and the value of co-creating knowledge, it is of paramount importance that TAE curricula and agricultural research adopt a food-systems perspective in the context of the bioeconomy. The breadth of the sustainability challenge (as it intersects with agrifood systems) requires greater interdisciplinary thinking than has been the norm and requires significant co-operation across the African continent among TAE institutions.

Solutions Need To Be Owned and Led by Africa

It is evident that Africa has relevant strategies in place that can be expected to yield high impact if implemented consistently and backed by appropriate political will and financial investment by African governments and institutions. African leaders at all levels must take the responsibility, and be held accountable, for establishing higher-education institutions capable of envisioning, conceptualizing, strategizing and crafting scientific and innovation models that solve the region's agricultural challenges and equip the next generation with the skills they need.

Endnotes

[1] Throughout this chapter, tertiary - or higher - agricultural education refers to post-secondary school (i.e. diploma and degree) studies in agriculture and related fields, such as aquaculture, fisheries, forestry, etc.
[2] In contrast to traditional business models based on secrecy, open innovation refers to practices which promote collaboration beyond individual institutions in order to foster an innovative, entrepreneurial culture.

References

African Capacity Building Foundation (2017) *Africa Capacity Building Report 2017: Building Capacity for Science, Technology and Innovation for Africa's Transformation*. Available at: https://www.africaportal.org/publications/africa-capacity-report-2017-building-capacity-in-science-technology-and-innovation-for-africas-transformation/ (accessed 22 February 2021).

African Development Bank Group (2016) *Jobs for Youth in Africa: Catalyzing Youth Opportunity Across Africa*. Available at: https://www.afdb.org/fileadmin/uploads/afdb/Images/high_5s/Job_youth_Africa_Job_youth_Africa.pdf (accessed 22 February 2021).

African Development Bank Group (2018) *Feed Africa: The High Five for Transforming Africa*. Available at: https://www.afdb.org/fileadmin/uploads/afdb/Documents/Generic-Documents/Brochure_Feed_Africa_-En.pdf (accessed 22 February 2021).

African Green Revolution Forum (2018) New report finds ten years on, African countries that embraced agriculture saw food production, GDP and nutrition all improve. (7 June) Available at: https://agrf.org/new-report-finds-ten-years-on-african-countries-that-embraced-agriculture-saw-food-production-gdp-and-nutrition-all-improve/ (accessed 22 February 2021).

ASSAf (Academy of Science of South Africa) (2017) *Revitalizing Agricultural Education and Training in South Africa*. Available at: http://research.assaf.org.za/handle/20.500.11911/85 (accessed 22 February 2021).

AUC (African Union Commission) (no date) *Science, Technology and Innovation Strategy for Africa 2024*: STISA–2024. Available at: https://au.int/sites/default/files/documents/29957-doc-stisa-published_book.pdf (accessed 22 February 2021).

AUC (2014a) *Malabo Declaration on Accelerated Agriculture Growth and Transformation for Shared Prosperity and Improved Livelihoods*. Available at: https://www.au.int/web/sites/default/files/documents/31247-doc-malabo_declaration_2014_11_26.pdf (accessed 22 February 2021).

AUC (2014b) *Agenda 2063 Linkages with the Sustainable Development Goals*. Available at: https://au.int/sites/default/files/documents/33126-doc-07_linkage_with_the_sdg.pdf (accessed 3 May 2021)

AUC (2015a) *Agenda 2063: The Africa We Want*. Available at: https://www.un.org/en/africa/osaa/pdf/au/agenda2063.pdf (accessed 22 February 2021).

AUC (2015b) *Continental Strategy for Africa 2016–2025*. Available at: https://au.int/sites/default/files/documents/29958-doc-cesa_-_english-v9.pdf (accessed 22 February 2021).

AUC (2018) *Inaugural Biennial Report of the Commission on the Implementation of the June 2014 Malabo Declaration on Accelerated Agricultural Growth and Transformation for Shared Prosperity and Improved Livelihoods and the 2017 Progress Report to the Assembly*. Department of Rural Economy and Agriculture. Available at: https://au.int/sites/default/files/documents/34698-doc-33640-rp-33640-wd-full_br_report_eng.pdf (accessed 3 May 2021).

AUC (African Union Commission), UNECA (United Nations Economic Commission for Africa), AfDB (African Development Bank) and UNDP (United Nations Development Programme) (2017) *Africa Sustainable Development Report: Tracking Progress on Agenda 2063 and the Sustainable Development Goals*.

Available at: https://www.afdb.org/en/documents/document/2017-africa-sustainable-development-report-tracking-progress-on-agenda-2063-and-the-sustainable-development-goals-executive-summary-99312 (accessed 28 May 2021).

Anderson, K. (2010) Globalization's effects on world agricultural trade, 1960–2050. *Philosophical Transactions of the Royal Society* 365(1554), 3007–3021.

Balz, A. Heil, E.A. and Jordan, I. (2015) Nutrient-sensitive agriculture: A new term or new concept. *Agriculture and Food Security* 4(6), 1–16.

Beegle, K. Christiaensen, L. Dabalen, A. and Gaddis, I. (2016) *Poverty in a Rising Africa: Africa Poverty Report*. World Bank, Washington, DC. Available at: https://openknowledge.worldbank.org/handle/10986/22575 (accessed 22 February 2021).

Benin, S. and Yu, B. (2013) *Complying with the Maputo Declaration Target: Trends in Public Agricultural Expenditures and Implications for Pursuit of Optimal Allocation of Public Agricultural Spending*. ReSAKSS Annual Trends and Outlook Report 2012. International Food Policy Research Institute. Available at: http://www.ifpri.org/publication/complying-maputo-declaration-target-trends-public-agricultural-expenditures-and (accessed 22 February 2021).

Benin, S. (2016) *Impacts of CAADP on Africa's Agriculture-Led Development*. IFPRI Discussion Paper 1553. International Food Policy Research Institute, Washington, DC.

Benson-Wahlén, C. (2018) African Statisticians Reflect on Data Needs for SDGs, Agenda 2063. IISD/SDG Knowledge Hub, 9 October. Available at: https://sdg.iisd.org/news/african-statisticians-reflect-on-data-needs-for-sdgs-agenda-2063/ (accessed 22 February 2021).

Brody, A. (2015) *Gender and Food Security*. BRIDGE Cutting Edge Programmes. In Brief. UK International Development Studies. Available at: http://www.wocan.org/sites/default/files/Gender%20and%20FS%20In%20Brief.pdf (accessed 22 February, 2021).

Bunch, S. and Mehra, R. (2008) *Women Help Solve Hunger. Why is the World Still Waiting?* International Center for Research on Women. Available at: https://www.icrw.org/wp-content/uploads/2016/10/Women-Help-Solve-Hunger.pdf (accessed 22 February, 2021).

Burney, J.A. Davis, S. J., and Lobell, D.B. (2010) *Greenhouse Gas Mitigation by Agricultural Intensification. Proceedings of the National Academy of Sciences of the United States of America (PNAS)* 107, 12052–12057. Available at: https://www.pnas.org/content/pnas/107/26/12052.full.pdf (accessed 22 February 2021).

Campbell, B.M. Vermeulen, S. Aggarwal, P.K. Corner-Dolloff, C. Girvetz, E. *et al.* (2016) Reducing risks to food security from climate change. *Global Food Security*. Available at: http://dx.doi.org/10.1016/j.gfs.2016.06.002 (accessed 22 February 2021).

Chakeredza, S. Temu, A.B. Saka, J.D.K. Munthali, D.C. Muir-Leresche, K. *et al.* (2008) Tailoring tertiary agricultural education for sustainable development in sub-Saharan Africa: Opportunities and challenges. *Scientific Research and Essay* 3, 326–332.

Chaminuka, P. Beintema, N. Flaherty, K. and Liebenberg, F. (2019) Public agricultural research and development spending in South Africa: Update. *Agricultural Economics Research, Policy and Practice in Southern Africa* 58(1), 7–20. Available at: https://doi.org/10.1080/03031853.2018.1550427 (accessed 22 February 2021).

Christiaensen, L. and Kaminski, J. (2016). *Structural Change, Economic Growth and Poverty Reduction: Micro-Evidence from Uganda*. African Development Bank Working Paper 229. Available at: https://www.afdb.org/fileadmin/uploads/afdb/Documents/Publications/WPS_No_229_UG_micro-level_structural_transformation_and_poverty__B.pdf (accessed 22 February 2021).

Connolly-Boutin, L. and Smit, B. (2016) Climate change, food security, and livelihoods in Sub-Saharan Africa. *Regional Environmental Change* 16(2): 385–399.

De Janvry, A. and Sadoulet, E. (2017) *Achieving the Sustainable Development Goal on Poverty Eradication*. FERDi Working Paper 190. Available at: https://are.berkeley.edu/esadoulet/wp-content/uploads/2018/10/achieving_ferdi.pdf (accessed 22 February 2021).

Development Initiatives (2018) *The Global Nutrition Report 2018: Shining a Light to Spur Action on Nutrition*. Available at: https://www.who.int/nutrition/globalnutritionreport/2018_Global_Nutrition_Report.pdf?ua=1 (accessed 22 February 2021).

FAO (Food and Agriculture Organization of the United Nations) (2011) *The State of Food and Agriculture 2010–2011. Women in Agriculture: Closing the Gender Gap for Development*. Available at: http://www.fao.org/3/i2050e/i2050e00.htm (accessed 22 February 2021).

FAO (2018a) *Transforming Food and Agriculture to Achieve the SDGs: 20 Interconnected Actions to Guide Decision-Makers*. Available at: http://www.fao.org/3/I9900EN/i9900en.pdf (accessed 22 February 2021).

FAO (2018b) *The State of World Forests: Forest Pathways to Sustainable Development*. In Brief. Available at: http://www.fao.org/policy-support/tools-and-publications/resources-details/en/c/1144279/ (accessed 22 February 2021).

FAO (2018c) *Leaving No One Behind: Empowering Africa's Rural Women for Zero Hunger and Shared Prosperity*. Available at: http://www.fao.org/3/CA1506EN/ca1506en.pdf (accessed 22 February 2021).

FAO and UNECA (2018) *Regional Overview of Food Security and Nutrition. Addressing the Threat from Climate Variability and Extremes for Food Security and Nutrition*. Available at: http://www.fao.org/fsnforum/resources/fsn-resources/regional-africa-overview-food-security-and-nutrition (accessed 22 February 2021).

Finch, C.R. and Crunkilton, J.R. (1999) *Curriculum Development in Vocational and Technical Education. Planning, Content, and Implementation* (5th ed.) Allyn and Bacon, Needham Heights.

Gbegbelegbe, S. Serem, J. Stirling, C. Kyazze, F. Radeny, M. (2017) Smallholder farmers in eastern Africa and climate change: A review of risks and adaptation option with implications for future adaptation programmes. *Climate and Development* 10(4), 289–306. Available at: https://ccafs.cgiar.org/resources/publications/smallholder-farmers-eastern-africa-and-climate-change-review-risks-and (accessed 22 February 2021).

Gengenbach, H. Schurman, R.A. Basset, T.J. Munro, W.A. and Moseley, W.G. (2018) Limits of the new green revolution for Africa: Reconceptualising gendered agricultural value chains. *The Geographical Journal* 184(2), 208–214. *Available at*: https://doi.org/10.1111/geoj.12233 (accessed 22 February 2021).

Glover, D. Sumberg, J. and Andersson, J. (2016) The adoption problem; or why we still understand so little about technological change in African agriculture. *Outlook on Agriculture* 45(1), 3–6.

Harvey, C.A. Rakotobe, Z.L. Rao, N.S. Dave, R. Rzafimahatratra, *et al.* (2014) Extreme vulnerability of smallholder farmers to agricultural risks and climate change in Madagascar. *Philosophical Transactions of the Royal Society*, 8 (369), 1639–1644. Available at: https://pubmed.ncbi.nlm.nih.gov/24535397/ (accessed 22 February 2021).

Hilderink, H. Brons, J. Ordonez, J. Akinyoade, A. Ieliveld, A. *et al.* (2012) *Food Security in Sub-Saharan Africa: An Explorative Study*. Netherlands Environmental Assessment Agency, The Hague.

ILO (International Labour Organization) (2018) ILOSTAT Database: Employment in Agriculture. Available at: https://data.worldbank.org/indicator/sl.agr.empl.zs (accessed 22 February 2021).

Intergovernmental Panel on Climate Change (2007). Summary for policymakers. In: Parry, M.L. Canziani, O.F. Palutikof, J.P. van der Linden, P.J. and Hanson, C.E. (eds.) (2007) *Climate Change: Impacts, Adaptation and Vulnerability*. Contribution of Working Group II to the Fourth Assessment Report of the Intergovernmental Panel on Climate Change. Cambridge University Press, Cambridge, UK.

Juma, C. (2015) *The New Harvest: Agricultural Innovation in Africa* (2nd ed.) Oxford University Press, New York.

Juma, C. and Serageldin, I. (2016) *Rebooting African Development: Science, Technology and Innovation Strategy for Africa* Belfer Centre for Science and International Affairs, Harvard Kennedy School, Harvard University, Cambridge, Massachusetts, USA.

Kirui, O.K. and Kozicka, M. (2018) *Vocational Education and Training for Farmers and Other Actors in the Agri-Food Value Chain in Africa*. ZEF Working Paper Series 164. University of Bonn, Center for Development Research, Bonn.

Kolavalli, S. Flaherty, K. Al-Hassan, R.M. and Baah, K.O. (2010) *Do CAADP Processes Make a Difference to Country Commitments to Develop Agriculture? The Case of Ghana.* IFPRI Discussion Paper 1006. International Food Policy Research Institute. Available at: http://www.ifpri.org/publication/do-caadp-processes-make-difference-country-commitments-develop-agriculture-case-ghana (accessed 22 February 2021).

Laborde, D. Lallemant, T. McDougal, K. Smaller, C. and Traore, F. (2018) *Transforming Agriculture in Africa & Asia: What Are the Policy Priorities?* Report for the International Institute for Sustainable Development and the Food Policy Research Institute. Available at: https://www.iisd.org/sites/default/files/publications/transforming-agriculture-africa-asia.pdf (accessed 22 February 2021).

Lalani, B. Dorward, P. Holloway, G. and Wauters, E. (2016) *Smallholder Farmers' Motivations for Using Conservation Agriculture and the Roles of Yield, Labour and Soil Fertility in Decision Making. Agricultural Systems*. Available at: https://core.ac.uk/download/pdf/74244323.pdf (accessed 22 February 2021).

Leavy, J. and Hossain, N. (2014) *Who Wants to Farm? Youth Aspirations, Opportunities and Rising Food Prices*. Institute of Development Studies Working Paper 439. Available at: https://opendocs.ids.ac.uk/opendocs/bitstream/handle/123456789/3550/Wp439r.pdf?sequence=4 (accessed 22 February 2021).

Lee, J. Gereffi, G. and Beauvais, J. (2010) *Global Value Chains and Agrifood Standards: Challenges and Possibilities for Smallholders in Developing Countries. Proceedings of the National Academy of Sciences of the United States of America (PNAS)* 109 (31), 12326–12331. Available at: https://www.pnas.org/content/pnas/109/31/12326.full.pdf (accessed 22 February 2021).

Liliana, H. (2005) *The Food Gap.The Impacts of Climate Change on Food Production: A 2020 Perspective*. Universal Ecological Fund, Alexandria, Virginia, USA. https://www.eenews.net/assets/2011/01/19/document_cw_02.pdf (accessed 22 February 2021).

Lipper, L. Thornton, P. Campbell, B.M. Baedeker, T. Braimoh, T. *et al.* (2014) Climate-smart agriculture for food security. *Nature for Climate Change* 4, 1068–1072.

Makombe, T. and Benin, S. (2018) *Is CAADP Having the Desired Impact on Agricultural-Led Development? International Food Policy Research Institute*. Available at: https://egyptssp.ifpri.info/2018/03/11/is-caadp-having-the-desired-impact-on-agricultural-led-development/ (accessed 22 February 2021).

Makombe, T., Tefera, W. and Benin, S. (2018) Tracking key CAADP indicators and implementation processes. In: *Boosting Growth to End Hunger by 2025: The Role of Social Protection*, Wouterse, F. and Taffesse, A.S. (eds.) Regional Strategic Analysis and Knowledge Support System (RESAKSS) Annual Trends and Outlook Report International Food Policy Research Institute, Washington, DC.

Maruping, A.M. (2017) *African Agenda 2063: What Progress since Adoption in 2015?* Available at: https://au.int/sites/default/files/newsevents/mediaadvisories/31932-ma-african_agenda_2063_where_are_we_-_towards_implementation_23_01_2017.pdf (accessed 22 February 2021).

McGrath, S. (2011) Where to now for vocational education and training in Africa? *International Journal of Training Research* 9, 35–48. Available at: https://eric.ed.gov/?id=EJ931057 (accessed 22 February 2021).

McGrath, S. (2012) Vocational education and training for development: A policy in need of a theory? *International Journal of Educational Development* 32(5), 623–631. Available at: https://doi.org/10.1016/j.ijedudev.2011.12.001 (accessed 22 February 2021).

McKinsey Global Institute (2010) *Lions on the Move: The Progress and Potential of African Economies*. Available at: https://www.mckinsey.com/featured-insights/middle-east-and-africa/lions-on-the-move (accessed 22 February 2021).

MIJARC/IFAD/FAO (2012) *Summary of the findings of the project implemented by MIJARC in collaboration with FAO and IFAD: 'Facilitating access of rural youth to agricultural activities'*. The Farmers' Forum Youth session, 18 February. Available at: http://www.ifad.org/farmer/2012/youth/report.pdf (accessed 22 February 2021).

Morton, J.F. (2007) *The Impacts of Climate Change on Smallholder and Subsistence Agriculture. Proceedings of the National Academy of Sciences of the United States of America (PNAS)* 104, 19680–19685. Available at: https://www.pnas.org/content/pnas/104/50/19680.full.pdf (accessed 22 February 2021).

NEPAD (2003) Comprehensive Africa Agriculture Development Programme. Available at: https://library.faraafrica.org/wp-content/uploads/2019/11/Comprehensive-Africa-Agriculture-Development-Programme.pdf (accessed 22 February 2021).

NEPAD (2015) *Agricultural Education and Skills Improvement Framework (AESIF) 2015–2025*. NEPAD, Pretoria, South Africa. Available at: http://www.psetresearchrepository.dhet.gov.za/sites/default/files/documentfiles/505.%20agriculture_education_and_skills_improvement_framework_aesif_-_web%20%283%29.pdf (accessed 22 February 2021).

Ngongalah, L. Emerson, W. Rawlings, N.N. and Musisi, J.M. (2018) *Research Challenges in Africa: An Exploratory Study on the Experience and Opinions of African Researchers*. Available at: https://www.biorxiv.org/content/biorxiv/early/2018/10/17/446328.full.pdf (accessed 22 February 2021).

Niang, I. Ruppel, O.C. Abdrabo, M.A. Esse, I A. Lennard, C. *et al.* (eds.) (2014) Climate Change: Impacts, Adaptation, and Vulnerability. Part B: Regional Aspects. Contribution of Working Group II to the Fifth Assessment Report of the Intergovernmental Panel on Climate Change. Cambridge University Press, Cambridge, Massachusetts, USA.

O'Brien, K. Leichenko, R.M. Kelkar, U. and Venema, H.D. (2004) Mapping vulnerability to multiple stressors: Climate change and globalization in India. *Global Environmental Change* 14, 303–313.

OECD (Organization for Economic Cooperation and Development) (2018) *Aid to the Education Sector*. Available at: http://www.oecd.org/dac/financing-sustainable-development/ (accessed 3 May 2021).

Oketch, M (2007) To vocationalise or not to vocationalise? Perspectives on current trends and issues in Technical and Vocational Education and Training (TVET) in Africa. *International Journal of Educational Development* 27 (2), 220–234.

Palmer, R. (2007) Skills for work? From skills development to decent livelihoods in Ghana's rural informal economy. *International Journal of Educational Development* 27, 397–420.

Pingali, P. (2012) *Green Revolution: Impacts, Limits and the Path Ahead. Proceedings of the National Academy of Sciences of the United States of America (PNAS)* 109(31), 12302–12308. Available at: https://www.pnas.org/content/pnas/109/31/12302.full.pdf (accessed 22 February 2021).

Psacharopoulos, G. (1991) Vocational education theory VOCED 101: Including hints for 'Vocational Planners'. *International Journal of International Development* 11(3), 193–199.

Roland, J. Buresh, P. Sanchez, A. and Calhoun, F. (eds.) (1997) *Replenishing Soil Fertility in Africa: SSSA Special Publication Number 51*. Soil Science Society of America and American Society of Agronomy. Madison, Wisconsin, USA. Available at: http://soil5813.okstate.edu/Africa/Replenishing_soil_fertility_in_africa.pdf (accessed 1 June 2021).

Ruel, M.T. Quisumbing, A.R. and Balagamwala, M. (2018) Nutrient-sensitive agriculture: What have we learned so far? *Global Food Security* 17(2), 128–153.

Signé, L. (2017) The quest for food security and agricultural transformation in Africa: Is the CAADP the answer? Africa in Focus, Brookings, 16 October. Available at: https://www.brookings.edu/blog/africa-in-focus/2017/10/16/the-quest-for-food-security-and-agricultural-transformation-in-africa-is-the-caadp-the-answer/ (accessed 22 February 2021).

Sissoko, F. (2017) Agenda 2030 monitoring: Contributions from UNECA on the sources of data for the indicators. Presentation at the ITUC-Africa ATUNDN Steering Committee Meeting, 20–21 June, Abudja, Nigeria. Available at: https://www.ituc-csi.org/IMG/pdf/atudn_sc_abuja_2017_agenda_2030_monitoring_uneca_sissoko.pdf (accessed 22 February 2021).

Spielman, D. Ekboir, J. Davis, K. Ochieng, C.M.O. (2008) An innovation systems perspective on strengthening agricultural education and training in Africa. *Agricultural Systems* 98, 1–9.

Swanepoel, F., Ofir, Z. and Stroebel, A. (2014) *Towards Impact and Resilience: Transformative Change in and through Agricultural Education and Training in Sub-Saharan Africa*. Cambridge Scholars Publishing, Newcastle upon Tyne UK.

Thornton, P.K., Jones, P.G., Ericksen, P.J., Challinor, A.J. (2011) Agriculture and food systems in sub-Saharan Africa in a 4 C+ world. *Philosophical Transactions of the Royal Society of London A* 369(1934), 117–136.

Tosun, J. and Leininger, J. (2017) Governing the interlinkages between the Sustainable Development Goals: Approaches to attain policy integration. *Global Challenges* 1(9), 1–12. Available at: https://doi.org/10.1002/gch2.201700036 (accessed 22 February 2021).

Trienekens, J. H. (2011) Agricultural value chains in developing countries: A framework for analysis. *International Food and Agribusiness Management Review* 14(2), 51–82.

UNECA (United Nations Economic Commission for Africa) (2017a) *2017 Africa Sustainable Development Report: Tracking Progress on Agenda 2063 and the Sustainable Development Goals*. Available at: https://repository.uneca.org/handle/10855/24022 (accessed 22 February 2021).

UNECA (2017b) *Integrating Agenda 2063 and 2030 Agenda for Sustainable Development into national development plans*. Available at: https://www.uneca.org/sites/default/files/PublicationFiles/integrating-agenda-2063-and-2030-agenda-for-sustainable-development-into-national-development-plans_en.pdf (accessed 22 February 2021).

UNESCO (United Nations Educational, Scientific and Cultural Organization) (2018) *One in Five Children, Adolescents and Youth Is Out of School*. UNESCO Institute for Statistics Fact Sheet 28. Available at: http://uis.unesco.org/sites/default/files/documents/fs48-one-five-children-adolescents-youth-out-school-2018-en.pdf (accessed 22 February 2021).

UNESCO (2020a) UNESCO Institute for Statics data on SDG 4.3.2: Gross attendance ratio for tertiary education, by sex and location. Available at: http://data.uis.unesco.org (accessed 3 May 2021).

UNESCO (2020b) News: New UIS data for SDG 9.5 on research and development (R&D). (29 June) Available at: http://uis.unesco.org/en/news/new-uis-data-sdg-9-5-research-and-development-rd (accessed 3 May 2021).

UNESCO (2021) *Continental Overview: Bridging CESA and SDG4 in Africa*. Africa Regional Report. Available at: http://uis.unesco.org/sites/default/files/documents/bridging_cesa_and_sdg4_in_africa-final.pdf (accessed 3 May 2021).

United Nations (2015) *Transforming Our World: The 2030 Agenda for Sustainable Development: A/RES/70/1*. Available at: https://sustainabledevelopment.un.org/content/documents/21252030%20Agenda%20for%20Sustainable%20Development%20web.pdf (accessed 3 May 2021).

Walker, K. and Hofstetter, S. (2016) *Study on Agricultural Technical and Vocational Education and Training (ATVET) in Developing Countries. Swiss Agency for Development and Cooperation*. Available at: https://www.shareweb.ch/site/Agriculture-and-Food-Security/focusareas/Documents/ras_capex_ATVET_Study_2016.pdf (accessed 22 February 2021).

WEF (World Economic Forum) (2017) *The Africa Competitiveness Report 2017. Addressing Africa's Demographic Dividend*. Available at: http://documents.worldbank.org/curated/en/733321493793700840/pdf/114750-2-5-2017-15-48-23-ACRfinal.pdf (accessed 22 February 2021).

World Bank (2007) *Agriculture for Development: World Development Report 2008*. Washington, DC.

World Bank (2008) *Accelerating Catch-Up: Tertiary Education for Growth in Sub-Saharan Africa*. Washington, DC.

World Bank (2013) *Africa's Pulse: An Analysis of Issues Shaping Africa's Economic Future 7*. Washington, DC.

World Bank (2018) Poverty and Equity Data Portal: Global Poverty Indicators. Available at: http://povertydata.worldbank.org/Poverty/Home (accessed 22 February 2021).

World Bank (2020) *Poverty and Shared Prosperity 2020: Reversals of Fortune*. Washington DC. Available at: https://www.worldbank.org/en/publication/poverty-and-shared-prosperity (accessed 3 May 2021).

Wouterse, F. and Taffesse, A.S. (eds.) (2018) *Boosting Growth to End Hunger by 2025: The Role of Social Protection*. Regional Strategic Analysis and Knowledge Support System (RESAKSS) Annual Trends and Outlook Report 2017–2018. International Food Policy Research Institute, Washington, DC. Available at: http://ebrary.ifpri.org/utils/getfile/collection/p15738coll2/id/132831/filename/133042.pdf (accessed 22 February 2021).

Part II

Sectoral and Institutional Context

2 Agricultural Food System Transformation and Its Implications for Tertiary Agricultural Education

John Lynam[1]* and Kay Muir Leresche[2]
[1]*Independent Scholar, based in Kenya;* [2]*Retired professor of agricultural and natural resource economics, based in South Africa*

2.1 Introduction

Agricultural economies in sub-Saharan Africa (SSA) have changed rapidly since the turn of the millennium. The drivers of this rapid structural change are overall economic growth, urbanization, new technologies and a growing private sector, prompted by market liberalization, democratization and investment in transport infrastructure. Accordingly, labour markets in these agricultural economies are also changing. Tertiary agricultural education (TAE) is a critical aspect of these changes, providing the human resources to support the change, while developing the new skills and competencies needed. In order to comprehend this changing demand for new skills, it is important to understand the process of structural change, which extends beyond the farm to include: (i) the set of services that support agricultural productivity and commercialization, including digital information and communications technology (ICT), (ii) increasing integration within a changing urban economy, (iii) growth in the non-farm rural economy (NFRE), (iv) changing consumption patterns (including growing domestic demand for agricultural products), (v) the impact of climate change and (vi) agro-industrial growth. All of this is encapsulated in what is termed 'the agrifood system' (FAO, 2017).

This chapter assesses the achievement of African development outcomes through market-led rural transformation within the context of food systems. Food systems, in turn, are often framed around guaranteeing good nutrition, which requires an understanding of supply-chain management, the impact of agro-processing on nutritional food value, the reduction of food waste and - ultimately - closer linkages between urban food delivery and distribution on the one hand, and high-quality supply from rural areas on the other. An agrifood-system approach expands these issues to incorporate: (i) structural change in urban food wholesaling and retailing, (ii) the organization of supply chains in relation to changing urban demand (e.g. dairy and horticulture), (iii) reduced seasonality of supply, (iv) processing to provide convenience without compromising nutrition and (v) the development of price differentials to address supply quality. Nutrition and equity outcomes significantly depend on how supply chains are organized and, especially, on successfully developing ways to aggregate smallholder produce and increase cost efficiencies. In many ways, this process draws on the smallholder experience from India. It involves

*Corresponding author: johnklynam@gmail.com

 Transforming Tertiary Agricultural Education in Africa (eds D. Kraybill, J. Lynam and Adipala Ekwamu)
DOI: 10.1079/9781789246544.0002

redesigning cooperatives and farmer associations, and developing information, service and marketing linkages between smallholders and growing urban markets.

2.2 Rural Development from a Food-System Perspective

Broadly defined, a food system includes all the functions in a value chain - from production to consumption, including marketing and processing - but set within a systems framework. The utility is in understanding how change in one function affects change in the overall system, understanding the exogenous (i.e. external) system drivers and analyzing how interventions in the system can meet overarching development objectives, including the impact on the natural environment. A food-system perspective expands the purview of rural development beyond just the agricultural sphere. FAO (2017, p. 3) makes this distinction in its definition of different spheres of transformation, defining rural transformation as capturing

> all aspects of agricultural transformation but also [including] the emergence of livelihood and income-generating opportunities in the rural non-farm sector. Improvements in access to services and infrastructure in rural areas lead to the expansion of remunerative off-farm employment and enterprises.

The recent literature on economic transformation in Africa has particularly focused on the RNFE and its role in meeting the challenges of employment and rural poverty (e.g. Barrett *et al.*, 2017). New ICT is expanding these opportunities. Employment in the RNFE is particularly important given the demographic 'bulge' of rural youth[1] soon to enter the workforce and 'urbanization without industrialization' in which urban employment primarily comprises services outside the formal economy.

The principal growth linkages which underlie the economic transformation of most African countries are found within the food system. Changes in urban demand towards higher-value products and convenience create growth linkages to the farm level, agro-processing and marketing. The other primary growth linkage is increased agricultural productivity, which has lagged but is critical to: (i) the development of the RNFE through the development of input supply systems, (ii) the sale of marketable surpluses, (iii) grain storage and (iv) small-scale processing for local markets. The engine of growth in the food system is balance between the pull of demand and the push of productivity. This balance leads to increasing rural employment associated with: (i) small-town development and reduced migration to urban areas, (ii) rising farm incomes which have consumption linkages to the small-town economy and (iii) improvements in nutrition. What is less clear is whether such a market-driven growth dynamic can have a major impact on rural poverty. In part this has to do with land and capital markets and whether smallholder farmers can: (i) maintain their position, given competition with larger commercial farms in both domestic and export markets and (ii) maintain their access to land, based on projections that the rural population will continue to grow in absolute terms for the next couple of decades.

2.3 Drivers of Change in the Food System

Since the turn of the millennium, the African food system has undergone quite rapid structural change in response to a number of concurrent drivers. These are briefly reviewed here, with the notion that they will continue to drive change for at least the next decade or so. These drivers vary across African countries, so the rate of food-system change will also vary significantly. The focus is on countries with the most dramatic change because they provide a template for change in the countries which are lagging.

Economic Growth, Urbanization and an Expanding Middle Class

Economic theory posits that as economies develop, resources - particularly labour - move out of the agricultural sector and into the urban sector where returns are higher. A pattern for many Asian countries is that labour moves into labour-intensive manufacturing, in many cases for export. Growth in the urban economy then attracts urban migrants, influencing the rate of migration and, at the same time, the search for labour-saving techniques in farming, usually associated with declining rural population (in absolute terms). The indicators of structural change in African economies present a very different pattern, however, involving (as mentioned above) 'urbanization without industrialization'. Very high rates of rural–urban migration have been associated with the creation of urban employment, primarily in the (informal) service sector. In turn, this has created an associated demand for services, boosted by high commodity prices and either mineral or agricultural exports, leading to robust economic growth in many African countries over the past decade and a half. With the associated requirement for urban infrastructure, construction plays a significant role in driving urban economies, but in Africa, much of this is occurring without effective planning. At the same time urban development is associated with rapid development of slums and disconnected neighbourhoods.

A recent World Bank report described such urbanization as a low development trap (Lall *et al.*, 2017). A United Nations Economic Commission for Africa report (UNECA, 2017) highlights the fact that African cities face low productivity; limited job creation, infrastructure and service gaps; weak linkages with rural areas; high levels of informality; increasing inequalities; growing environmental damage and vulnerability to climate change, as well as weak institutional systems and capacities. Nevertheless, economic growth has produced an urban middle class which is now large enough to create major changes in food-consumption patterns (AfDB, 2011; Tshirley *et al.*, 2015). The middle class has grown at 3.1% over the past several decades and constitutes around 350 million people. This segment of the population has the potential to produce major consumption linkages with either domestic agricultural growth or food imports. Linkages to imports are particularly occurring in resource-dependent economies (UNECA, 2017).

These trends offer both opportunities and challenges for food systems in African countries. On the one hand, rapidly changing food-demand patterns - particularly a shift to higher-value commodities and, as Tshirley *et al.* (2015) describe, towards more perishable goods - have potential impacts on farm production patterns and income levels, on agro-processing and on the structure of food wholesaling and retailing. Supermarkets have been increasing as a form of food retailing, but the level of penetration is yet to be determined, particularly the continuing dependence in low-income neighbourhoods on local shops due to the relative cost of transport. Moreover, food in African cities is expensive - 35% more than comparable cities outside Africa (Nakamura *et al.*, 2016). The increased linkages and greater flow of goods is providing opportunities for the increased processing and sale of indigenously-consumed products in urban supermarkets. Recognition by the middle class of the health value and social acceptability of consuming traditional products - such as dried mopane worms, baobab leaves and juice, and indigenous vegetables (like nightshades) - is growing (Muhanji *et al.*, 2011).

Market Liberalization and Restructuring of Input and Output Markets

The 1990s structural adjustment programmes, which focused on macroeconomic stability, and market liberalization programmes, which focused on reducing the role of government in markets, laid the basis for expanding private-sector investment in the agricultural sector. A lag arose between the public sector's withdrawal from markets (often in the form of agricultural marketing boards) and the development of efficient and integrated agricultural markets led by private-sector investment. The market-development process got a major boost from the spike in world food prices around 2005, 'enabling thousands of small-, medium- and large-scale private firms to rapidly respond to profitable incentives' (Jayne *et al.*, 2018, p. 4). In effect, that process continues, often with government re-entry into markets - such as through fertilizer subsidy programmes, which have become prevalent over the past decade. Studies provide evidence of a continuing lack of market integration caused by credit limitations, search and transport costs and imperfect competition, suggesting further gains could be made in improving market efficiency and reducing marketing costs (Bergquist, 2017). Looking forward, improved information flows resulting from the use of mobile phones, blockchain technology and big data provide real opportunities to integrate these markets more effectively (Aker and Fafchamps, 2015; Dillon and Dambro, 2017).

Investment in Road and Port Infrastructure

Africa has the lowest road density of any region globally, and despite the increased investment in principal road corridors, road density actually declined during 1990–2011 (World Bank, 2017). This reflects declines in the lower-income countries and significant increases in the emerging middle-income countries. Moreover, only about 16% of the road network is paved. Transport costs are a major contributor to the eventual price of a bulky input like fertilizer, for example, and distance to markets affects the ability of farmers to commercialize their crops. Thus, transport infrastructure has a major impact on farmers' ability to participate in markets, access inputs, locate processing facilities and absorb assembly and bulking costs (e.g. Dorosh *et al.*, 2011 and Iimi *et al.*, 2015). A recent approach to maximizing the utility of existing road infrastructure is that of agricultural growth corridors, in which governments create agricultural investment incentives along principal road corridors, usually linked to ports (Picard *et al.*, 2017). The focus tends to be on agro-industry, often linked to investment in large commercial farms. The idea is to achieve low-cost links to major urban markets and export points, to reduce costs through agglomeration and to develop growth poles which will have spillover impacts. The jury is out on how effective these will be, but the approach is indicative of the impacts of investment in transport infrastructure on the food system (Serraj *et al.*, 2015).

Advances in Information and Communications Technology

Rapidly changing ICT is providing real opportunities for smallholders to maintain competitiveness in growing urban markets. The much greater transparency and improvement in information flows - which particularly creates logistical efficiencies - make it easier to institutionalize aggregation of commodity outputs, inputs and services, especially through farmer cooperatives. This in turn enables a sharing economy which is able to more evenly spread capital costs. These advances also open up new areas of employment for youth, who are in a better position to understand, adopt and adapt new technologies to local situations.

ICT has enhanced the production, marketing and dissemination of knowledge. Mobile phones are able to disseminate market information efficiently (Aker and Fafchamps, 2015) and are increasingly effective in delivering extension services (Misaki *et al.*, 2018). Advances in ICT provide opportunities for the traceability required to differentiate prices related to quality, and to participate in fair-trade and organic-produce value chains. ICT also facilitates direct linkages among small groups and provides access to more reliable information on weather, diseases and markets. These advances have shown their potential in a number of areas where they have affected smallholder farmers' ability to access better marketing, finance, insurance, extension and weather services. Examples of these benefits exist across Africa, but they remain isolated and have yet to be translated into increases in rural productivity. Investment in connectivity, literacy and enabling policies is essential to integrate smallholder farmers into an evolving food system through the use of ICT.

2.4 Changing Structure of Food Demand

As discussed above, growing urban populations, a rising middle class and economic growth are driving major changes in food demand, such that domestic and regional markets are becoming principal competitors with traditional export markets - e.g. the substitution by farmers in central Kenya of horticulture for coffee[2]. As Reardon *et al.* (2015, p. 63) note, 'urban food markets have become the dominant market for farmers' in all three regions of SSA. Urban food purchases are 48% of all food purchases in East and Southern Africa, and closer to three-quarters of all purchases in West Africa (Reardon *et al.*, 2015). Growth of food markets has been rapid:

'over the past four decades, African food markets have expanded as much as eight-fold' (Reardon *et al.*, 2015, p. 63).

A significant part of the relatively recent growth in domestic food markets is the increasing reliance of rural populations on food purchases. In four countries in East and Southern Africa, rural populations were found to purchase 40–49% of their overall food basket (Dolislager, 2017). Moreover, for very poor rural households, the purchased share is higher than the next quintile, and then rises further with increasing income. This is supported by literature which shows that smallholders are both sellers and buyers of crops, particularly staples. A distinct seasonal price pattern is present, but because of lack of storage, credit and hence arbitrage, farmers generally 'sell low and buy high' (Burke *et al.*, 2017, p. 1). A dominant policy question is how smallholders will respond to changing market conditions and, particularly, greater market integration, which will potentially allow them to move out of subsistence production and specialize instead in higher-value crops. For the smallholder farmer, changing food markets catalyze a potential response on both the food-demand and food-supply sides (discussed in the next section).

Nevertheless, urban markets remain the principal driver of changing food-consumption patterns, facilitated by changes in food retailing. As Tshirley *et al.* (2015, p. 635) point out for East and Southern Africa, 'the middle class is still only 27 per cent of the population but controls fully 54 per cent of the total expenditures and 48 per cent of the food expenditures'. Dolislager (2017, p. 2) summarizes the implications as follows:

> Diets are changing in three ways: food is becoming more purchased, more perishable, and more processed. These changes drive a multitude of effects on the agrifood system.

These dietary changes have implications for farmer production patterns, but even more so for what might be called the middle of the agrifood system - namely, the capacity requirements needed for processing, storing and marketing perishable commodities, and the changes in wholesaling and retailing needed to respond to the growing middle class. In turn, these changes have implications for employment and the new sets of skills needed to support these structural changes, all of which have implications for curricula and training in TAE institutions.

2.5 Generating a Production and Productivity Response

Farm-Level Transformation

In order to adapt to the changing demand for agricultural products and the differences in urban requirements, it is essential that smallholders increase their marketable surpluses and shift to higher-value production. This diversification of farming systems is what de Janvry and Sadoulet (2017) term a rural transformation, requiring the development of integrated value chains for both cash crops and staples. At the same time, this shift requires greater staple-crop productivity, or what they term the Green Revolution component, which - together with the rural transformation component - form the basis of structural transformation of the rural economy.[3] Increasing staple yields requires improved technologies, but farmers' adoption of those technologies requires:

> securing property rights over land to induce investment, improving infrastructure for access to water and markets, making financial institutions accessible to the poor for credit and insurance, improving extension services to inform farmers and help them decide, and engaging in policy reforms to reduce price distortions and excessive taxation on agriculture (de Janvry and Sadoulet, 2017, p. 4).

Thus, synergy exists between farmers' adoption of new technologies and farm diversification and growth in the NFRE; neither can precede the other.

These transformation processes require a dynamic private sector, which introduces a critical policy issue: the impact on both poverty and employment of a rural economy which is becoming more commercial, particularly in relation to the growing youth cohort. As rural populations in Africa continue to rise in absolute terms, and urban employment growth is primarily generated by self-employment in the service sector (McCullough, 2017), rural employment growth remains critical to employment growth in the overall economy. Moreover, due to high rural population densities and declining farm sizes in many countries, growth in the NFRE is seen as the principal means of generating rural employment. Using living standards measurement study (LSMS) data for four countries[4], McCullough (2017) analyzed employment statistics in the early 2010s, with relatively-consistent findings. In particular, over 80% of the rural population was directly self-employed in agriculture. For wage employment, the bulk was in agriculture (15% of rural population in three countries, but 33% in Malawi). The next-largest sector providing rural wage employment was the service sector, the highest being in Uganda (11% of the rural population). Wage earners in this sector tended to be younger, although the majority of youth (more male than female) were employed in agriculture. Comparisons among surveys indicate a decline in the share of workers employed in agriculture (despite an increase in absolute terms), and an increase in the share employed in the NFRE (Yeboah and Jayne, 2017) Nevertheless, growth in the NFRE is from a relatively small base. Although the trends are in the right direction, the key is whether the NFRE is expanding at a sufficient rate to absorb the rapidly-increasing youth cohort.

Employment and poverty alleviation, at least for the medium-term future, are inextricably linked to the income-generating potential of small-scale farms. Productivity growth in this sector also drives growth in the NFRE (Haggblade *et al.*, 2007). Market-driven, smallholder development, however, is dependent on the structure of urban supply chains and the ability of smallholders to both diversify and increase staple productivity. Given the requirement for continuity of supply for urban markets (in part mediated by development of storage capacity), the high cost of logistics in the aggregation process and the increasing demand for quality standards, many supply chains focus on larger commercial farms, often operating with irrigation. These, however, have little impact on both employment and poverty, unless they develop in size, such as in the case of Kenya's horticultural sector. While such larger farms may lead in the development of commercial supply chains, the policy question is whether they lead to progressive integration of smallholders into these supply chains or whether they marginalize the smallholder sector through competition for land, with the possibility of compensation through wage employment. Thus, a smallholder strategy for agricultural transformation must address three principal issues: (i) whether smallholders can lead in urban-market development, (ii) the mechanism for diversifying smallholder farming systems and (iii) the potential for a smallholder green revolution for food staples.

Debates over Smallholder-Led Growth and Increasing Land Pressure

Debates about a smallholder development strategy have broadened over the past decade to focus on structural transformation of the agricultural sector. On the one hand is what might be termed 'the pragmatic argument'. This is best represented by Collier and Dercon (2014), but is also represented by several government policies (e.g. Nigeria). It basically asserts: (i) that a structural transformation of smallholder agriculture will not be rapid enough to meet the changes occurring in the African food system, (ii) that larger commercial farms are already meeting this challenge and should be recognized as another possible pathway, (iii) that commercial farms also create significant levels of employment (Reardon *et al.*, 2009) and (iv) that the potential exists to integrate commercial farms and smallholder producers as a means of increasing smallholder links to markets. The strategy is also associated with the development of agricultural growth corridors (Kuhlmann *et al.*, 2011). These initiatives promote private-sector investment, particularly organized around agribusiness clusters along corridors which link primary rail and roads to ports. The Southern Agricultural Growth Corridor of Tanzania is one of two initial pilot projects. It focuses on agricultural development, most often involving large commercial farms linking the southern highlands to domestic markets - particularly in Dar es Salaam - and to export markets.

The other side of the debate focuses on equitable or inclusive growth, whereby agriculture is seen as the key to reducing poverty, generating employment and balancing rural–urban migration. Achieving these outcomes is deemed to require a policy focus on smallholder farmers (IFAD, 2016; de Janvry and Sadoulet, 2017; and

Hazell, 2017). The key is smallholders' ability to respond to emerging market incentives, which in turn requires efficient access to input and output markets, often through cooperative arrangements, effective extension systems and available credit and insurance services. The design of the institutions providing such services incorporates inclusive procedures which do not discriminate against smallholders, although this may come at an additional cost. Microcredit, index-based crop and livestock insurance, farmer field school extension, warehouse receipt systems (Coulter and Onumah, 2002) and cooperative marketing arrangements are all examples of programmes designed to reach smallholders (Hazell and Rahman, 2014). New technology provides better opportunities to support this approach. The task, however, is to incorporate these arrangements into a coherent policy framework which drives smallholder development.

Evidence from the recent experience in commercialization supports both of these strategies, although a smallholder strategy is more generalizable across a range of value chains. On the one hand, Neven *et al.* (2009) find that, with the increase in supermarkets for urban retailing in Kenya, well-resourced medium-sized farms dominate as suppliers of horticultural products rather than rainfed smallholders. The authors argue that a critical capital threshold is necessary for market entry. On the other hand, Andersson *et al.* (2013) find that smallholders (of less than 1 hectare) in a peri-urban farming area around Nairobi can participate in contractual arrangements with supermarkets and achieve significantly-increased incomes compared with supplying traditional markets. Notably, however, participating smallholders had significantly higher assets than non-participants, and participation turnover rates were high due to a number of constraints. This suggests the presence of a leading cohort of smallholders who are able to participate in supplying higher-value markets. Hazell *et al.* (2017) developed a typology of smallholders from farm-level surveys in Ghana, Ethiopia and Tanzania (Table 2.1). Of the participating smallholders, in Ghana and Tanzania 30–40% were commercial farmers selling over half their crop production, whereas in Ethiopia the smallholder share was only 11%. Half to two-thirds of households had diversified their livelihood strategies, whereby off-farm sources represented more than one-third of their total incomes. This suggests that a critical mass of smallholders do respond to market incentives, but this is more difficult in markets where the quality and timing in supply chains are essential. Moreover, a majority of households also participate in the NFRE. Thus, evidence strongly supports the viability of a smallholder-led agricultural transformation (Ogutu and Qaim, 2019).

Most countries, however, are at an early stage of agricultural market development. As markets move to greater efficiency, the issue is whether smallholders will continue to maintain a competitive advantage or whether medium-sized commercial farms will begin to crowd the smallholders out through greater ability to access land or to provision high-requirement supply chains, such as for supermarkets or export markets. Recent research has demonstrated that, at the early stage of an agricultural transformation, farm-size distribution is significantly changing in several countries. Jayne *et al.* (2016, p. 197) surveyed Kenya, Tanzania, Ghana and Zambia and found the following:

> Among all farms below 100 hectares in size, the share of land on small-scale holdings under five hectares has declined except in Kenya. Medium-scale farms (defined here as farm holdings between 5 and 100 hectares) account for a rising share of total farmland, especially in the 10–100 hectare range where the number of these farms is growing especially rapidly. Medium-scale farms control roughly 20% of total farmland in Kenya, 32% in Ghana, 39% in Tanzania, and over 50% in Zambia.

The declining share of smallholder land is hypothesized to be due to land consolidation or changes in land allocation under customary tenure (or both). As de Janvry and Sadoulet (2017, p. 6) note, the 'key issue is the pacing of land consolidation and mechanization with labour absorption in the rural non-farm economy and rural–urban migration'. Currently in SSA, these two rates appear to be out of balance, and a strategic question arises as to how to increase the productivity, incomes and growth of smallholder agriculture.

Sustainable System Intensification

A smallholder strategy encompasses diversification, rising productivity and increased

Table 2.1. Typology of smallholders (less than 4 hectares) by livelihood strategy

	Share of crop production sold (%)		
Non-farm income as a share of total income (%)	Low (<5%)	Medium (5–50%)	High (>50%)
Ghana (GLSS 2012–2013; sample size = 7,743)			
Low	8.1	14.5	22.6
High	38.9		15.9
Ethiopia (ESS 2013–2014; sample size = 3,000)			
Low	17.2	32.0	6.8
High	39.4		4.6
Tanzania (NPS 2012–2013; sample size =2,855)			
Low	5.1	15.8	14.6
High	49.9		14.6

Source: Hazell *et al.* (2017, Annex 1, Table 1.1).

commercialization of farming systems. These three pathways also improve the resilience of smallholder systems threatened by climate change. Moreover, as smallholder farming systems have lost the possibility of maintaining soil fertility through traditional fallowing (Binswanger-Mkhize and Savastano, 2017), the key to achieving these objectives is to improve nutrient management or what has been termed 'sustainable intensification' of these systems. Because of the high relative cost of fertilizer, the critical role of soil organic matter in managing tropical soils (Marenya and Barrett, 2008) and the diversity of smallholder soils, integrated soil-fertility management approaches are critical to sustainable increases in productivity (Vanlauwe *et al.*, 2015) and to moving smallholders out of poverty (Tittonell and Giller, 2013).

These conditions on smallholder transformation can be framed within a research and development approach organized around systems. Such approaches mark a significant change from the past and increase the complexity of both research and extension in Africa. Nevertheless, productive soils increase the impact of investments in improved varieties, chemical fertilizers and disease management. System approaches such as agroforestry, integrated crop–livestock systems, integrated disease and pest management and integrated soil-fertility management are based on enhancing ecological mechanisms which improve system productivity - namely, agroecological approaches (Nelson and Coe, 2015). In addition, system diversification enhances agroecological intensification. African smallholder systems evolved around adaptation to local ecologies, but the move to permanent systems over a relatively short period of time has meant that trial-and-error adaptation has not been possible. Pathways to system intensification need to be based on farmers 'learning by doing' but within strong programmatic support to enhance access to information, inputs and output markets.

Increasing Staple Productivity in Smallholder Systems

Food staples take up, by far, the greatest percentage of cultivable land and are almost universally produced by smallholders - although smallholders are not the only producers of staples. A green revolution of African staples has yet to occur in SSA. As with the strategic direction of the Alliance for a Green Revolution in Africa (AGRA), staple productivity can be increased through a smallholder strategy. AGRA pursues an input-led strategy, primarily focused on developing private-sector delivery systems for improved varieties and fertilizer. Managing these inputs in rainfed smallholder farming systems is more complex than it may first appear. Time to maturity and potential for adaptation are important issues in varietal selection under changing seasonal distribution of rainfall. Also, nitrogen and phosphorous fertilizers are often unresponsive in acidic soils or soils with low soil organic matter and other nutrient deficiencies. Using LSMS data in three countries, Sheahan and Barrett (2014) found that farmers do not vary their use of chemical fertilizer or organic inputs based on their perception of a plot's soil quality or erosion status. Moreover, the authors also found that farmers rarely combine the use of chemical fertilizer and organic inputs, suggesting little understanding of integrated soil fertility management.

For most smallholders, chemical fertilizer is a relatively recent technology, and optimizing its use is still a work in progress. Moreover, given nutrient-depletion rates in most smallholder systems, getting the right mix and balance of nutrients is critical for profitable fertilizer use. In Nigeria, where the rate of fertilizer use is higher than in most African countries (Sheahan and Barrett, 2014), response rates for nitrogen application are low, fertilizer costs are high and a majority of farmers do not get a profitable response to application (Liverpool-Tasie *et al.*, 2015). Non-profitable application may reflect a virtually-minimal yield if fertilizer is not used - given the poor nutrient status of soils, as well as the high cost of buying maize. Evidence to date suggests that access to improved varieties and fertilizer, without other required inputs, will not be a pathway to large increases in smallholder yields of cereal and root-crop staples.

It might be argued that the traditional high-yielding variety and fertilizer technology of the green revolution is a necessary but by no means sufficient condition for increasing smallholder staple productivity across Africa. Moreover, long-term trials under experimental conditions - which usually start with highly fertile, undisturbed soils - show a positive response in terms of yields over time, but when integrated with organic inputs the yields are higher and more sustained (Batiano *et al.*, 2012). Where nutrient levels are initially limited, as is common on smallholder fields, the organic inputs act to balance nitrogen and phosphorus and sustain yields over time. An agroecological approach to extension is more likely to provide the sufficiency condition to enable sustainable increases in staple crop yields in smallholder systems, thereby ensuring profitable and sustainable productivity increases. A good example of an agroecological approach is push–pull technology[5] in East Africa, which results in the reduction of fertilizer use over time and large improvements in the efficiency of fertilizer use (Kassie *et al.*, 2018).

2.6 Structural Changes in Market Supply Chain and Growth in the Non-Farm Rural Economy

The strategic core to a smallholder-led, rural transformation in SSA is increasing market participation of smallholder farmers. Achieving this goal, however, requires investment in and organization of a range of critical services to support smallholders in increasing their participation in competitive markets. A broad characterization of smallholder livelihood systems in SSA is: (i) that subsistence needs still form a significant part of production systems, (ii) that market participation depends on periodic surpluses sold in small lots in local markets and (iii) that household income includes a significant component derived from non-farm sources. Increasing smallholders' marketed production share depends on increasing their production efficiency, market integration and market access (in terms of both inputs and outputs), along with the provision of private-sector services, complemented by the public sector. As detailed above, however, not all smallholders have the land, human and financial capital and other resources needed to move effectively to a more commercial orientation. One study estimates that only about one-third of smallholders have the necessary resources (Tshirley *et al.*, 2017, drawing on data in Table 2.1). This raises an important question for policymakers about whether the NFRE or expanding urban employment will be able to absorb the youth coming from non-commercial farms.

Development of the NFRE - which requires investment in infrastructure - generates employment and more efficient financial flows, linking the rural economy with the larger economy. This process is largely market driven and depends on locational advantages (i.e. agroecological potential, road infrastructure and distance to processing plants or urban markets), costs of aggregating supply (which can be made more feasible with new technology) and levels of prior investment in infrastructure to support service provision (storage, agro-dealer distribution networks, aggregation points, banking, and so on). Product markets eventually evolve towards greater efficiency and integration, reducing marketing margins and increasing spatial arbitrage. For perishable products or ones for which processing is time-sensitive, marketing arrangements may evolve towards contracts between farmers and processors or supermarkets. Areas of increasing agricultural productivity are associated with growth in the NFRE, and the two tend to be mutually reinforcing. Lower output prices arising from increasing productivity make the economy more competitive. A threshold of market infrastructure is necessary to provide incentives for farmers to invest in diversification away from subsistence crops and to cover the additional cost of improving productivity. Rising smallholder commercialization is associated with a process of rural innovation, which sustains the growth process.

Much of the policy discussion of alternative futures and the integration of smallholders centres around how supply or value chains are organized. Byerlee and Haggblade (2013) identify five alternative models which balance transaction costs, economies of scale, market standards and locational advantages in the organization of the supply chain. As with the example of supermarkets discussed earlier, as smallholders move away from trading on spot markets, potential arises to increase incomes either from higher prices or from improved efficiency and lower supply chain costs. A principal means of reducing transaction costs in the aggregation of supply from dispersed smallholders or the distribution of bulky inputs (such as fertilizer) is collective action by producer associations or cooperatives (Badiane and Ulimwengu, 2017). If they are appropriately organized, producer associations can act as a mechanism of inclusivity, integrating poor farmers (Fischer and Qaim, 2012). The pool of smallholders participating in contract farming is still relatively small, although the income benefits are significant (Ton *et al.*, 2018). It is the better endowed smallholders that participate in contract farming. Such contracts tend to involve input provision, farmer associations, and price premiums in crops other than staples (Ton *et al.*, 2018). Thus potential income gains to smallholders arise from organizational innovations rather than dependence on spot markets, but broadening the scope of these innovations and ensuring inclusive engagement is a challenge in market-led rural transformation. Better information flows, transparency and the use of information and communications technology and mobile phones will assist in this transformation.

As analysed above, employment in the NFRE is growing quickly, but from a very small base. There is reason for concern about whether employment creation can keep pace with the expansion in medium-scale farms and the impact that is having on access to land. Nagler and Naude (2017) found that, while 44% of rural households participate in the NFRE, it only generates 15% of household incomes on average. They also found that 95% of rural enterprises employ fewer than five workers. The differences in performance are significant, and many enterprises cease operations due to lack of finance or profitability, or because of livelihood shocks. A Darwinian process moves rural enterprises in the direction of increasing their size and efficiency. While noting the recent increase in entry by large agribusinesses, many analysts argue that small and medium-sized enterprises are the key to integrating smallholders into markets. For example, Tshirley *et al.* (2017, p. 54) argue that:

> investments in specialized training, building of relationships between firms and farmers, preferential credit access and other programme or project activities can be important and should be pursued. But the return to investment in these focused activities will be vastly greater if the policy setting is conducive, and if infrastructure is in place to reduce costs and risks in the system.

African agricultural transformation will be based on simultaneous and reinforcing development of small and medium-sized enterprises and sustainable intensification of smallholder farms.

2.7 The Role of Tertiary Agricultural Education in a Transforming African Food System

Building the skills of both educated and uneducated youth to work along the agricultural value chain has the potential to solve the problem of youth unemployment and, at the same time, increase agricultural productivity. Rapid technological change requires a culture of continuous learning and an agricultural education and training system focused on both how to learn and what to learn. Such a system uses new methods to position its graduates to lead change, to be adaptable and to work with people across the food system. Universities should enhance the skills of graduates in ways which can respond to rapidly changing technological, environmental and structural changes. They must be innovative and able to adapt their learning to complex and changing food systems. Surveys by the IUCEA (2015) and other organizations across Africa and the world repeatedly show that although technical skills are valuable, employers place even greater importance on 'soft skills'[6] (Kalufya and Mwakajinga 2016; Ngalomba 2018). It is important to prepare graduates to continually adapt and master complicated technologies by providing them with the skills needed to work in a complex and changing ecosystem and to ensure that they are included in networks sharing scientific, technical and market information.

A rapidly changing food system highlights the transformational role of the NFRE. This sub-sector has the potential of attracting the growing youth demographic and, as outlined above, technical and organizational innovation give them greater opportunities for employment and contribution to rural transformation. At the same time African farmers could capture a greater share of global markets and make more inroads through organic and fair trade opportunities, provided they are serviced by professionals who have been trained to access new and changing information on production, management and marketing to meet demand. Farmers need to be aware of international standards, of how to build links across the globe and of how to become integrated into international markets.

The development of the NFRE requires significant investment in human capital, in terms of both technical training and higher agricultural education. Business skill development is a critical need in the early stages of market development. As firms grow in terms of scale, specialization and complexity, demand rises for a range of skills and competencies which can only be provided at the university level. Demand is also rising for degrees in agribusiness, food processing, agricultural engineering, agricultural economics, agricultural finance and - more generally – entrepreneurship (see Ch. 12, this volume). Expansion of universities based in rural areas is needed to enable youth to obtain these skills and apply them in local contexts.

Institutional divisions between teaching, research and extension need to be overcome. More effort needs to be placed on programmes which integrate universities into communities, on the one hand, and along the full value chain, on the other. TAE institutions need to be active in preparing their graduates and farmers to cope with the vagaries of climate change, to benefit from the opportunities offered by new technologies and to collaborate to influence policy and strengthen their links to markets. The new information and communications technologies should be much more effectively used by universities and colleges to achieve integration into a continuous, lifelong learning paradigm for students, faculty, institutions and all stakeholders in the rural economy and food system.

2.8 Conclusion

A major policy issue is how to 'nudge' market-led, rural transformation to reduce poverty, decrease malnutrition and improve sustainable management of the natural resource base. In the search for cost efficiencies in an increasingly market-oriented economy, there tends to be a move towards specialization and concentration, and those actors who cannot compete must search for alternative livelihoods. This is evidenced in the increasing competitive advantage of medium-scale farms, and the increasing number of those farms, at a time when rural populations continue to expand. Urban households are increasingly investing in medium-scale farms and displacing smaller-scale rural households. Movement out of farming is a feature of structural transformation of the economy, as occurred in Latin America over the past 50 years or in China more recently. However, in those parts of the world, structural transformation was based on the development of labour-intensive manufacturing. In contrast, formal employment in the manufacturing sector is small in African economies, and the major source of employment, by far, is in the low-skilled, informal sector. Rural-to-urban migration flows are enormous and are overwhelming urban infrastructure. This pressure could be alleviated, and growth could be more equitable, through the development of small and medium-sized towns and the NFRE. However, inclusive, market-based growth does not occur without government investment, and commitment to integrating smallholder farmers into markets is critical to success.

Endnotes

[1] The demographic 'youth bulge' commonly occurs once a country achieves success in reducing infant mortality, but mothers still have a high fertility rate. As a result, a large share of the population comprises children and young adults (Lin, 2012).
[2] This section relies almost wholly on papers produced by authors at Michigan State University, particularly Dolislager (2017), Reardon *et al.* (2015) and Tshirley *et al.* (2015).
[3] Agricultural transformation is currently used to encompass these processes, although often with multiple meanings.
[4] McCullough, (2017) analyzed 2010–2011 data for Tanzania, Malawi and Uganda, and 2013–2014 data for Ethiopia. At that time these countries had not experienced the high growth in the NFRE which occurred in Kenya or Senegal; nevertheless, they are probably representative of most of SSA countries.
[5] 'Push–pull' technology was developed by the International Centre of Insect Physiology and Ecology as an approach to controlling stem borers using chemical ecology, whereby a surrounding grass would attract (pull) stem borers to the grass to lay eggs, and a legume inter-planted in the maize would repel (push) them from the maize. Over time, the system evolved to control striga, aflatoxin and fall army worm, while also improving soil organic matter.
[6] Such skills include, in particular, self-direction, continuous learning, better communication, decision-making, critical thinking and problem-solving skills, as well as social intelligence, new media literacy and virtual collaboration skills.

References

AfDB (African Development) Bank (2011) *The Middle of the Pyramid: Dynamics of the Middle Class in Africa*. Market Brief, April 20. Available at: https://www.afdb.org/fileadmin/uploads/afdb/Documents/Publications/The%20Middle%20of%20the%20Pyramid_The%20Middle%20of%20the%20Pyramid.pdf (accessed 1 March 2021).

Aker, J.C. and Fafchamps, M. (2015) Mobile phone coverage and producer markets: Evidence from West Africa. *World Bank Economic Review* 29(2), 262–292.

Andersson, C. I. M. Kiria, C.G. Qaim, M. and Rao, E.J.O. (2013) *Following Up On Smallholder Farmers and Supermarkets*. GlobalFood Discussion Papers 23, Universität Göttingen, Research Training Group 1666, GlobalFood, Göttingen. Available at: https://www.econstor.eu/bitstream/10419/90618/1/GlobalFood_DP23.pdf (accessed 1 March 2021).

Badiane, O. and Ulimwengu, J. (2017) Business pathways to the future of smallholder farming in the context of transforming value chains. Ch. 2 in *Africa Agriculture Status Report 2017: The Business of Smallholder Agriculture in Sub-Saharan Africa*. Issue 5. Alliance for a Green Revolution in Africa, Nairobi. Available at: https://agra.org/wp-content/uploads/2017/09/Final-AASR-2017-Aug-28.pdf (accessed 1 March 2021).

Barrett, C. Christiaensen, L. Sheahan, M. and Shimeles. A. (2017) *On the Structural Transformation of Rural Africa*. Policy Research Working Paper 7938. World Bank, Washington, DC.

Batiano, A. Waswa, B. Abdou, A. Bado, B.V. Bonzi, M.E. *et al.* (2012) Overview of long term experiments in Africa. In Bationo, A. Waswa, B. Kihara, J. Adolwa, I. Vanlauwe, B. *et al.* (eds.) *Lessons Learned from Long-Term Soil Fertility Management Experiments in Africa*, Springer Science, Dordrecht, Germany.

Bergquist, L.F. (2017) *Efficiency, Competition, and Welfare in African Agricultural Markets*. University of California, Berkeley. Available at: https://escholarship.org/uc/item/1hb5j87b (accessed 1 March 2021).

Binswanger-Mkhize, H.P. and Savastano, S. (2017) Agricultural intensification: The status in six African countries. *Food Policy* 67, 26–40.

Burke, M. Falcao Bergquist, L. and Miguel, E. (2017) *Selling Low and Buying High: An Arbitrage Puzzle in Kenyan Villages*. Stanford University. Available at: https://web.stanford.edu/~mburke/papers/MaizeStorage.pdf (accessed 1 March 2021).

Byerlee, D. and Haggblade, S. (2013) *African Food Systems to 2030: Toward Inclusive Business Models. Center on Food Security and the Environment*. Paper for the Stanford Symposium Series on Global Food Policy and Food Security in the 21st Century, Stanford University, Stanford, California, USA.

Collier, P. and Dercon, S. (2014) African agriculture in 50 Years: Smallholders in a rapidly changing world? *World Development* 63, 92–101.

Coulter, J. and Onumah, G. (2002) The role of warehouse receipt systems in enhanced commodity marketing and rural livelihoods in Africa. *Food Policy* 27, 319–337.

De Janvry, A. and Sadoulet, E. (2017) *Achieving the Sustainable Development Goal on Poverty Eradication*. FERDi Working Paper 190. Available at: https://are.berkeley.edu/esadoulet/wp-content/uploads/2018/10/achieving_ferdi.pdf (accessed 22 February 2021).

Dillon, B. and Dambro, C. (2017) How competitive are crop markets in sub-Saharan Africa? *American Journal of Agricultural Economics* 99(5), 1344–1361.

Dolislager, M. (2017) Food Consumption Patterns in Light of Rising Incomes, Urbanization and Food Retail Modernization. Evidence from Eastern and Southern Africa, unpublished PhD dissertation, Michigan State University, East Lansing, Michigan, USA.

Dorosh, P. Wang, H. You, L. and Schmidt, E. (2011) Road connectivity and its impact on crop production. *PositionIT* 32 (Aug/Sept).

FAO (Food and Agriculture Organization of the United Nations) (2017). *The State of Food and Agriculture: Leveraging Food Systems for Inclusive Rural Transformation* FAO, ROME.

Fischer, E. and Qaim, M. (2012) Linking smallholders to markets: Determinants and impacts of farmer collective action in Kenya. *World Development* 40(6), 1255–1268.

Haggblade, S., Hazell, P. and Reardon, T. (2007) *Transforming the Rural Non-Farm Economy*. International Food Policy Research Institute, Washington, DC.

Hazell, P. and Rahman, A. (eds.) (2014). *New Directions for Smallholder Agriculture*. Oxford University Press, Oxford, UK.

Hazell, P. (2017) Why an inclusive agricultural transformation is Africa's way forward. Ch. 1 in *Africa Agriculture Status Report 2017: The Business of Smallholder Agriculture in Sub-Saharan Africa*. Issue 5. Alliance for a Green Revolution in Africa, Nairobi. Available at: https://agra.org/wp-content/uploads/2017/09/Final-AASR-2017-Aug-28.pdf (accessed 1 March 2021).

Hazell, P. Wood, S. Bacou, M. and Dey, B. (2017) Operationalizing the typology of smallholder households. Annex 1 in *Africa Agriculture Status Report 2017: The Business of Smallholder Agriculture in Sub-Saharan Africa*. Issue 5. Alliance for a Green Revolution in Africa, Nairobi. Available at: https://agra.org/wp-content/uploads/2017/09/Final-AASR-2017-Aug-28.pdf (accessed 1 March 2021).

IFAD (International Fund for Agricultural Development) (2016) *Rural Development Report 2016: Fostering Inclusive Rural Transformation*. IFAD, Rome.

Iimi, A., You, L., Wood-Sichra, U. and Humphrey, R.M. (2015) *Agriculture Production and Transport Infrastructure in East Africa : An Application of Spatial Autoregression*. World Bank, Washington, DC.

IUCEA (Inter Universities Council for East Africa) (2015) *Harmonization of Higher Education in East Africa: Status Update July 2015*. IUCEA, Kampala.

Jayne, T.S., Chamberlin, J. Traub, L. Sitko, N. Muyanga, M. *et al.* (2016) Africa's changing farm size distribution patterns: The rise of medium-scale farms. *Agricultural Economics* 47(supplement), 197–214.

Jayne, T.S. Chamberlin, J. and Benefica, R. (2018) Africa's unfolding economic transformation. *Journal of Development Studies* 54(5), 1–11.

Kalufya, N. and Mwakajinga, L. (2016) Employability of graduates from higher education institutions in Tanzania. *Institute of Social Work Journal* 1(2), 51–68.

Kassie M., Stage J., Diiro G., Muriithi B., Muricho G. *et al.* (2018) Push-pull farming system in Kenya: Implications for economic and social welfare. *Land Use Policy* 77, 186–198.

Kuhlmann, K. S. Sechler and J. Guinan. (2011) *Africa's Development Corridors as Pathways to Agricultural Development: Regional Economic Integration and Food Security in Africa*. The Aspen Institute. Available at: https://www.aspeninstitute.org/wp-content/uploads/files/content/docs/pubs/TransFarmAfrica_DevelopmentCorridors_7%20JulyFINAL.pdf (accessed 1 March 2021).

Lall, S.V., Henderson, J.V. and Venables, A.J.(2017). *Africa's Cities: Opening Doors to the World*. World Bank, Washington, DC.

Lin, J. (2012) Youth bulge: A demographic dividend or a demographic bomb in developing countries? World Bank Blogs, Let's Talk Development. Available at: https://blogs.worldbank.org/developmenttalk/youth-bulge-a-demographic-dividend-or-a-demographic-bomb-in-developing-countries (accessed 3 March 2021).

Liverpool-Tasie, L.S.O. Omonona, B.T. Sanou, A. and Ogunleye, W. (2015) *Is Increasing Inorganic Fertilizer Use in Sub-Saharan Africa a Profitable Proposition? Evidence from Nigeria*. Policy Research Working Paper 7201. World Bank, Washington, DC.

Marenya, P. and Barrett, C. (2008) Farmers' perceptions of soil fertility and fertilizer yield response in Kenya. *Agricultural Economics* 40, 561–572.

McCullough, E.B. (2017) Labor productivity and employment gaps in Sub-Saharan Africa. *Food Policy* 67, 133–152.

Misaki, E. Apiola, M. Gaiani, S. and Tedre, M. (2018) Challenges facing sub-Saharan small-scale farmers in accessing farming information through mobile phones: A systematic literature review. *Electronic Journal Information Systems Developing Countries* 84(2), 1–12.

Muhanji, G. Roothaert, R.L. Webo, C. and Stanley, M. (2011) African indigenous vegetable enterprises and market access for small-scale farmers in East Africa. *International Journal of Agricultural Sustainability* 9(1), 194–202.

Nagler, P. and Naude, W. (2017) Non-farm entrepreneurship in rural sub-Saharan Africa: New empirical evidence. *World Development* 67, 175–191.

Nakamura, S., Harati, R., Lall, S., Dikhanov, Y., Hamadeh, N., *et al.* (2016) *Is Living in African Cities Expensive?* World Bank, Washington, DC.

Nelson, R. and Coe, R. (2015) Agroecological intensification of smallholder farming. In Herring, R.J. (ed.) *The Oxford Handbook of Food, Politics, and Society*. Available at: http://www.oxfordhandbooks.com/view/10.1093/oxfordhb/9780195397772.001.0001/oxfordhb-9780195397772-e-006 (accessed 1 March 2021).

Neven, D. Odera, M.M. Reardon, T. and Wang, H. (2009) Kenyan supermarkets, emerging middle-class horticultural farmers, and employment impacts on the rural poor. *World Development* 37(11), 1802–1811. Available at: https://doi.org/10.1016/j.worlddev.2008.08.026 (accessed 1 March 2021).

Ngalomba, S. (2018) Employers need graduates to be taught vital soft skills. *University World News*. 19 October. Available at: https://www.universityworldnews.com/post.php?story=20181015070551149 (accessed 1 March 2021).

Ogutu, S.O. and Qaim, M. (2019) Commercialization of the small farm sector and multidimensional poverty. *World Development* 114, 281–293. Available at: https://doi.org/10.1016/j.worlddev.2018.10.012 (accessed 1 March 2021).

Picard, F. Coulibaly, M. and Smaller, C. (2017) *The Rise of Agricultural Growth Poles in Africa*. Investment in Agriculture Policy Brief 6. International Institute for Sustainable Development, Winnipeg, Manitoba, Canada. Available at: https://www.iisd.org/system/files/publications/rise-agricultural-growth-poles-in-africa.pdf (accessed 1 March 2021).

Reardon, T. Neven, D. Makokha, M. Honglin, O. and Wang H. (2009) Kenyan supermarkets, emerging middle-class horticultural farmers, and employment impacts on the rural poor. *World Development* 37(11), 1802–1811.

Reardon, T. Tschirley, D. Minten, B. Haggblade S., Liverpool-Tasie, S. *et al.* (2015) Transformation of African agrifood systems in the new era of rapid urbanization and the emergence of a middle class. Ch. 4 in Badiane, O. and Makombe, T. (eds.) *Beyond a Middle Income Africa: Transforming African Economies for Sustained Growth with Rising Employment and Incomes*. RESAKKS Annual Trends and Outlook Report. Available at: http://ebrary.ifpri.org/utils/getfile/collection/p15738coll2/id/130121/filename/130332.pdf (accessed 1 March 2021).

Serraj, R. Byers, B. Engel, P. Bizzotto Molina, P. Rampa, F. and Sayer, J. (2015) *Agricultural Growth Corridors and Agricultural Transformation in Africa: Research Needs for Impact, Implementation and Institutions*. Available at: http://www.fao.org/3/bp142e/bp142e.pdf (accessed 1 March 2021).

Sheahan, M. and Barrett, C.B. (2014) *Understanding the Agricultural Input Landscape in Sub-Saharan Africa: Recent Plot, Household, and Community-Level Evidence*. Policy Research Working Paper 7014. World Bank, Washington, DC.

Tittonell, P.A. and Giller, K.E. (2013) When yield gaps are poverty traps: The paradigm of ecological intensification in African smallholder agriculture. *Field Crops Research* 143, 76–90.

Ton, G. Vellema, W. Desiere, S. Weituschat, S. and D'Haese, M. (2018) Contract farming for improving smallholder incomes: What can we learn from effectiveness studies? *World Development* 104, 46–64.

Tschirley, D., Reardon, T. Dolislager, M. and Snyder, J. (2015) The rise of a middle class in East and Southern Africa: Implications for Food System Transformation. *Journal of International Development* 27(5), 628–646.

Tshirley, D. Reardon, T. Haggblade, S. Jayne, T.S. Liverpool-Tasie, S. *et al.* (2017) Engaging the agribusiness sector in inclusive value chain development: Opportunities and challenges. Ch. 3 in *Africa Agriculture Status Report 2017: The Business of Smallholder Agriculture in Sub-Saharan Africa*. Issue 5.

Alliance for a Green Revolution in Africa, Nairobi. Available at: https://agra.org/wp-content/uploads/2017/09/Final-AASR-2017-Aug-28.pdf (accessed 1 March 2021).

UNECA (United Nations Economic Commission for Africa) (2017) *Economic Report on Africa 2017: Urbanization and Industrialization for Africa's Transformation*. Available at: https://repository.uneca.org/handle/10855/23723 (accessed 22 February 2021).

Vanlauwe, B. Descheemaeker, K. Giller, K.E. Huising, J. Merckx, R. *et al.* (2015) Integrated soil fertility management in sub-Saharan Africa: Unravelling local adaptation. *Soil* 1, 491–508. Available at: https://doi.org/10.5194/soil-1-491-2015 (accessed 1 March 2021).

World Bank (2017) *Africa Pulse: An Analysis of Issues Shaping Africa's Economic Future*. Vol. 15. World Bank, Washington, DC.

Yeboah, F. and Jayne, T. (2017) Africa's evolving employment trends: Implications for economic transformation. *Africagrowth Agenda Journal* 14(1), 18–22. Available at: https://www.canr.msu.edu/fsp/publications/peer-reviewed-publications/Yeboah_and_Jayne_2017-_AGA_article.pdf (accessed 1 March 2021).

3 Positioning Tertiary Agricultural Education within a Changing Policy and Institutional Context

John Lynam[1]* and Eusebius Mukhwana[2]
[1]*Independent Scholar, based in Kenya;* [2]*Kenya National Qualifications Authority*

3.1 Introduction

Tertiary Agricultural Education (TAE) institutions[1] perform a range of functions that complement their principal focus on training and human resource development. These include: research, outreach and community development, policy and advocacy and - increasingly - product development and public–private partnerships. Upstream and downstream from TAE institutions are other kinds of institutions which also provide services to agriculture, mostly specializing in what could be termed an 'organizational ecosystem' within the sector. Faculties of agriculture face a dilemma in that these subsidiary functions are necessary to ground faculty in a rapidly changing sector and to provide graduates with the skills needed in the labour market. At the same time, universities have limited financial resources with which to support these functions. Their ability to fulfil their mandate of delivering development-oriented scientific training and research on limited budgets can be enhanced through closer linkages with other faculties and organizations within the sector in what is commonly referred to as an 'agricultural innovation system' (AIS). However, such linkages require coordination, resources to support transaction costs and the realization of mutual benefits. Because these conditions are often absent, institutional linkages have proven to be more the exception than the rule, and this is even more the case in the context of a rapidly-changing policy and institutional environment.

This chapter presents an historical perspective on institutional developments in support of the agricultural sector, highlighting the principal drivers of institutional change. Agriculture's institutional ecosystem has varied across time and throughout Africa, as well as in relation to its colonial origins. The discussion principally focuses on the main trends, occasionally noting regional differences. The timeline is divided into three periods: the next section lays out the structure of the institutional ecosystem, Section 3.3 describes the post-independence period to the mid-1980s and Section 3.4 analyzes the succeeding period to approximately 2005, which was dominated by structural adjustment and market liberalization programmes. Section 3.5 reviews the rapid changes which occurred after 2005, framed in terms of the institutional linkages necessary to achieve the subsidiary functions outlined above most effectively and, hence, TAE's role in an evolving AIS. The implications for TAE are highlighted throughout the chapter.

*Corresponding author: johnklynam@gmail.com

 Transforming Tertiary Agricultural Education in Africa
(eds D. Kraybill, J. Lynam and Adipala Ekwamu)

DOI: 10.1079/9781789246544.0003

3.2 The Institutional Ecosystem Supporting Agriculture, Pre-1950s To The 1990s

The institutional ecosystem supporting African agriculture has its origins in the initial designs from the colonial period, and in an agrarian economy facing multiple market failures. By the 1950s, colonial administrations had recognized the critical role of the African smallholder in maintaining the stability of the agricultural economy. Research by economic anthropologists indicated that smallholders would respond to price incentives, but domestic demand was highly constrained because urban markets were very small and local rural markets were easily saturated. Export orientation required an efficient assembly and bulking capacity in areas where transport infrastructure was limited and expensive. This led to the design of institutions supporting the marketing function, with the future prospect that marketing would shift from public control to competitive private markets, which did not occur until the structural adjustment of the 1990s. Hence, the early focus of public management was the design of marketing institutions and associated arrangements.

Another primary function in the development of the institutional ecosystem was the provision of public goods, particularly research and its complement, agricultural extension services. Technologies supporting growth in agricultural productivity were viewed as essential, particularly given the challenges of maintaining soil fertility and dealing with an array of indigenous pests and diseases on which little research had been done. The design of research and extension organizations has gone through several structural changes in the post-independence period. Funding for these public goods primarily relied on international aid funds, and many of the design changes came from outside Africa. Limited local participation meant that, often, the implementation of research and extension policies and programmes was not deeply rooted in or sufficiently adapted to the local context.

The principal constraint on the functioning of agricultural institutions was the lack of adequately-trained personnel to staff these organizations. University development was in an embryonic state during the initial independence period, and the cadre of trained personnel was limited in relation to needs. For decades, university graduates were virtually guaranteed positions in the public sector, and faculties focused on the skills needed to fill these positions. As the end point in the educational pipeline, universities had to cope with poor preparation of entrants as a result of under-investment in primary and secondary education. The size of agricultural faculties, and their curricula, depended on the supply and degree of preparedness of potential university students, the resources allocated to tertiary education and the changing job market for graduates.

Eicher (1999) argues for a systems approach to what he calls the 'agricultural research triangle', namely agricultural research, extension and university institutions. A systems approach, he argues, addresses the interplay between the services offered by these three institutions. Weakness in one constrains the provision of services by the other two, so Eicher calls for planned and sequenced investment in the core set of institutions driving agricultural productivity growth. The idea of the destabilizing effect of unbalanced development is key to understanding the capacity of these institutions to support agricultural growth and transformation. This balance, or lack thereof, is influenced by government investment priorities and structural changes in the agricultural market economy, which - as will be seen - has been a significant factor in the performance of the institutional ecosystem.

3.3. Public-sector Dominance, Independence to the mid-1980s

The Colonial Legacy at Independence

The principal dimensions of the colonial legacy with respect to agriculture can be defined in terms of two issues: (i) the development of an institutional ecosystem to support agriculture and (ii) the delayed development of universities and training of indigenous agricultural scientists. The institutional ecosystem of the African colonies varied significantly between those of France and those of Great Britain. In the post-WWII period, France developed a centralized, federated system across its colonies in West Africa, while Britain supported separate administrative capacities in each of its colonies, although there were regional approaches to agricultural research in the British colonies in East and West Africa. The institutional framework in place during the independence period became the template for administering the agricultural sector by the new governments, with the principal issue being how to staff these institutions.

In the British colonies, up until the post-WWII period, each colony was expected to support its own governmental expenditure. Colonies that could not meet this threshold were considered to be 'lacking responsible government' (Morgan, 1964, p. 8). The finances of these colonies were under the control of the British Treasury. With the exception of Southern Rhodesia, most colonies in sub-Saharan Africa (SSA) fell under this rubric, which meant that all expenditure by the colonial governments came under the scrutiny of the colonial office. For agriculture, the Colonial Agricultural Service staffed the agricultural departments in the colonies with British graduates who spent one year at Cambridge and one year at the Imperial College of Tropical Agriculture in Trinidad. The departments of agriculture were headed by a director of agriculture, who reported to the governor and oversaw both research and field staff. Field officers were assigned at either the provincial or the district level, with district agricultural officers being at the frontline (Masefield, 1972). Before independence in the 1950s, governance in the colonies shifted to a ministerial system based on local elections (in Kenya, for example, this took place in 1954 [Trapman, 1974]), ultimately leading to the creation of ministries of agriculture (MOAs). Departments of agriculture also supported an agricultural research system in each of the colonies. Extension had also become more formalized in the post-WWII period, particularly in large settlement and agricultural development projects focusing on soil conservation. Settlement and soil conservation projects, however, often involved coercive measures which caused farmer animosity towards government extension programs, thereby undermining the efficacy of these initiatives.

A critical institution was the 'commodity marketing board'. These differed between the settler colonies in East and Southern Africa and the smallholder-based export economies of West Africa. A principal activity of the settler community was cereal production, but in a context of limited urban demand, small-volume markets, price volatility and competition from smallholder production. The depression of the 1930s led local governments to pass acts to create marketing boards to set prices and regulate supplies. These acts led to (Jayne and Jones, 1996, p. 5):

> (1) the establishment of restrictions on grain movement from African towns, mines and other demand centers where African production could otherwise undercut European grain production; (2) the creation of monopoly state crop buying stations in European farming areas without similar investments in African farming areas; and (3) assured prices for European farmers (typically above export parity) that were financially sustained through 'rake-off' taxes on maize sales by Africans to licensed private traders operating as agents of the boards.

West African economies were based on smallholder production of both non-tradable food staples, e.g. cassava and yams, and export production of cocoa and palm oil and kernels. In 1942, the West African Cocoa Control Board, established in 1940 to manage cocoa purchases and exports, was extended to include palm products and groundnuts. In 1947, commodity boards were established in each of the West African colonies to manage export products (Williams, 1985).

The last major institution to be developed in the British colonies was university colleges. The Asquith Commission of 1945 recommended that higher-education institutions (HEIs) be created in the colonies, reflecting the need for an educated elite in getting ready for self-rule. Prior to the Commission, early 'university' education focused on vocational training, with pass rates determined by the number of new staff which needed to be recruited to support the British agricultural officers. From an early period, agriculture was associated with inferior career opportunities (Nwauwa, 1993), which has continued to this day. This was compounded by the stress in British higher education on liberal arts training (i.e. natural and social sciences and arts and humanities) as superior to vocational training (Sicherman, 2005), which - even until today - has disadvantaged technical and vocational education and training (see Ch. 13, this volume). The first universities evolved from these technical colleges, and given the focus on academic rigor, discouraged the development of faculties of agriculture (for the case of Makerere University in Uganda, see Sicherman, 2005).

As a result of the Asquith Commission, university colleges were founded in Nigeria (University College of Ibadan), the Gold Coast (University College of the Gold Coast) and Uganda (Makerere University College/University of East Africa), Southern Rhodesia/Nyasaland and Sudan. These university colleges were affiliates of the University of London, which designed the curriculum, set entry requirements and awarded the degrees. This went against a 1938 commission report, which recommended a curriculum adapted to African conditions. These colleges had relatively small class sizes in relation to the need for trained graduates. In the period 1952–1963, university colleges in anglophone Africa produced about 150 graduates in agriculture compared with only four in francophone Africa (Eisemon *et al.*, 1985). One critique of the Asquith Commission's approach focused on over-specialization and lack of relevance, lack of linkages to technical institutes and too much focus on fundamental research (Ashby, 1966).

The dominant characteristic of the former francophone colonies was that institutions continued to be populated by French expatriate staff well into the 1980s. Little effort was made at developing HEIs before independence, and there was only one faculty of agriculture in all of francophone Africa. Whereas the British relied on 'indirect rule', delegating authority to chiefs and elites, the French relied on direct rule through expatriate staff in the management of all institutions. This applied even to farmer co-operatives, which were compulsory in francophone colonies and managed by French staff (Develtere, 2008). Similarly, the French developed a network of agricultural research stations, which were operated at a federal or regional level but were managed through research centres based in France. The research centres focused exclusively on export crops; an institute focusing on food crops was not created until 1960 at the time of independence of the French colonies. These research centres based in France each operated a network of research centres across the francophone region in West Africa.

Management of the Agricultural Economy by the Ministry of Agriculture

In anglophone countries, MOAs carried out many functions critical to development of the agricultural sector, and they generally had one of the largest budgets among the government ministries. The marketing boards, however, were maintained as separate entities or parastatals. Each board handled a different commodity and was given the authority to set prices and organize supply. Kenya, for example, had 12 separate agricultural boards, one of which supported irrigation (Trapman, 1974), while others regulated coffee, tea, sisal, cotton, cereals and milk. The boards all carried out similar functions but were orientated to serving smallholder farmers, particularly in the former settler colonies. This was done by expanding state buying stations into smallholder areas and, with links to MOA departments, providing subsidized inputs and credit (Jayne and Jones, 1996). The MOAs had departments for research and extension, agricultural credit and infrastructure development, all of which were organized at national, provincial and district levels, but with decision-making centralized at the national level. MOAs in francophone countries had less of an implementation role and more of a planning and coordination role. As Moris (1994, p. 39) notes:

> [The MOA] serves in a technical and advisory capacity, planning the project portfolio for publicly financed activities in the agricultural sector, which, once they have become part of the national plan and subsequently the annual budget, are usually implemented by an array of parastatal institutions dealing with individual export crops (groundnuts, cotton, cereals, and sugarcane).

These projects were usually implemented by region, with significant variation in project implementation depending on institutional capacity, which up to the 1980s depended significantly on expatriate staff.

The critical changes in the relevant institutional 'architecture' after independence are reflected in the role and structure of the MOA. The functions carried out by the MOA over this period have reduced significantly over time, which has affected how these functions are coordinated and funded in the wider ecosystem.

Much of the funding of agricultural services and functions is dependent on annual budgetary allocations, primarily through MOAs. So as many of those functions shifted out of the MOAs, the tendency towards underinvestment in the overall agricultural institutional framework developed. In many cases government budgets were supplemented by international aid funds, but aid priorities which supported agricultural projects through to the mid-1990s shifted away from agriculture to rural education and health. At the same time funding of agricultural faculties comes through ministries of education, which has often served to isolate those faculties from principal policy initiatives in the agricultural sector and to hinder the maintenance of balance in the aforementioned agricultural knowledge triangle (agricultural research, extension and university institutions).

Fragmentation in Agricultural Research

Significant agricultural research capacity had been developed in most countries at independence, but it was majorly under-resourced. As Roseboom *et al.* (1998, p. i) note, 'In 1961, public (agricultural research) systems in 33 of 48 African countries employed fewer than 25 full-time equivalent (FTE) researchers'. This also included a number of regional agricultural research centres into which most of the countries were integrated during the colonial period. In the anglophone countries, this included: (i) the East African Agricultural and Forestry Organization (EAAFRO), based in Kenya; (ii) regional institutes for palm oil, cocoa and rice in West Africa and (iii) the Empire Cotton Growing Corporation's central station at Namulongue in Uganda. In West Africa, the regional institutes were nationalized at independence. EAAFRO continued to function until the collapse of the East African Community in 1977, at which point the research stations were integrated into the science division of the Kenyan MOA and the Namulonge station into the Ugandan MOA. The collapse of regional structures meant that many countries lost access to research support for important crops in their economy and lost access to the institutional memory that had been built over time. In many cases, the newly independent countries inherited a research infrastructure that did not undertake research relevant to local smallholder needs.

Research capacities were now incorporated in various MOA departments in many countries, resulting in fragmentation and poor coordination of research efforts. Moreover, to focus on stakeholders, priorities shifted away from export crops to food crops. This often led to competition among research departments for funding to build capacities, which were sometimes duplicated. The macroeconomic mismanagement and stagnant growth of the 1980s and the inefficiencies of the MOAs led to the restructuring of agricultural research. Many countries created national agricultural research institutes (NARIs), which were autonomous parastatals with centralized agricultural research management and budgeting, although in most cases via MOA budgets.

The development of agricultural research institutions in francophone Africa took a different tack. At independence in 1960, the federal system of agricultural research created by the French disappeared, as each newly-independent country ostensibly assumed responsibility for its own research. However, given the shortage of personnel, research centres in the former French colonies continued to be managed by eight research institutes in France up until the mid-1970s, based on agreements with France at independence. As Roseboom *et al.* (1998, p. 18) describe:

> A continuation of French ownership or long-term rights of access to the local research infrastructure throughout Africa formed part of the cooperative agreements that France signed with nearly all the French African territories that gained independence during the late 1950s and early 1960s.

In 1970, these France-based centres were brought under one administrative structure, Groupe d'Etudes et des Recherches pour le Développement de l'Agronomie Tropical, which evolved into CIRAD (the French agricultural research and international cooperation organization), and became the principal organization for French aid in agriculture. Moreover, by the mid-1970s, most francophone countries had taken over full management of the research centres in their respective countries. Because of this history, institutional development was a much slower process in French West Africa than in the former British colonies, especially in terms of the 'Africanization' of these institutions and their ability to focus on local problems.

Integrated Rural Development, and Training and Visit Extension

Agricultural extension in the anglophone countries was administered by MOAs and focused on food-crop production by smallholders. Different methods were employed, including farmer training centres, model farms, demonstration trials and learning from progressive farmers. Ruthenberg (cited in Lele, 1975, p. 64) summarized these different approaches as either the 'take it or leave it' approach or the contract-farming method. The first essentially offered advice or techniques to farmers, leaving it up to them to decide whether to adopt. This avoided the coercion used during the colonial period, which was usually associated with cash crops (e.g. tea or cotton) and was based on contractual forms of participation. MOA extension agents were the front line in credit and subsidized-input programs, which targeted cash-crop production (Mukhwana, 2000). Contract farming arrangements were also used in the francophone countries, particularly in cotton and groundnut programmes, often organized through farmer cooperatives.

Robert McNamara's 1973 speech in Nairobi shifted the development agenda to poverty alleviation, which, in Africa, meant rural poverty. This led to the World Bank's support for integrated rural development programmes, or IRDPs (Lele, 1975; Yudelman, 1976). These programmes initially built on colonial-period rural-development projects, particularly in anglophone countries, which integrated various services in particular regions and relied on MOAs, especially extension departments (Hodge, 2007). Extension personnel were still relatively few in relation to the smallholder population, and not well-trained. IRDPs increased the training of extension personnel and expanded extension capacity. As the term 'integrated' implies, technical advice was combined with credit and input-delivery programmes. These programmes, however, were found to be too complex for the institutional capacities in place in that period, and investment in IRDPs declined and eventually ceased (World Bank, OED, 1992).

The decline in support to IRDPs coincided with the World Bank's new initiative, training and visit (T&V) extension. The Bank had supported this extension model in Asia in the mid-1970s and piloted the approach in Kenya in 1981, leading to a 1983 loan supporting a complete reorganization of extension. The Bank would go on to support T&V projects in 30 African countries (Bagchee, 1994). Extension in these countries was organized around a single model consisting of (i) a hierarchical management structure, (ii) a rigid schedule of visits to pre-selected contact farmers, (iii) a focus on a few dominant crops, (iv) periodic interaction with research staff and (v) a significant expansion in and training of extension personnel (Anderson *et al.*, 2006). The T&V approach dominated extension departments within MOAs in Africa through the 1980s and into the early 1990s, when World Bank support for it ended. The financial sustainability of the T&V approach was increasingly in question, and its use declined when the World Bank withdrew its support (Mukhwana, 2000).

Expanding National Universities and Government Management

An enormous deficit remained in the ability of newly-independent nations to staff and manage the complex institutions necessary for agricultural growth and development - including faculties of agriculture themselves. At independence, five higher-education faculties provided a degree in agriculture in the anglophone countries, one in francophone West Africa and one in the Congo. The former university colleges affiliated with the University of London were re-constituted as national universities. In East Africa, Makerere University was reconfigured in 1963 as one of three colleges constituting the University of East Africa, joining more recently-established colleges in Nairobi and Dar es Salaam. The vast majority (90%) of the academic staff in the newly-constituted university were expatriates (Court, 1991). The University of East Africa ceased to exist in 1977 when the East African Community collapsed, and its units in each of the three countries became national universities.

During the decolonization in the 1960s, American foundations (particularly Ford, Rockefeller and Carnegie) supported the development of higher education in SSA (Livsey, 2017). Support was also provided by the United States Agency for International Development (USAID) and the World Bank. These programmes had two prongs: higher-degree training of Africans in northern universities, and capacity-strengthening programmes for new national universities. The result was a significant expansion of African universities in the 1960s and 1970s, and particularly those offering agricultural and veterinary degrees through faculties of agriculture (Fig. 3.1). Nevertheless, enrolment in agricultural degree programmes was not high. In 1979–1981, of all tertiary students in 21 countries, only about 6% and 7% were enrolled in BSc- and MSc-level agricultural degree programmes, respectively. For 19 countries outside Nigeria and South Africa, which had by far the largest number of students, this amounted to only 4,102 students at the BSc level and 493 students at the MSc level, compared with 3,931 students at both levels in Nigeria (Beintema *et al.*, 1998).

USAID was an early donor to the development of African TAE, particularly based on adapting the land grant model which had developed in the United States. One of the earliest investments was in 1953, when funding was given to Oklahoma State University to develop an agricultural university at Alemaya in Ethiopia. This was followed by support to Michigan State University to apply the land-grant model in the creation of the University of Nigeria, at Nsukka in eastern Nigeria, at independence in 1960 (Livsey, 2017). Ahmadu Bello University in the northern region and the University of Ife in west also received support. Altogether, 11 universities in Africa received USAID support, each for a decade, which is long compared with the standard length of donor aid projects (Alemneh, 2014). The land-grant model integrated education, research and extension, with each reinforcing the other two. At Ahmadu Bello University in Nigeria, a research institute (at Zaria) was integrated with the university and a 'research-liaison' (extension) department was created, all supported by the provincial Northern MOA (Eicher and Haggblade, 2013). However, most countries failed to achieve this functional integration, mainly because the administration of research and extension were separated under the MOAs on the one hand, and the education ministries on the other (Eicher and Haggblade, 2013).

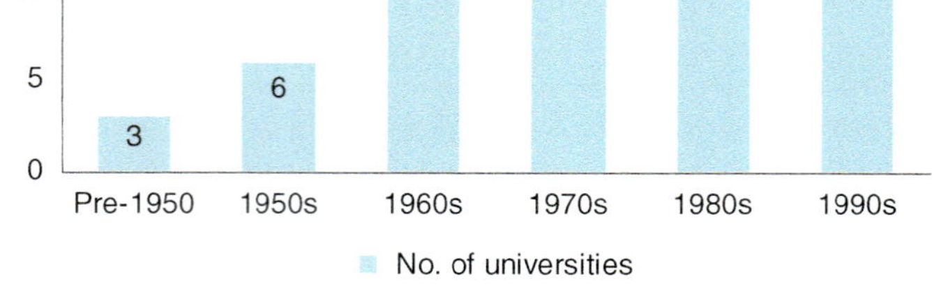

Fig. 3.1. Number of African universities established with a faculty of agriculture or veterinary science, pre-1950s–1990s
Source: Beintema *et al.* (1998, Fig. 1).

3.4. Structural Adjustment and Market Liberalization, mid-1980s to mid-2000

Market Liberalization and the Demise of Marketing Boards

The 1980s in SSA was a period of stagnant economic growth, declining per capita food production and struggling institutions. Expansion of primary education and health services stretched government budgets, and countries tended to run large budget deficits and maintain overvalued exchange rates. Overstaffing overwhelmed the budgets of public institutions, leaving little for capital investment and operations. Government-controlled farm prices and regulations on the movement, sale and trade of produce dampened agricultural investment incentives. By the mid-1980s, this period of 'African pessimism' led to the formulation by the World Bank and International Monetary Fund (IMF) of policy reforms designed to reverse these trends. These structural adjustment and market liberalization policies were implemented as conditions of loans (Woomer and Mukhwana, 2004). African countries depended on these loans because private foreign investment was minimal. Policy reforms from the mid-1980s to the turn of the century were targeted towards achieving macroeconomic stability, which translated into control over deficit spending, liberalization of foreign exchange controls, relaxation of trade restrictions and liberalization of markets, especially agricultural markets. These reforms led to a major restructuring of the institutional ecosystem supporting the agricultural sector.

Agricultural policy reform was implemented around three policy and institutional changes underlying the market liberalization process in the agricultural sector. The first focused on liberalization of agricultural output and input prices. This required the elimination of subsidies and constant pricing across the country, which was also tied to liberalization of foreign exchange markets. Second was the removal of myriad regulatory controls put in place for public management of agricultural markets, such as restrictions on internal movement of agricultural produce. The final reform focused on the institutions themselves, particularly restructuring or eliminating marketing boards and privatizing processing parastatals. (Kherallah *et al.*, 2002). The result of these policy reforms would be to disengage the public sector from participating directly in agricultural markets, thereby encouraging growth of private-sector participation and investment (Woomer and Mukhwana, 2004).

This reordering of functions between the public and private sectors in the agricultural economy would provide the rationale for redesigning public institutions to support agricultural development. Marketing boards were either dismantled or turned into strategic grain reserves, such as the Agricultural Development and Marketing Corporation in Malawi. Significant parts of many MOAs lost their roles in a liberalized market economy, particularly those departments providing services which the private sector could provide. As might be expected with such a quick shift to reliance on a liberalized market economy, a significant lag resulted in the entry of the private sector into the agricultural economy. This was especially the case because lack of infrastructure, uncertainty as to whether the government would fully extricate itself from market interventions and the high costs of integrating smallholders into the market economy all resulted in higher returns to investment in the urban sector (Mukhwana *et al.*, 2004). Stepping into the breach was civil society, whose role significantly expanded with the surge in democratization across the continent (Bates and Block, 2013). Non-governmental organizations (NGOs), farmer associations and cooperatives, microcredit facilities and other forms of collective action filled the institutional void left by the market-liberalization process.

Redefining Extension

The World Bank played a dominant role in African agriculture during the 1990s. At the same time as it was imposing structural adjustment policies, the Bank was also increasing its loans for T&V extension. In the end, the two were incompatible: T&V extension required a significant expansion in the number of extension personnel, and their mobility, leading to significant increases in recurrent budgets. This occurred at the same time as other parts of the Bank were supporting macroeconomic reform, which demanded budgetary restraint. A uniform T&V extension model had been rolled out in 30 countries, beginning in 1983, but by 1998 all support to T&V extension was terminated. A 1997 report by the Bank's office of evaluation found the following:

> [T]he Bank had erred in the extent to which it has promoted the T&V extension management system in relatively uniform packages of investments and extension practices in large state and national programs . . . the increased flexibility in African T&V projects was not likely to resolve the problems encountered and therefore the T&V design was unlikely to be the most appropriate approach for improving extension in many African countries (cited in Anderson *et al.*, 2006, p. 21).

The rapid withdrawal left a vacuum in agricultural extension capacity. This then prompted a period of experimentation, primarily through NGOs, with agricultural extension methods, including the farmer field school methodology piloted by the Food and Agriculture Organization of the United Nations, and new forms of media, such as local language radio and video. Farmer participation, involving group methods and farmer associations, formed the cornerstone of most of the new approaches. Collective action by farmers was viewed as a precondition for cost-effective integration of smallholders into expanding agricultural markets (see Ch. 2, this volume). The new pluralism in agricultural extension rejected a uniform methodology delivered by a single agency and argued for multiple methods to be offered by a range of providers, including an expanding private sector. The potential downside of this approach was possible bias in the coverage of the agricultural population, with a tendency for under-coverage of marginal areas or those with high rates of poverty. The experimentation included methods linking smallholders to markets. Civil society responded with integrated approaches intended to fill the vacuum in both extension services and private-sector market development (Woomer and Mukhwana, 2004).

Restructuring Ministries of Agriculture and their Changing Role

During this period, the role of MOAs fundamentally changed, and many of the former functions were shifted to other institutions. Research moved to the NARIs, for example, and market coordination moved to the private sector. Financial support for extension collapsed with the withdrawal of the World Bank from T&V extension, and pilot projects devolved extension to districts as decentralization took hold. The implementation functions of the MOA were progressively devolved, and the role of the ministry was relegated to regulation, facilitation and the provision of public goods. Budgets for agriculture were pared to critically-low levels - in some cases, to 3% of the overall government budget[2]. The African Union meeting in Maputo in 2003 issued a declaration stating that government spending on agriculture should reach a target of 10% of overall public expenditure. In many ways this was an opportunity for the MOA to reimagine itself and redefine its role, but in most cases the MOA retrenched and was unable 'to respond to such diverse constituencies, claims and demands, operating in an increasingly democratic but also decentralized setting. As technical line ministries they [were] often poorly equipped to do this' (Cabral and Scoones, 2006, p. 30).

At the turn of the century, the World Bank and IMF realized that structural adjustment policy reform was having a negative impact on poverty rates which, in 1999, led them to require borrowing countries to prepare a poverty reduction strategy. Targeted investments in rural development became a principal means of poverty reduction, generating debate as to whether rural development strategies should be primarily focused on agriculture or be extended to health and education services. It was easier for governments to direct annual budgeting to health and education than to the more complex relationship between agriculture and poverty. A review by the World Bank (2004, p. 13) noted that there was a

> disconnect between [Poverty Reduction Strategy Paper (PRSP)] priorities and follow-up of the rural themes through Poverty Reduction Strategy Credits (PRSCs), other instruments, and other donors. This disconnect is most pronounced for rural finance, rural private sector development, food security, risk and vulnerability, livestock, decentralization and governance issues. On the other hand, rural education, health and water supply and

> sanitation, issues related to [natural resource management], the incentive system, as well as agricultural support services, are followed-up in one way or another in all countries where these had been prioritized in the PRSP.

The marginalization of the MOA and the lack of public-sector spending on agriculture was eventually viewed as counterproductive given the sector's importance to the economy and the weight of rural voters in the democratic process. This led to the development of agricultural plans, which were viewed as critical to justifying the sector's strategic importance in the policy process. In Uganda, the Plan for the Modernization of Agriculture was put in place in 2001, with the overall goal of reducing poverty but 'through a profitable, competitive, sustainable and dynamic agricultural and agro-industrial sector' (Oxford Policy Management, 2005, p. 2). In Kenya, the Strategy for the Revitalization of Agriculture, developed in 2004, vested implementation in an interministerial agricultural sector coordination unit. Such strategies were designed to support development led by the private sector through the integration of smallholders into an expanding market economy. These country-level initiatives were reinforced by the African Union's 2003 launch of the Comprehensive Africa Agriculture Development Programme (CAADP), which supported an Africa-wide process for developing investment plans for the sector.

A Return to Regional Approaches in Research

After rapid growth in funding of agricultural research in the 1960s (6.8% per annum), the rate of annual spending growth

> declined from 2% in the 1970s to 1.3% in the 1980s, and to only 0.8% in the 1990s. Excluding Nigeria and South Africa, total public agricultural R&D spending in Africa actually declined by 0.3% per year in the 1990s (Beintema and Stads, 2004, p. 241).

Although the early reforms in agricultural research were in part supported by donor funding, this dependence increased until the turn of the century. In the early 1980s, the donor share of expenditure averaged 34%, increasing to 43% in 1991 - or 49%, excluding Nigeria (Pardey *et al.*, 1995). By 2000, donor funding accounted for an average of 35% of spending (Beintema and Stads, 2004). Although donor funding was volatile, it accounted for much of the operating and capital budgets of NARIs. Therefore, the termination of large programmes by the World Bank and other donors had a destabilizing effect on research.

Their relatively large investments in agricultural research gave donors a vested interest in improving its effectiveness, especially through re-structuring. The trend throughout the 1990s was to create NARIs, which consolidated research capacities under a unified management structure with both budgetary and decision-making autonomy. Such reorganization resulted in a decline in fragmentation (Pardey *et al.*, 1995). By 1991, 28 of 48 countries had adopted the NARI model (Roseboom *et al.*, 1998). To address the poor performance of agricultural research during the 1990s and the increasing move towards decentralization, another institutional reform process was widely adopted at the turn of the century:

> A major new dimension in the overall focus of the reform agenda . . . is the shift from centralization to decentralization. Moreover, the new reform agenda forces agricultural research to be more outward looking, client oriented, and impact driven. It is not sufficient to produce good science: research organizations are being urged to ensure that their knowledge and technology are being applied, preferably by resource-poor, subsistence-oriented farmers in hitherto neglected areas (Chema *et al.* 2003, p. 23).

Decentralization was designed to meet local farmers' needs and to partner with the complex set of rural actors involved in extension and integration of smallholders into markets. The disarray in extension created pressure for NARIs to move further downstream in piloting innovative farmer-participation methodologies.

At the turn of the century, half the countries in SSA had less than 100 full-time equivalent researchers (Beintema and Stads, 2004). The size distribution of NARIs in SSA was heavily skewed. Seven countries had NARIs of what could be considered a robust size to manage the diversity in crops and ecologies, while NARIs in the remaining countries were unable to build the research capacity necessary to address all critical aspects of agriculture. More collaborative research structures were needed, and these built on the commodity-research networks managed by the CGIAR centres and the Southern African Centre for Cooperation in Agriculture Natural Resources Research and Training, which was created in 1984 to coordinate agricultural research in the Southern African Development Community countries. With donor support, the Association for Strengthening Agricultural Research in Eastern and Central Africa was created in East Africa in 1994 and an existing French network, the West and Central African Council for Agricultural Research and Development, was extended to anglophone and lusophone countries in West Africa. These sub-regional organizations (SROs) were managed by the executive directors of the NARIs in the member countries. The SROs did not have research capacity of their own but rather utilized competitive agricultural research grants to identify projects with regional impact (Roseboom and Flaherty, 2016, p. 42).

Fiscal Autonomy and Change in Universities and Tertiary Agricultural Education

Institutional change in TAE in this period was driven by overall changes at the university level. Universities by the mid-1980s were suffering from the same malaise as other institutions. They depended on government education budgets to meet their expenditures, and these competed with the increasing demand for primary and secondary education. Student costs were covered by government grants, and graduates were guaranteed a place in the public service. Because government controlled the purse, there was often political interference in the management of universities to contain student political activity. 'Real spending per student in Africa declined from an average of $6,300 in 1980 to $1,500 in 1988' (Beintema *et al.*, 1998, p. 15). Adequate staffing of faculty with graduate degrees was a challenge given the limited pool of talent, especially as staff salaries fell by 30% during 1980–1988 (Beintema *et al.*, 1998). Limited budgets, low salaries and increasing demand for university education significantly reduced institutional effectiveness. Structural adjustment programmes and reordering of spending priorities resulted in reduced reliance on government budgets, loss of full support to students and termination of assured student employment.

The university sector became more responsive to market forces to meet pressured demand for university degrees. Tuition fees became a key source of university finances and, at the same time, sparked a significant expansion in private universities. The investments countries had made in secondary education resulted in increasing demand for university education. Given the participation of youth in the democratic process, public universities expanded in number and size.

> Kenya had only two public universities in 1984 and Ethiopia only two up to 1991. By 2007, Ethiopia had 22 and, by 2013, Kenya also had 22, of which 15 were added in that year alone. In Kenya in 1983, university enrolment was 6,800 students. This grew to 60,000 by 2002 and to 200,000 by 2012 (Lynam, 2017, p. 33).

Such rapid expansion had numerous effects. Firstly, faculties of agriculture focused almost exclusively on teaching to accommodate the expanded enrolment. There were staff shortages, increased class sizes and recruitment of MSc-qualified lecturers. Quality of education became a key concern. Secondly, many experienced researchers in the NARIs transferred to universities to become teachers, draining research capacity. Finally, the concentration on teaching at the expense of research, and the focus on first-degree students, caused a drain on graduate teaching capacity (Mukhwana *et al.*, 2017).

At the turn of the century, faculties of agriculture had gone, firstly, through a phase of consolidation and, secondly, through a phase of expansion. Another cycle of consolidation was needed and, given the number of competing faculties, differentiation of institutions and programmes was needed. Some faculties were now located in rural areas, away from the capital city, and could focus on particular ecologies. Some had the potential to develop strong graduate programmes; others targeted the needs of the expanding private sector. The focus of governments and foreign aid donors was shifting back from rural social services to investments in agriculture (Kane and Eicher, 2004, p. 34):

> Between 1991 and 2001, the share of aid to agriculture in Africa fell from 19 to 10 percent, while that of social services, especially health and education, increased from 32 to 56 percent. As agriculture lost support among development organizations, both in absolute and relative terms, poverty in Africa has persisted, and unlike other regions that have made major advances in agricultural productivity to improve the livelihoods of the poorest, Africa has regressed.

Agriculture was poised to move back onto the development agenda, but this required strong, interacting institutions.

3.5 Positioning Tertiary Agricultural Education in a Dynamic Institutional Context

This historical survey demonstrates the dynamic changes in the institutional ecosystem supporting agriculture. It outlines how individual institutions responded to different drivers, and the interactions of market liberalization, democratization and decentralization. Over this period of time, the functions performed by the various institutions significantly shifted. The key question for this chapter, therefore, is how TAE should position itself within this institutional framework moving forward. To assess this requires an understanding of how functions are distributed within this ecosystem, with TAE institutions having primary responsibility for building human-resource capacity, but with other institutions contributing from the vantage point of their specialization. Another feature of this ecosystem is how the different institutions in the agricultural sector interact with regard to the larger agricultural development agenda. This interaction, while not necessarily coordinated, forms what is now termed an AIS and is based on collective institutional self-interest (World Bank, 2012). Over the past 50 years, institutions often acted with little coordination. Today, given its mandate of education, research, extension and policy, TAE has the potential to foster interaction among the institutional actors in the AIS (Spielman *et al.*, 2008). This section presents an assessment of the role of TAE within these functional domains and in relation to the other institutions operating within them.

Tertiary Agricultural Education in the Context of Education and Human-Resource Development

TAE institutions are the major providers of the knowledge, training and skills necessary to drive the agricultural growth process in Africa. Faculties of agriculture face the challenge of meeting the needs of a highly-differentiated labour market, which is increasingly driven by the requirements of the private sector. Few other institutions apart from TAE institutions are so important in training the agricultural labour force, although institutions providing training for specialized skills, such as business management, are also of relevance to agriculture.

As this volume details, the TAE sector is undergoing significant change, driven by market forces, namely: (i) a differentiated labour market, (ii) a tuition-paying student population which can choose courses and faculty and (iii) an increasing number of competing sources of supply. The rapid expansion in public and private universities has created a huge increase in demand for PhD-qualified teaching staff, which is not being met by supply. This limits pedagogy, innovation in the development of curricula and graduate courses, quality assurance and research capacity. The enhanced competitive environment should eventually lead to improved quality and differentiation across faculties of agriculture, whereby some faculties focus on research and graduate education. To a certain extent, this differentiation process is occurring, facilitated by regional network approaches (see Ch. 14, this volume) and drawing on academic expertise where it exists in the region. The current need is for differentiation of faculties such that a few specialize in providing high-quality graduate education. In many faculties of agriculture, class sizes are too small to justify the investment needed to develop high-quality graduate programmes. The World Bank programme, Partnership for Skills in Applied Sciences, Engineering and Technology, has promoted this process by supporting centres of excellence (World Bank, no date). However, such faculties must meet the criteria to justify the title of 'centre of excellence', and that comes from (i) critical changes in the internal development of graduate programmes, (ii) the pursuit of excellence and (iii) staff involvement in other critical functions beyond just teaching (which is where staff time is currently focused) to include research, community development and policy.

Tertiary Agricultural Education within Research Systems

In the design of the colonial research system, research was considered essential for degree training, but degree training was not considered essential for research. Independent research institutes were created with little linkages to the, then, embryonic university system. This separation of the research and training functions characterized the evolution of the two institutions until the turn of the century. With the critical focus on graduate programmes, research capacity in faculties of agriculture has become the necessary means of keeping teaching staff up-to-date with current knowledge, and to ensure the quality of MSc and PhD programmes. A strong argument currently exists for a more-effective division of labour between national agricultural research institutes and faculties, along with improved synergy between the two. Such a division of labour could be founded on the organizational basis of the research: namely, disciplines in universities, and crop, livestock and natural resource management programmes in NARIs. Examples of this could be the development of molecular-breeding techniques in faculties for application in the crop-breeding programmes of the NARIs. Similar examples exist in the area of soils research, plant pathology and entomology. Faculties would have a particular comparative advantage in agricultural economics, food technology and agricultural engineering. Moreover, graduate students are a highly effective means of carrying out dedicated research - assuming appropriate supervision and material support. Multiple students could be organized around a multifaceted research problem, whether in the NARIs or in the faculties themselves. All of this argues for strong research partnerships to maintain the skills of academic staff and develop strong graduate programmes.

Research in faculties of agriculture is currently driven by funding, and partnerships are usually organized around grants. Sources of such funding vary by country but include (i) national competitive research funds, usually administered by NARIs, (ii) the grant programmes of the SROs or the Regional Universities Forum for Capacity Building in Agriculture (RUFORUM), (iii) a few international research-support programmes, such as those provided by foundations and (iv) partnerships with CGIAR centres or northern universities through the grants they have received. Faculty staff compete with the NARIs for funding from all these sources. At the same time, faculties are hiring staff from NARIs and are increasing their competitiveness. Success in attracting research funds requires networking, technical skills, writing skills and the inclination to devote time to developing and executing grants over and above teaching responsibilities.

Some faculties and universities are now branding themselves as research universities with strong graduate programmes. This requires developing institutional research capacities recognized by the broader scientific community, not just the capacities of individual scholars. Examples of this are the plant-breeding and molecular biology programmes at Makerere University and the University of Ghana, in collaboration with the West Africa Centre for Crop Improvement, which has invested in laboratories, greenhouses, experiment farms, skilled staff, partnerships with similar institutions, advanced curricula and competitive recruitment of graduate students. Faculties of agriculture across East and Southern Africa are developing new or expanded research and teaching programmes in dryland agriculture, integrated soil-fertility management, livestock production systems and other topical areas. The locus of competitiveness in these centres of excellence is the academic department, facilitated by partnerships and funding. Going forward, research will be a significant thrust of faculties of agriculture seeking to strengthen their graduate programmes. The research will be led by strong departments whose programmes complement, rather than displace, the research undertaken in NARIs (see Ch. 8, this volume).

The Role of Tertiary Agricultural Education in Extension and Community Development

Faculty participation in extension should reinforce teaching and student education. Extension is one of several means of locating student research in ongoing field sites, thereby developing synergies between research and extension. Effective extension involves (i) an understanding of smallholder systems, (ii) collaboration with farmer groups and community development organizations and (iii) partnership with other rural institutions. Extension encompasses processes of rural innovation, which some academic departments are now organized to address. An example is the RUFORUM–supported collaborative PhD programme in rural innovation at Makerere University, Egerton University and Sokoine University of Agriculture, where faculties test and evaluate new methodologies, such as innovation platforms (Dror *et al.*, 2016). Properly designed and led, research and extension promoting rural innovation can be a point of interaction with other actors in rural development, and a bridge to farming communities.

The Role of Tertiary Agricultural Education in the Policy Arena

For much of the post-independence period, faculties of agriculture have been marginalized in the agricultural policy process. The absence of universities in policy formulation and analysis is partly because diverse faculties could not speak with one voice, as compared with the MOA. Also, most faculties of agriculture are located outside the capital city, and departments of agricultural economics - where much of the policy expertise has resided - could not coordinate and respond quickly to emerging policy issues. Moreover, MOAs have their own policy-research units which support ministerial policy processes. This began to change with market liberalization in the late 1990s and with the support of the African Capacity Building Foundation, which funded the creation of dedicated policy-research institutes in several African universities. Examples are the Economic Policy Research Centre at Makerere University, a part of which works on agricultural policy, and Tegemeo Institute of Agricultural Policy and Development at Egerton University. Some research institutes take a long view of policy issues by investing in their own surveys and data collection, and by conducting independent research on policy-relevant issues. The institutes struggle, however, to find the funding required to maintain collection of their own core data. Most agricultural policy research institutes draw on graduate students to assist in undertaking research, but student involvement should be increased. The institutes should develop clearly-delineated research priorities and help universities build their capacity to address those priorities. Research institutes also organize seminars for policy-makers on their latest research, thus becoming a more proactive voice in the policy process. A question going forward is whether an inter-university capacity is possible, thereby strengthening agricultural policy research across the faculties, or whether developing such capacity is part of the institutional differentiation process.

3.6 Conclusion

Growth and development of the agricultural sector is supported by a range of institutions which provide several critical functions and deliver a range of services. Most of these are public institutions, but in historical terms the number of private institutions has increased, and the role of civil society organizations has expanded. TAE institutions are a critical component of this institutional architecture; they provide a key function in training the human-resource capacity which feeds into these institutions and, increasingly, supports private-sector expansion. To prepare students for these roles, TAE institutions must take on some of the functions provided by these other institutions, particularly in the areas of research, extension and policy. These additional functions strengthen teaching and skills development, and at the same time provide points of interaction with the other institutions supporting the agricultural sector. In many cases, these additional functions and institutional linkages are latent, given heavy teaching loads and scarce financial resources. Faculties of agriculture can play a pivotal role in the development of a functional AIS because they provide the knowledge and skills for those who will staff other institutions within the AIS. Understanding the changing and dynamic nature of the institutional ecosystem of agriculture is a key factor in TAE institutions' ability to position themselves to fulfil that role.

Endnotes

[1] In the context of this chapter, 'institutions' refer, principally, to formal organizations. Alternatively, the more specific use of the term refers to the norms, laws, and regulations that govern economic activity, as is particularly used in institutional economics.

[2] In a public expenditure review of agricultural spending in Kenya in 2004 involving the Ministry of Agriculture, Ministry of Livestock and Fisheries Development and the Ministry of Cooperative Development and Marketing, the findings included: (i) a 'Sharp decline in overall public expenditure in agriculture as a percentage of total government spending: 10% after Independence, an average of 7.5% in 1980–1989, 3% in 1990–2000 and 3.1% in 2002/03" and (ii) "Recurrent expenditure dominated by salary payments [accounting for] 70–90% of total expenditure in the three ministries. . ..' (cited in Cabral and Scoones, 2006, p. 21).

References

Alemneh, T. (2014) Agricultural higher education in sub Saharan Africa: Partnerships and the land-grant model. In Swanepoel, F. Stroebel, A. and Ofir, Z. (eds.) *Towards Impact and Resilience: Transformative Change In and through Agricultural Education and Training in Sub-Saharan AJC&PABfrica.* Cambridge Scholars Publishing, Newcastle upon Tyne, UK.

Anderson, J.R. Feder, G. and Ganguly S. (2006) *The Rise and Fall of Training and Visit Extension: An Asian Mini-drama with an African Epilogue.* World Bank Policy Research Working Paper 3928. Available at: https://www.researchgate.net/publication/23549956_The_Rise_and_Fall_of_Training_and_Visit_Extension_An_Asian_Mini-Drama_with_an_African_Epilogue (accessed 3 March 2021).

Ashby, E. (1966) *Universities: British, Indian, African. A Study in the Ecology of Higher Education.* Harvard University, Press, Cambridge, Massachusetts, USA.

Bagchee, A. (1994) *Agricultural Extension in Africa.* World Bank Discussion Papers 231. Available at: http://documents1.worldbank.org/curated/en/479471468742860560/pdf/multi-page.pdf (accessed 3 March 2021).

Bates, R. and Block, S. (2013) Revisiting African agriculture: Institutional change and productivity growth. *Journal of Politics* 75(2), 372–384. Available at: https://www.journals.uchicago.edu/doi/abs/10.1017/S0022381613000078 (accessed 3 March 2021).

Beintema, N. Pardey, P. and Roseboom, J. (1998) *Educating Agricultural Researchers: A Review of the Role of African Universities.* EPTD Discussion Paper 36. International Food Policy Research Institute, Washington, DC. Available at: https://www.ifpri.org/publication/educating-agricultural-researchers (accessed 15 February 2021).

Beintema, N. and Stads, G.-J. (2004) Sub-Saharan African agricultural research: Recent investment trends. *Outlook on Agriculture* 33(4), 239–246. Available at: https://www.asti.cgiar.org/pdf/Outlook.pdf (accessed 3 March 2021).

Cabral, L. and Scoones, I. (2006) *Narratives of Agricultural Policy in Africa: What Role for Ministries of Agriculture?* Future Agricultures Consortium, Brighton, UK. Available at: https://assets.publishing.service.gov.uk/media/57a08c39ed915d3cfd00124c/MoA_paper.pdf (accessed 3 March 2021).

Chema, S. Gilbert, E. Roseboom, J. (2003) *A Review of Key Issues and Recent Experiences in Reforming Agricultural Research in Africa.* ISNAR Research Report 24. International Service for National Agricultural Research, The Hague.

Court, D. (1991) The development of university education in sub-Saharan Africa. In Altbach, P.G. (ed.) *International Higher Education: An Encyclopedia*. Vol. I, St. James Press, Chicago and London, pp. 329–347.

Develtere, P. (2008) Cooperative development in Africa up to the 1990s. In Develtere, P. Pollet, I. and Wanyama F. (eds.) *Cooperating Out of Poverty: The Renaissance of the African Cooperative Movement.* International Labour Organization, Geneva.

Dror, I. Cadilhon, J. Schut, M. Misiko, M. and Maheshwari, S. (eds.) (2016) *Innovation Platforms for Agricultural Development*. Routledge, New York.

Eicher, C. (1999) Institutions and the African farmer. *Issues in Agriculture* 14. CGIAR Secretariat, Washington, DC. Available at: https://cgspace.cgiar.org/bitstream/handle/10947/5763/issues14.pdf?sequence=1 (accessed 3 March 2021).

Eicher, C. and Haggblade, S. (2013) *The Evolution of Agricultural Education and Training: Global Insights of Relevance for Africa.* MAFS Working Paper 4. The Modernizing African Food Systems Consortium, East Lansing, Michigan, USA. Available at: https://ideas.repec.org/p/ags/midasp/11816.html (accessed 3 March 2021).

Eisemon, T.O. Davis, C.H. and Rathgeber, E.M. (1985) Transplantation of science to anglophone and francophone Africa. *Science and Public Policy* 12, 191–202.

Hodge, J.M. (2007) *Triumph of the Expert: Agrarian Doctrines of Development and the Legacies of British Colonialism.* Ohio University Press, Athens, Ohio, USA.

Jayne, T.S. and Jones, S.P. (1996) *Food Marketing and Pricing Policy in Eastern and Southern Africa: Lessons for Increasing Agricultural Productivity and Access to Food*. Food Security International Development Policy Syntheses 11337, Department of Agricultural, Food, and Resource Economics, Michigan State University, East Lansing, Michigan, USA. Available at: https://ideas.repec.org/p/ags/midiwp/54694.html (accessed 3 March 2021).

Kane, S. and Eicher, C.K. (2004) *Foreign Aid and the African Farmer*. Staff Paper 2004-13. Department of Agricultural Economics, Michigan State University, East Lansing, Michigan, USA. Available at: https://ideas.repec.org/p/ags/midasp/11602.html (accessed 3 March 2021).

Kherallah, M. Delgado C., Gabre-Madhin, E. Minot, N. and Johnson, M. (2002) *Reforming Agricultural Markets in Africa: Achievements and Challenges.* Johns Hopkins University Press, Baltimore, Maryland, USA.

Lele, U. (1975) *The Design of Rural Development: Lessons from Africa.* Johns Hopkins University Press, Baltimore, Maryland, USA.

Livsey T. (2017) Multilateral negotiations: Nigerian universities, the United States, and the Cold War. In *Nigeria's University Age*. Cambridge Imperial and Post-Colonial Studies Series. Palgrave Macmillan, London.

Lynam, J.K. (2017) Assessing the CGIAR's approach to capacity development in R&D in sub-Saharan Africa. Annex B in *Evaluation of Capacity Development Activities of CGIAR: Evaluation Reports and Reviews.* Vol. III, Issue Papers. Independent Evaluation Arrangement of CGIAR, Rome. Available at: https://cas.cgiar.org/sites/default/files/pdf/CD-Eval-Report-Vol-III-Issue-Papers.pdf (accessed 3 March 2021).

Masefield, G.B. (1972) *A History of the Colonial Agricultural Service.* Clarendon Press, Oxford.

Morgan, D.J. (1964) *Colonial Development: A Factual Survey of the Origins and History of British Aid to Developing Countries.* British Aid-5. Overseas Development Institute, London. Available at: https://www.odi.org/sites/odi.org.uk/files/odi-assets/publications-opinion-files/8077.pdf (accessed 3 March 2021).

Moris, J. (1994) *Extension Alternatives in Tropical Africa.* Agricultural Administration Unit, Occasional Paper 7. Overseas Development Institute, London.

Mukhwana E. J. (2000) Food security and the impact of agricultural development on small scale farmers in Kenya. Overseas Development Institute, London. *Agricultural Research and Extension Network Newsletter* 41, 21–26.

Mukhwana, E. Gabre-Madhini, E. and Njuki, S. (2004) Sustainable smallholder marketing initiatives in Africa. Presented at the International Food Policy Research Institute workshop in Entebbe, Uganda.

Mukhwana, E.J. Kande, A and Too, J. (2017) Transforming university education in Africa: Lessons from Kenya. *African Journal of Rural Development* 2(3), 341–352.

Nwauwa, A.O. (1993) The British establishment of universities in tropical Africa, 1920–1948: A Reaction against the spread of American 'radical' influence. *Cahiers d'étudesafricaines* 33(130), 247–274.

Oxford Policy Management (2005) *Evaluation Report: The Plan for the Modernization of Agriculture.* Oxford Policy Management, Oxford, UK.

Pardey, P.G., Roseboom, J. and Beintema, N. (1995) *Investments in African Agricultural Research.* EPTD Discussion Paper 14. International Food Policy Research Institute, Washington, DC. Available at: http://citeseerx.ist.psu.edu/viewdoc/summary?doi=10.1.1.58.5348 (accessed 3 March 2021).

Roseboom, J. Pardey, P. and Beintema, N. (1998) *The Changing Organizational Basis of African Agricultural Research*. EPTD Discussion Paper 37. International Food Policy Research Institute, Washington, DC. Available at: https://www.ifpri.org/publication/changing-organizational-basis-african-agricultural-research (accessed 3 March 2021).

Roseboom, J. and Flaherty, K. (2016) The evolution of agricultural research in Africa: Key trends and institutional developments. In Lynam, J. Beintema, J. Roseboom, J. and Badiane, O. (eds.) *Agricultural Research in Africa: Investing in Future Harvests.* International Food Policy Research Institute, Washington, DC. Available at: https://www.ifpri.org/publication/agricultural-research-africa-investing-future-harvests (accessed 3 March 2021).

Sicherman, C. (2005) *Becoming an African University: Makerere 1922–2000.* Africa World Press, Trenton, New Jersey, USA.

Spielman, D.J. Ekboir, J. Davis, K. and Ochieng, C.M.O. (2008) An innovation systems perspective on strengthening agricultural education and training in sub-Saharan Africa. *Agricultural Systems* 98(1), 1–9. Available at: https://doi.org/10.1016/j.agsy.2008.03.004 (accessed 3 March 2021).

Trapman, C. (1974) *A Case Study of Kenyan Agricultural Development.* Overseas Development Institute, London.

Williams, G. (1985) Marketing without and with marketing boards: the origins of state marketing boards in Nigeria. *Review of African Political Economy* 12(34): 4–15. Available at: https://doi.org/10.1080/03056248508703647 (accessed 3 March 2021).

Woomer, P.L and Mukhwana, E.J (2004) Working with smallholder farmers to improve maize production and marketing in Western Kenya. *Uganda Journal of Agriculture* 9, 491–500.

World Bank (no date) Partnership for skills in Applied Sciences, Engineering and Technology (PASET). Available at: https://www.worldbank.org/en/programs/paset (accessed 3 March 2021).

World Bank. (2004) *A Review of Rural Development Aspects of PRSPs and PRSCs, 2000–2004*. World Bank, Washington, DC. Available at: http://www.hubrural.org/IMG/pdf/wb-review-rural-dev-in-prsp-prsc.pdf (Accessed 16 April 2021).

World Bank (2012) *Agricultural Innovation Systems: An Investment Sourcebook.* Agricultural and Rural Development, World Bank, Washington, DC. Available at: http://hdl.handle.net/10986/2247 (accessed 3 March 2021).

World Bank, OED (Operations Evaluation Department) (1992) *Evaluation Results for 1990–Rural Development Revisited*. World Bank, Washington, DC.

Yudelman, M. (1976) The role of agriculture in integrated rural development projects: The experience of the World Bank. *Sociologia Ruralis* 16(4), 308–325.

4 Global and African Trends in Tertiary Education

Aldo Stroebel[1]*, Melody Mentz-Coetzee[2] and Frans Swanepoel[2]
[1]*National Research Foundation, South Africa;* [2]*Centre for the Advancement of Scholarship, University of Pretoria, South Africa*

4.1 Introduction

Education, including tertiary education, is considered integral to achieving global development goals. As noted in Ch. 1, this volume, education is fundamental to the achievement of several of the Sustainable Development Goals (SDGs) and is directly linked to Goal 4, to 'Ensure inclusive and equitable quality education and promote lifelong learning opportunities for all' (United Nations, 2015, p. 21). Tertiary education benefits both individuals and societies. For individuals, tertiary qualifications are linked to improved employment opportunities and quality of life (OECD, 2018a). For societies, having a higher proportion of skilled and professional workers in the labour force increases economic growth through the generation of knowledge products and innovations which improve competitiveness (Pillay, 2010a; World Bank, 2017). In Africa, education systems and economic prospects are inextricably linked, and investments in higher education deliver a rate of return of 21% - the highest in the world. A one-year increase in average tertiary education levels eventually yields as much as a 12% increase in gross domestic product (GDP) (AAI, 2015).

This chapter examines several global trends in tertiary education, drawing parallels, where possible, with trends in African tertiary education, in particular tertiary agricultural education (TAE). The chapter also highlights key challenges in the broader African tertiary education system as the ecosystem within which TAE is situated.

4.2 Tertiary Education in the Context of Global Mega-Trends

Regardless of context, tertiary education is influenced to varying degrees and in different ways by social, economic and political trends and changes. At least five global 'mega-trends' will have a direct impact on tertiary education globally and in Africa. However, while it is likely that these trends will affect tertiary education across the globe, it is also highly likely that in several cases their impact in Africa may differ from their impact in the Global North. The five trends identified for discussion are: (i) population size and age shifts, (ii) increasing urbanization, (iii) rapid technological transformation, (iv) shifting geopolitics, including climate change and (v) competition for scarce resources.

*Corresponding author: aldo.stroebel@nrf.ac.za

DOI: 10.1079/9781789246544.0004

Population Size and Age Shifts

The medium variant[1] of the 2017 United Nations projections predicts that the world population will continue to grow and will not peak until the end of the century. Under this model, the global population will rise to over 8.6 billion by 2030 - an increase of approximately 1.1 billion from mid-2017. Furthermore, more than half of the expected population growth to 2050 is expected to come from Africa, translating into an additional 1.3 billion people. At this point, it is estimated that one in every four people on earth will live in Africa - an increase from the current 17% (UN DESA, 2017). In addition to growth in the absolute size of the population, the age distribution of populations is set to shift across the world. The Global North will have an older population than ever before, and aging will occur in other parts of the world as well, such that by 2050, nearly a quarter or more of the populations of all regions of the world, except Africa, will be aged 60 and above (UN DESA, 2017). Aging populations will have a significant impact on countries, placing strain on several systems - including pension schemes, health care, and social services. An older populace is also linked to slower economic growth as the size of the labor force decreases (World Bank, 2015, 2016; Lee, 2016).

In contrast, Africa is experiencing a 'youth bulge' (Ch. 2, this volume). African children under age 15 years accounted for 41% of the population in 2017, and people aged 15–24 years accounted for an additional 19% (UN DESA, 2017); thus, 60% of Africans were under the age of 24 years as of 2017. Furthermore, the proportion of the population aged 25 to 59 years (the age cohort accounting for most people of working age) is projected to continue to grow for many decades - from 35% in 2017 to 45% by 2090 (UN DESA, 2017). Furthermore, just in the period 2015–2035, the working-age population in Africa is projected to rise by approximately 450 million people (WEF, 2017a). This youthful population is, simultaneously, a potential asset and a potential risk. It could be an asset if Africa can capitalize on the demographic dividend - which is when a population is 'characterized by more people of working age and fewer dependents (children and elders) [which] can boost economic growth simply because a larger share of the population is productive' (WEF, 2017a, p. 3). However, this would require viable employment opportunities and other pathways to prosperity. The absence of such opportunities signals increased risk because rapid population growth (particularly in countries with weak institutions) is a very strong predictor of social or political instability: youth with poor prospects are more likely to protest, engage in criminal activity or join insurgent movements than youth with adequate employment opportunities (Goldstone, 2002; Cincotta *et al.*, 2003; Urdal, 2006; World Bank, 2013; Goldstone *et al.*, 2014; Walker, 2015).

The implication of these demographic trends is that Africa has a growing opportunity to address some of its most pervasive economic and social challenges by leveraging its youth population, but this will require intentional shifts in the design and implementation of policies and strategies to mitigate the risks associated with failing to provide livelihood opportunities. Undoubtedly, these population trends have significant implications for the region's tertiary education systems. As African countries seek to evolve into knowledge economies, an increasing number of skilled knowledge workers will be needed and, for individuals, attaining a tertiary education will become increasingly essential in securing employment and achieving prosperity. These national and individual needs will intensify the demand for access to tertiary education over the coming decades.

Demographic trends also have direct implications for TAE. A large proportion of Africa's workforce - both younger and older generations - continues to rely on the agrifood system for their livelihoods, and 40% of African youth work in the agricultural sector (WEF, 2017a). The 2017 African Competitiveness Report (WEF, 2017a) estimates that, given the projected demographic patterns, and under existing policies, 75% of new entrants in the labour market will work in either microenterprises or agriculture. Given this trajectory, TAE clearly remains of critical relevance to Africa; yet, youth are often uninterested in agricultural careers or see agriculture as a low-status employment option (Leavy and Hossain, 2014). This situation requires TAE to critically reflect on its current status and to reposition itself for optimal impact in the future.

Increasing Urbanization

More than half of the global population live in urban areas, and this proportion is projected to rise to two-thirds by 2050 (United Nations, 2018). In Africa, the proportional size of urban populations is lower than the global average, with the percentage of the population living in urban areas approaching 40%. This is, however, projected to increase to almost half the population by 2025 and 70% by 2050 (ADF, 2016). Sustainable cities are a key development challenge in Africa because their infrastructure frequently does not meet the standards required for a growing population. The large populations of urban poor living in informal settlements on the edges of cities are especially vulnerable, with a lack of access to safe drinking water and proper sanitation posing extensive hazards to their everyday health and safety (Dodman *et al.*, 2017; UNAS, 2019).

Increasing urbanization also has implications for food security at the national level (Szabo, 2016). Research suggests that different aspects of food security are influenced differently by urbanization. For example, increased urban populations create an increased demand for food, and although urban populations tend to have better access to food, this does not necessarily translate into access to food of nutritious or safe quality. In particular, Africa's urban poor are at a heightened risk of food insecurity (Szabo, 2016; IPCC, 2019). In turn, food insecurity and lack of access to safe drinking water have detrimental impacts on the health burden in urban areas, making them vulnerable to several other stressors (Ashbolt, 2004; Battersby, 2012; IHME, 2013).

Urbanization patterns need to be understood in the broader context of Africa's changing demographics and the youth bulge (discussed above), with more than 60% of unemployed individuals being under the age of 35 (Honorati and Johansson de Silva, 2016) and youth unemployment being higher in urban areas (Mo Ibrahim Foundation, 2015). Positively, urbanization is also associated with higher levels of education, even after controlling for the level of development of countries, and urbanization leads to enhanced incentives to invest in education (Bertinelli and Zou, 2008). In addition, more recent research has shown urbanization to have a positive effect on enrolment in education in Africa (Arouri *et al.*, 2014); it is thus likely that rapid urbanization will also contribute to future increased demand for tertiary education in Africa's cities.

Rapid and Radical Technological Transformation

The so-called Fourth Industrial Revolution (4IR) - which includes the 'Internet of Things'[2], big data, machine learning, virtual reality, artificial intelligence and robotics, as well as 3D printing - is predicted to have a disruptive impact on all aspects of society. What distinguishes 4IR from technology trends to date is that it is expected to bring about the merging of the physical, digital and biological worlds through rapid technological transformation on a scale and at a breadth and depth which is unprecedented and highly unpredictable (Schwab, 2017). Holding both great promise and potential peril, 4IR technologies are likely to be highly disruptive and may force organizations, sectors and countries to rethink their development trajectories. In many ways, Africa has essentially missed the opportunities of the second and third industrial revolutions (Harvey, 2017). Africa's structural transformation from low-productivity informal agriculture has translated mostly into low-productivity, urban-based retail services which have not contributed to increased economic growth to the same extent that industrial manufacturing has done in other regions. 4IR is perhaps Africa's opportunity to accelerate economic transformation on the wave of these new technologies (Cilliers, 2018). In particular, 4IR may hold potential to help solve one of Africa's most persistent challenges - namely, malnutrition (Development Initiatives, 2018). The World Economic Forum, in collaboration with McKinsey & Company (WEF, 2019), identified 12 'categories' of emerging technology innovations with the potential to transform food systems, positively affecting their sustainability, inclusivity, efficiency and health impacts. The capacity of African countries to adapt, adjust and adopt the changes of 4IR is, however, a concern (WEF, 2019), and it raises questions about the extent to which the region will be able to successfully capitalize on the opportunity. While radical new technologies are by no means a simple solution to such deeply entrenched challenges as malnutrition, their inevitable arrival and potential to bring about change should not be disregarded in African policy, agricultural development and education.

In the context of this chapter, one of the most pertinent potential impacts of 4IR will be its effect on employment. Some predict that increased sophistication in robotics and artificial intelligence will lead to large-scale job losses and even to the disappearance of many low-skilled jobs (WEF, 2016). Analysts have attempted to determine the extent to which jobs in Africa will be disrupted by 4IR. Estimates of the proportion of all work activities susceptible to automation are as much as 41% in South Africa, 44% in Ethiopia, 46% in Nigeria and 52% in Kenya. Ultimately, the extent to which 4IR changes the face of employment will depend to a large degree on the sector, with some jobs and industries at higher risk than others. It is also likely that losses will be moderated by comparatively low labour costs and offset by new job creation in sectors related to information and communications technology (ICT). However, ICT-intensive jobs in Africa are not likely to be in the lower-skilled delivery of digital products, but in digital design, creation and engineering.

One of the implications of the changing nature of employment is that tertiary education under the 4IR will need to adapt both its content and its modes of delivery. In the first instance, as connectivity improves across Africa, ICT has particularly impressive potential to radically increase access to cutting-edge educational materials - e.g., on handheld digital devices or through such 4IR technologies as virtual reality (Cilliers, 2018). From the perspective of content, tertiary education must incorporate digital literacies into its curricula in order to accelerate the acquisition of digital skills which match the way people will work and collaborate (WEF, 2017b). At the same time, rapid evolution in technology will mean that content becomes obsolete more quickly than before. Thus, tertiary education must shift towards empowering students to adapt to constant and rapid change (as opposed to focusing on teaching content which rapidly becomes obsolete) and must find ways to produce lifelong learners who are creative and emotionally-intelligent critical thinkers, capable of constant adaptation to change (WEF, 2017b). Achieving this will be challenging because, for the most part, primary and secondary school systems have yet to refocus their curricula to incorporate the development of critical thinking skills, creativity and emotional intelligence. This complex objective therefore falls to the tertiary education system. Nevertheless, given that 4IR will profoundly affect both agriculture and education, TAE needs to respond to changes in both spheres in order to ensure its relevance and effectiveness.

Climate Change and Competition for Resources

Climate change is already a reality. The Fifth Assessment of the Intergovernmental Panel on Climate Change has shown that global climate change is damaging crops and undermining food-production capacity in much of the world, and particularly in poorer countries (IPCC, 2014). As a sector, agriculture will be heavily influenced by climate change; yet, at the same time, food systems contribute between 21% and 37% of total net anthropogenic greenhouse gas emissions and must therefore help seek solutions to reducing global warming (Vermeulen *et al.*, 2012; IPCC, 2019).

Sub-Saharan Africa (SSA) is particularly vulnerable to the impacts of climate change, compounded by strong dependence on rainfed agriculture and natural resources, high levels of poverty, low levels of human capital, low levels of preparedness for the effects of climate change and poor infrastructure in rural areas (IFAD, 2011). In the absence of interventions, agricultural yields in SSA could fall by as much as 50% by 2050; in addition, temperatures are already close to or beyond thresholds at which further warming reduces yields (Cline, 2008; Serdeczny *et al.*, 2016). Since agriculture is fundamental to the elimination of poverty and inequality in Africa, inaction is not an option (World Bank, 2008; AGRA, 2016). Addressing climate change and making agriculture more adaptive and climate-smart is, therefore, fundamental to further growth. In this context, climate-smart agriculture becomes vital from a food-security and development perspective. This approach to agriculture can help reduce the negative impacts of climate change and increase the adaptive capacity of farming communities to long-term climatic trends, as well as to increasing weather variability (FAO, 2010).

It is possible that climate change will negatively affect the participation of rural communities in education, but the scale of the potential decrease has not been extensively researched. A study in rural Ethiopia suggests that, as the effects of climate change disrupt the daily lives of rural communities, educational participation may decline (Randell and Gray, 2016). Tertiary education institutions have a responsibility to equip graduates with the knowledge and skills needed to adapt to a world directly and indirectly affected by climate change. No student should leave their institution without an understanding of what climate change is, how it is relevant to their core discipline and how it affects their everyday life. This imperative implies that institutions will need to critically evaluate their academic offerings through a climate change perspective and adapt, as needed, to ensure graduates understand the risks and how to address them in their professional lives (Dyer and Andrews, 2011).

The integration of climate change into curricula is essentially a call for context-relevant, local knowledge to receive greater prominence in African institutions. Climate change will have differential impacts across countries, and related curricula should not only include broad content of global relevance, but also country-specific knowledge. This need to contextualize knowledge presents an opportunity for African higher education institutions to focus attention on finding local and national solutions to this global challenge, and to give students the opportunity to work on solving real-world problems of direct relevance (Hewitson, 2010). The complex nature of climate change is also an opportunity for higher-education institutions to expose students to the principles of trans-, inter- and multidisciplinarity from the first year of study, promoting these approaches in graduate (i.e., master's and doctoral) studies and research.

The impact of climate change on agriculture is still unfolding, and this will have profound implications for TAE over time. In the first instance, as discussed above, curricula will need to be revisited to include relevant content, such as climate-smart agriculture. Nevertheless, curricula will also need to remain responsive to new knowledge of climate change and how to respond to it. Essentially, this need for responsiveness and agility is a key opportunity to ensure that TAE curricula are increasingly informed by research.

Globalization and Shifting Geopolitics

Globalization has strongly influenced higher education in recent decades, sometimes producing contradictory outcomes - including creating, rather than reducing, inequalities within systems and among institutions (as is the case in several other sectors). Criticisms of globalization and the rise of nationalism - as evidenced in the United Kingdom and United States most recently - may strongly influence the extent to which higher-education systems focus on 'internationalization' in these countries in the future. Other geopolitical trends, such as the growth of China, will likely also play an important role in global higher-education patterns, possibly through the emergence of more powerful collaborations among Asian nations (Van der Wende, 2017). The 'internationalization' of higher education, although linked to globalization, focuses on integrating international, intercultural or global perspectives into the purpose, function or delivery of tertiary education (Knight, 2003), including through (i) the exchange and mobility of staff and students, (ii) the inclusion of international content in curricula and (iii) partnerships for research, teaching or engagement.

From an African perspective, activities focused on internationalization can be structured at the subregional level (e.g., within the Southern African Development Community [SADC]), at the pan-African level or beyond Africa. Given the severe capacity constraints of African tertiary education systems (ACBF, 2017), internationalization and related partnerships can play a key role in enhancing capacity, if designed and implemented appropriately (Hagenmeier *et al.*, 2017). As discussed earlier in this chapter, the public funding available for tertiary education in Africa is inadequate to fully meet the needs of the system, and a diversified funding base is required. International partnerships not only offer a key opportunity to mobilize resources, but also a key opportunity to share resources, such as expertise and infrastructure.

4.3 Contextualizing Global Higher-Education Trends in the African Context

Massification of the Higher-Education System

Rates of enrolment in higher education have significantly increased in the past four decades. The number of students enrolled in tertiary education now exceeds 200 million globally, compared with 47 million in 1980. A range of factors have driven this massive expansion. On the one hand, the global emergence of the knowledge economy has necessitated higher numbers of skilled and knowledgeable workers who require one or other form of post-secondary education (Balán, 2015). On the other hand, individuals in these economies increasingly acknowledge the private benefit of a tertiary qualification, which has led to increased demand for educational opportunities. As an example, in the countries of the Organisation for Economic Co-operation and Development (OECD), adults with a tertiary degree 'are 10 percentage points more likely to be employed, and will earn 56% more on average than adults who only completed secondary education' (OECD, 2017, p. 23). In the coming decades, rates of participation in global tertiary education will continue to rise, particularly in emerging and developing countries. Enrolments are projected to exceed 660 million by 2040, representing 10% of the world's population aged 15 to 79 years, compared with 4% in 2012. The result is that, by 2040, most countries will have a participation rate in excess of 60% (Calderon, 2015).

In Africa, despite massive growth in enrolments from 200,000 about 40 years ago to 9.5 million in 2012 (ICEF, 2015), less than 10% of young people are enrolled in higher education, compared with the global average of 26%. In the African context, where most countries are still aiming to become knowledge-driven economies, tertiary education has a critical role to play (AAI, 2015) by developing human capital, creating knowledge and disseminating knowledge for application. In developing economies, higher education can yield significant public benefit, including: (i) greater tax revenues, as individuals earn higher incomes, (ii) increased savings and investment and (iii) a more entrepreneurial and technology-literate citizenry. Higher education can also improve health, contribute to reducing population growth and strengthen governance (Pillay, 2011). These factors are all likely to play a role in driving increased participation in higher education in the future.

The Declaration for Higher Education in Africa identified a target of 50% enrolment by 2063 (TrustAfrica, 2015). As the number of young Africans continues to rise (as previously discussed), an increasing number of students will seek access to the tertiary education system. Achieving a target enrolment of 50% will thus mean that a substantially larger absolute number of students will need to be accommodated. While Africa undoubtedly needs to significantly increase its participation rate in tertiary education, several challenges arise from this expansion. As enrolment numbers continue to grow, funding is unlikely to increase proportionally, which will affect teaching loads, student–teacher ratios and the quality of teaching, while exerting more pressure on already-limited institutional infrastructure. Inadequate numbers of PhD-qualified staff directly impedes institutions' teaching capacity and hence their ability to increase graduate student numbers. This has serious implications for the quality of graduate qualifications (see Section 4.4).

Managing and responding to growth in tertiary enrolments is a responsibility of both national governments and tertiary education institutions. Case studies, such as those discussed by Mohamedbhai (2014), illustrate the diversity of strategies which can be undertaken in response to massification. Evidence-informed planning, coordination and monitoring at both levels is needed to manage enrolments in line with institutions' (human, financial and infrastructural) capacity. Improved and integrated higher-education data and information-management systems at national and institutional levels are urgently required to support this level of planning. Strategies on how to (i) expand public (and private) enrolments responsibly, (ii) leverage the use of ICT and (iii) more efficiently use current resources will need to be informed by relevant data and be more widely documented, shared and adopted among institutional and national stakeholders.

Increased Demand on Limited Public Funding for Tertiary Education

As the tertiary education sector continues to grow, demand on public funding increases, yet there are limits to how much funding governments are capable of investing. As a result, the sector - and in particular individual institutions - must do more with a relatively constant, or even lower, level of public funding. At the same time, the per student costs for delivering education are rising, and governments cannot keep pace with these, so the sector faces increasing pressure (Altbach *et al.*, 2010). In some cases, government spending on tertiary education is declining. For example, in OECD countries, total spending between 2010 and 2015 decreased in more than two-thirds of the countries surveyed. In turn, students' financial contributions to their own education are rapidly increasing and will likely continue to rise (OECD, 2017). As of 2015, 75% or more of students in Australia, the United Kingdom and the United States benefitted from public loans, scholarships or grants, and open-admission systems to public and private tertiary institutions (or both) can be found in more than half the OECD countries (OECD, 2017). However, this funding model is changing as governments - even in Europe, where free education has historically been most accessible - become more reliant on non-tax revenues to support tertiary education (World Bank, 2010).

As of 2010, African countries' contribution to education as a percentage of GDP was already above the world average, including an allocation of nearly 20% of this amount to higher education. As a result, this share cannot reasonably be increased significantly (Saint, 2015). Although spending on tertiary education has increased in absolute value in many African countries, public funding has not kept pace with the rapidly- growing enrolments, and spending per student has significantly declined in a large number of countries since 1995 (World Bank, 2008). Despite this, in 2006, African countries' average public spending per university student was US$2,000 per year - more than twice as much as non-African developing countries invest in tertiary education (World Bank, 2010; Eicker and Stroebel, 2014). However, as the number of students in African tertiary education is projected to continue to rise, significant demand will be placed on available public funds, such that the necessary level of funding per student cannot be maintained.

In the context of TAE, delivering quality education - including adequate time devoted to farm-based or industry-based experiential learning, practical training and laboratory work - is a costly endeavour. On average, it is more costly than degree programmes which do not require high levels of practical training or equipment. When funds are limited, experiential learning opportunities and lab work are often side-lined, which has implications for the quality of TAE and may affect the costs for individuals to access high-quality TAE opportunities.

The Rise of Private Tertiary Education

Both of the above-mentioned trends - massification and public funding constraints - have contributed to a significant increase in the number and type of private tertiary education institutions across the globe. In 2010, the best estimate was that up to 30% of global enrolments were in private higher education (Levy, 2010). In line with global trends, private tertiary education enrolments have risen significantly in Africa, with an estimated 22% of students being privately enrolled in 2006 (Levy, 2010). However, the proportion of students enrolled in private higher education in Africa varies significantly by country (Levy, 2010). Given that public institutions and public funding cannot meet the increasing demand for access, private-sector development may help to diversify the sector, increase funding and absorb part of the projected increase in the number of students expected to enter higher education in the future. However, private tertiary education has some challenges in terms of quality; for it to make a positive impact in the broader sector, it needs to be governed by an adequate regulatory framework to ensure quality and appropriate monitoring, thereby enhancing its credibility (Levy, 2010).

Enrolment Patterns by Discipline

Over time, the profile of enrolments by major field of study has remained relatively stable globally, with enrolments in science, technology, engineering and mathematics (STEM) accounting for less than one-third of total enrolments. In OECD countries, the most popular degrees are business, administration and law. A low proportion of students are interested in science, mathematics or statistics (6%). In these countries, only 5% of newly-enrolled students are interested in an ICT degree, and while the demand for ICT degrees remains low, the demand for ICT skills in the workplace is high. Furthermore, the number of students interested in teaching qualifications does not match the high demand for teachers in OECD countries (OECD, 2017). A similar trend with STEM enrolments is noted in Africa. It is estimated that in 2004, just 28% of tertiary students were enrolled in science and technology fields (World Bank, 2008). Social sciences and humanities students accounted for 47% of graduations in 2005, and education students accounted for a further 22% (World Bank, 2008). For more than a decade, the Higher Education Research and Advocacy Network in Africa (HERANA) has worked with a group of leading national higher education institutions in Africa. Data from HERANA institutions for the period 2008–2011 indicate that enrolments in STEM fields in what are considered Africa's flagship universities[3] were similar to those reported in previous studies (Fig. 4.1). (See Ch. 5, this volume, for a specific discussion of TAE enrolments.)

Continued Gender Disparities in Tertiary Education

SDG target 4.3 envisages that, by 2030, there will be 'equal access for all women and men to affordable and quality technical, vocational and tertiary education, including university' (United Nations, 2015, p. 21). Significant progress has been made in increasing access to higher education for women globally. In 2013, of the 6 million students across OECD countries who graduated with a bachelor's degree, more than half were women (OECD, 2015). Nevertheless, this overall parity masks some stark disparities by field of study (Fig. 4.2). Women are overrepresented in the fields of education, health and welfare and social sciences, as well as in the arts and humanities; they are significantly underrepresented in the fields of ICT and engineering.

In the African context, Agenda 2063 envisages gender equality in all spheres of life, and the Declaration for Higher Education in Africa (TrustAfrica, 2015) identifies a target for gender parity in tertiary enrolments by 2025. Gross enrolment rates for women in SSA rose from 3.7% in 2000 to 7% in 2013. Yet, in 2013, for every 100 male students enrolled there were still only 72 female students (Darvas *et al.*, 2017). Similar to the global trend, female enrolments in STEM disciplines in Africa show greater inequalities than total enrolments. Data from the HERANA universities illustrate that, at Africa's leading

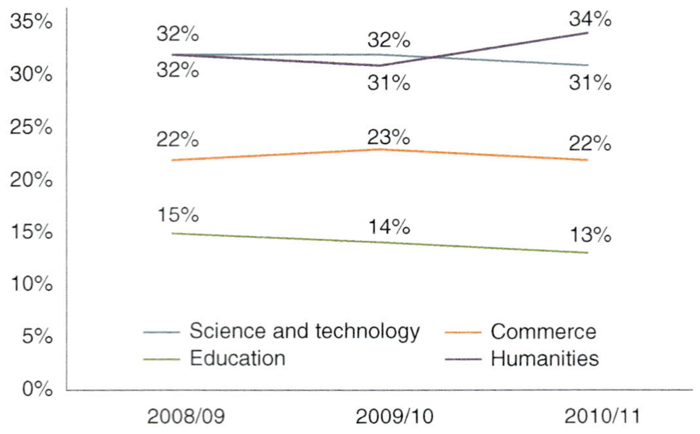

Fig. 4.1. Enrolment by field of study at HERANA institutions, 2008–2011
Source: Constructed by authors from Bunting *et al.* (2014).

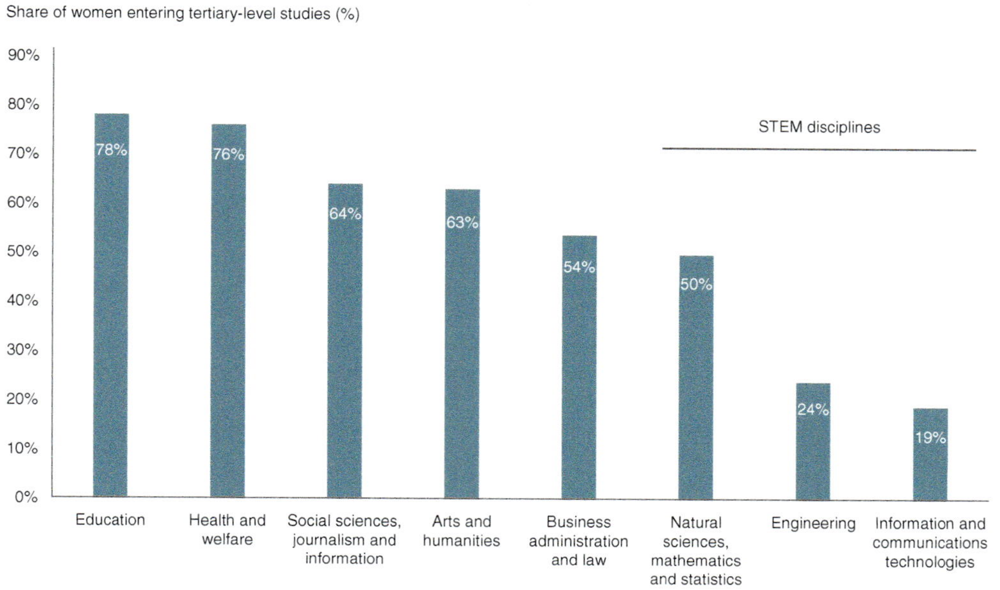

Fig. 4.2. Gender-disaggregated enrolment by field of study, 2015
Source: Adapted by authors from OECD (2017, p. 26).
Note: STEM = science, technology, engineering and mathematics.

institutions, enrolments of female students in science and technology were lower than 50% in all cases and lower than 30% in more than half the institutions (Fig. 4.3). Furthermore, in most cases, women remain overrepresented in education, as well as in the humanities and social sciences (see Ch. 5, this volume, for a discussion of the gender profile of TAE enrolments).

Technology-Mediated Education

Even before 4IR technologies have their potential disruptive impact in education, current technologies are already central to higher education globally, as a means to enhancing or replacing traditional models of learning and teaching - e.g., through online-only offerings. Opportunities for online-only distance education are likely to increase in the future, and several private international universities have positioned themselves as online providers. Massive open online courses (MOOCs) became one of the most high-profile aspects of technology use in teaching in recent years, with several leading international universities providing free courses, open to all participants via such platforms as Coursera and edX. Several years after MOOCs arrived in the public sphere, both enthusiasm and fears about their potential impact on higher education have subsided. MOOC enrolment is high, but retention and completion rates are low (less than 5% on some of the leading MOOC platforms) (University of Oxford, 2015).

While online learning environments offer massive potential to widen access, to date the greatest use of technology in tertiary education has been the integration of technology into traditional methods of teaching (CHE, 2014). Many universities now utilize an online learning management system, such as BlackBoard or Moodle. These platforms have contributed to the rise in blended approaches to teaching, combining online and face-to-face delivery. In Africa, a significant body of knowledge exists on the potential of, and critical need for, massive expansion in online distance-learning options to substantially increase access. Even globally, however, uptake of this has been slow (CHE, 2014). Affordability and inadequate infrastructure remain barriers to ICT adoption and use in SSA, and these barriers do not appear to be rapidly disappearing (Baller *et al.*, 2016). The global COVID-19 pandemic was a pervasive disruptor

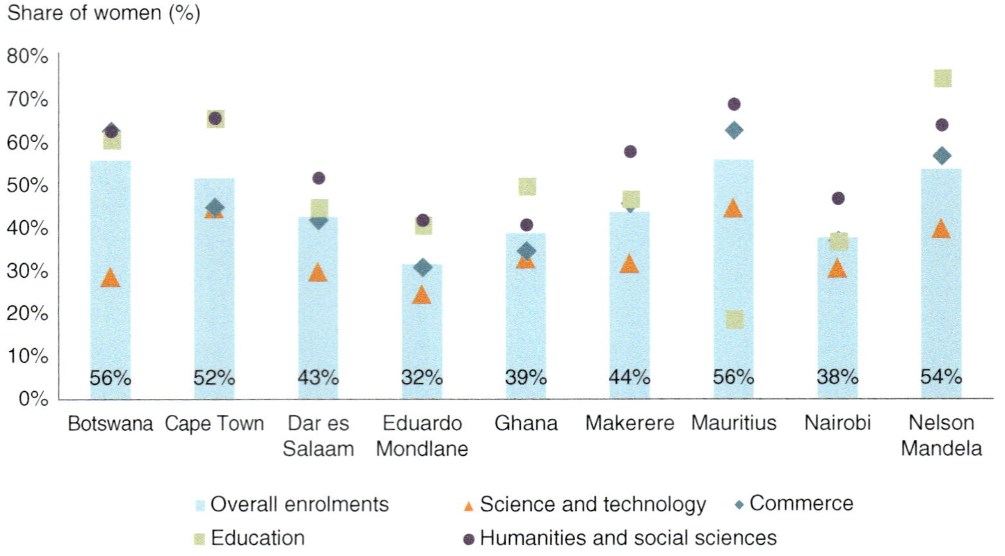

Fig. 4.3. Gender-disaggregated enrolments by field of study at HERANA institutions, 2010/2011
Source: Constructed by authors from Bunting *et al.* (2014).
Note: HERANA = Higher Education Research and Advocacy Network in Africa.

in higher education, which necessitated the rapid adoption of online technologies for teaching and learning. While the experience brought many opportunities for rethinking models of teaching and learning using technology, it also posed many challenges (Paterson, 2021). Africa already faced challenges in adopting technology in teaching before COVID-19, and these continued and, in some cases, were exacerbated. For example, a 2020 survey by the International Association of Universities found that Africa's share of HEIs which completely cancelled teaching or postponed/cancelled exams was higher than in any other region of the world (Marinoni *et al.*, 2020). This reality starkly highlights inequalities between students with access to resources and technology, and those without (Martin-Barbero, 2020).

The global market for mobile learning is predicted to grow at a compound annual rate of 20.6% between 2021 and 2026, increasing to $64.3 billion by 2026 (360 Research Reports, 2021). With the success of African mobile innovations in other sectors (e.g., M-Pesa in the financial sector), the question of the role and value of mobile learning in the African context needs to be explored. Mobile penetration is on the rise in the region, with 557 million mobile subscribers as of 2015 (46% of the population). Some examples of mobile educational innovation can already be found - e.g., Kenya's Daystar University has Daystar Mobile, a programme where students can earn a bachelor's degree in education primarily through their smartphones (Adams Becker *et al.*, 2017).

The reality is that current cohorts of students engage with technology in qualitatively different ways from earlier generations, and by implication, teaching staff engage with and think about technology in very different ways from their students. This has profound implications for education, which are not yet fully understood and have not yet been capitalized on. Changes in the way in which education is and should be delivered are challenging educators and institutions around the world. To benefit from emerging technologies and remain relevant to the world of work where graduates will be employed, the need to capacitate teaching staff in tertiary environments to employ effective technology-enhanced pedagogies is clear. These pedagogies will deliver learner-centred approaches, integrate experiential learning for employability, equip students with skills for self-employment and respond to demand-driven curricula (Almpanis, 2015).

4.4 Africa's Most Pressing Challenges in Tertiary Education

It is evident from the discussion above that the tertiary education system in Africa will need to increase access significantly, change its enrolment profile (in terms of gender and disciplines) and develop skills-driven and workplace-relevant curricula delivered through pedagogies which stimulate adaptability and deep, lifelong learning. All of this will need to take place in the midst of a rapidly transforming technological landscape which, if leveraged correctly, can be a significant enabler of change. Nevertheless, several inherent challenges need to be addressed to bring about these system-level changes, and a number of pervasive historical challenges in African tertiary education may hinder the achievement of the necessary transformation. These challenges are discussed below.

Consequences of Expanded Enrolments and the Need for Massification

It is clear that enrolments in African tertiary education must increase, which will be beneficial to regional and national agendas, but may pose significant institutional-level challenges. Increasing student numbers at Africa's already-limited number of universities negatively affects the quality of teaching and learning as class sizes grow rapidly, inadequate levels of funding are spread more thinly across a larger number of students and the carrying capacity of institutions is exceeded (Mohamedbhai, 2014; Girmaw, 2016). The solution is not as simple as increasing the number of institutions, whether public or private. As the number of institutions grows, so does the demand for qualified staff, institutional funding and physical infrastructure. Staff qualification levels at institutions pose a major challenge. Even at Africa's flagship institutions, the proportion of staff with a PhD qualification ranges from 17% to 63%, with most of the institutions having approximately 40–50% of staff with a PhD degree (Bunting *et al.*, 2014).

As has been argued previously, private tertiary education and technology-mediated options will need to play a role in increasing access to tertiary education but will require renewed approaches because, to date, implementing these options has not proven to be a quick solution. Both non-profit and for-profit private education systems in the United States, Turkey, Singapore, Malaysia, Kuwait, South Korea and Chile have played significant roles in dealing with the increased demand for higher education (Thwala, 2017). Private higher education, thus, can and should be seen as a viable option (or indication of the failure) to supplement, rather than replace public higher education. As previously discussed, regulatory frameworks which govern accreditation and quality assurance of (public and private) tertiary education will play an important part in ensuring the credibility and role of private higher education institutions in Africa.

Fully online distance education has been slow to take off across the globe. Among the reasons for this are the perception that it is of lower quality, and, for the institutions administering the programmes, that it is hard to achieve economies of scale because online programmes are inherently labour-intensive, requiring the involvement of both lecturers and information-technology technicians (CHE, 2014). This is perhaps one area where partnerships between African tertiary education stakeholders and development partners can be leveraged to pool the necessary resources to make these online offerings a viable and sustainable element of an expanded tertiary system. Finding viable ways to increase access without compromising quality needs to remain a priority in African higher education because gross enrolment rates must increase in order to realize the developmental benefits of the expansion of tertiary education.

The Need to Increase the Number of PhD Graduates in Africa

The reality that Africa needs to produce more PhD graduates is now widely acknowledged (Mentz-Coetzee *et al.*, 2020). Africa currently has only 198 researchers per million people, compared with 455 per million in Chile and more than 4,500 per million in the United Kingdom and United States. To achieve the world average ratio, Africa needs another million new PhD graduates (Christoffels, 2018). In addition to bolstering the capacity for research and innovation in the region, these graduates will have an important role to play as the number of (public and private) institutions increases (Swanepoel, 2019). Yet, scaling up the production of Africa's PhD graduates is significantly constrained by several factors. In 2014, only a minority of the 1,500 public and private universities across Africa offered graduate programmes (Hayward and Ncayiyana, 2014). Limited funding, specifically in terms of funding allocated to the development of graduate education, in turn inhibits the development and implementation of more graduate programmes (ANIE, 2018). Offering graduate programmes requires that institutions have an adequate number of staff with doctoral-level qualifications who can supervise and mentor graduate students and (as noted above) very few African universities are in this position (Hayward and Ncayiyana, 2014; ANIE, 2018). Questions have also been raised about the quality of PhD programmes in Africa. ANIE (2018, p. 23) – reporting on a joint study in six countries of SSA - found there were 'significant quality concerns . . . with evidence showing that standards of provision [were] slipping in the context of rapid increases in enrolments and diffusion across more institutions'. The report provides several examples of how collaborative (North–South and South–South) partnerships offer opportunities to address some of the challenges associated with escalating the number of PhD graduates in Africa, including through joint programmes (offered by two or more institutions), sandwich programmes (which include an industry placement) and scholarship schemes. Leveraging the critical mass and convening power in established centres of excellence should also be explored as a mechanism for developing and supporting high-quality PhD programmes (see Ch. 15, this volume). A multiplicity of approaches should be adopted and implemented to ensure that adequate numbers of PhD students can be trained without undermining the quality of training.

Funding

Funding for tertiary education in Africa is a longstanding challenge, and public spending on higher education as a proportion of the education budget (and of GDP) varies substantially among countries (Pillay, 2010b). However, SSA countries have gradually increased public spending on education by more than 6% each year, and African countries devote a substantial proportion of their government budgets to the education sector despite relatively low GDPs and a host of competing development challenges. Globally, public education expenditure accounts for 4.7% of the world's $18 trillion GDP per capita. Africa devotes 5% of its total GDP of $1.5 trillion to public education expenditure, which is the second-highest percentage after North America (AAI, 2015). However, despite these high spending shares, the sector is chronically underfunded, and, as was noted in the discussion on global trends in funding for tertiary education, it is unlikely that these budgets can be radically increased.

As funding to institutions decreases or remains stagnant, students are expected to increase their financial contributions. However, finances already present a major challenge for students to access high-quality tertiary education in Africa. The recent and protracted protests by university students in South Africa over fees demonstrate the rising tensions around the cost of higher education and who should pay for it (Wangenge-Ouma, 2012; Langa *et al.*, 2016; Jogee *et al.*, 2018). Expanding access to higher education for students whose families do not have the means to pay for it is necessary to disrupt the cycle of poverty. However, the extent to which entirely fee-free tertiary education is feasible in the African context under the projected continued expansion needs to be carefully considered (Oketch, 2016). Innovative approaches, thoughtful models and sustainable partnerships need to be explored as alternatives to address the complex challenges associated with student fees.

In summary, government spending alone will not be adequate to meet all the financing challenges in African tertiary education. Furthermore, in resource-poor contexts, such as SSA, there is a limit to the amount of funding which can come from individuals and their families through fees. A diversified funding base is thus necessary to ensure a high-quality, sustainable tertiary education sector (see Ch. 8, this volume). Critical analysis and investigation is needed to understand (i) the dynamics of the paradox that education requires higher than average spending and (ii) the continued scale of the challenges African institutions face. Simply advocating for increased funding in the sector, without understanding and taking steps to address the causes of the current inefficiencies, will not yield the desired improvements. While increased funding for tertiary education is Africa is most certainly necessary, efforts to address this issue by both institutions and governments should be coupled with efforts to radically improve efficiency in the use of current resources.

Graduate Unemployment and Opportunities for Entrepreneurship

As previously discussed, one of the factors strongly driving the increased demand for higher education is the benefit of holding a tertiary qualification, which typically affords individuals access to improved employment and social mobility. For example, as much as 84% of graduates in OECD countries are employed, and only 15% are not in education, employment or training (OECD, 2017).

Comprehensive data on graduate employment (or unemployment) is not readily available in Africa; however, evidence shows that, in some countries at least, graduates with tertiary qualifications (i) earn significantly more than their peers without qualifications and (ii) have lower levels of unemployment (Bloom *et al.*, 2015; Business Tech, 2017). In other cases, however, including in North Africa, graduate unemployment is high (ILO, 2017a,b). Particularly in the formal employment sector, job opportunities have not kept pace with population growth (WEF, 2017a). Limited job opportunities may mean that tertiary education does not immediately lead to employment and economic benefit for African graduates (NBS, 2018), but in the long-term, they will remain more likely to be in demand, particularly if they gain increasingly sought after skills or qualifications. Lack of opportunities could negatively contribute to the loss of skilled professionals as the global demand for talent intensifies, and populations in the global north age (increasing the demand for skilled workers). Currently, the African Union estimates that 70,000 skilled professionals emigrate from Africa each year (African Union, 2018). In the context of Africa's capacity challenges, drastic increases in this migration should be minimized.

At the same time, employers bemoan the lack of skills on the part of graduates and identify inadequately-skilled workforces as a major constraint to their businesses (WEF, 2017a). This presents an opportunity and a challenge for teaching and learning in African tertiary education. Curriculum transformation, in terms of both content and mode of delivery, will be fundamental to the ability to train workplace-ready graduates in the future. However, educational institutions are often not well-positioned to bring about the needed changes to ensure relevance to the workplace. Partnerships between the private sector and higher education are one of the most successful models for catalyzing this needed teaching and learning reform (Mourshed *et al.*, 2013). Equally important, partnerships with the private sector remain one of the most feasible opportunities for students to acquire real-world experience and develop skills through internships, work placements and experiential learning (see Ch. 16, this volume).

Furthermore, graduates could be better equipped to explore opportunities for self-employment or entrepreneurship if tertiary education curricula were intentional about developing skills for these alternatives (see Ch. 12, this volume). Graduate unemployment (and the likelihood that it may increase in the future) thus presents an opportunity to orient tertiary education towards empowering students with the skills they need to be job creators. Historically, preparing youth for careers in entrepreneurship has not been the focus of higher education, and if these institutions hope to empower their graduates with the necessary skills, partnerships with government and the private sector will be required. As a supplement to formal teaching and learning, the private sector can provide hands-on practical exposure to students, and industry leaders can serve as mentors to emerging entrepreneurs. Investment capital, through seed funding or low-interest loans, is another vehicle for the private sector to work with government and higher education to develop the potential of Africa's young entrepreneurs. Initiatives such as the Tony Elumelu Foundation Entrepreneurship Program (which committed $100 million to create 10,000 start-ups across Africa [Boer, 2015]) and the African Development Bank's ENABLE Youth programme (which seeks to create 300,000 agribusinesses and 8 million jobs in 30 African countries over five years [AfDB, 2017]) provide positive models of these potential partnerships.

The African agricultural sector, allied with industry, manufacturing and processing capacity, can lead to strong and sustainable economic development and the creation of wealth across the economy. If well harnessed, the sector holds great promise through the growth of agribusiness industries for the creation of millions of job opportunities, especially for youth (AfDB, 2016). This will require a new generation of young 'agri-preneurs' who will need to be trained through TAE. A pressing imperative most certainly exists for TAE and relevant stakeholders to develop responsive curricula in partnership to equip students with workplace-relevant skills (in addition to academic skills) and prepare them for entrepreneurial opportunities in the sector.

Despite their potential, to date, partnerships among tertiary education, governments and the private sector have not been widely adopted. Incentives are limited in academic institutions, and disincentives and competing priorities persist in the macro-environment. Cross-sectoral partnerships can be time-consuming and potentially costly to implement because they require significant coordination. To overcome these obstacles, partnerships should be designed to ensure they deliver mutual benefits to all parties; they should also be afforded the necessary human and financial investment needed to ensure their effectiveness and sustainability (see Ch. 16, this volume).

Quality Assurance

Rapid growth in the number of students and higher-education institutions, as well as the increased focus on employability, have made quality and quality assurance central topics in the higher-education community (HAQAA Initiative, 2017). The increase in the number of national quality-assurance agencies in Africa attests to this increased focus, with a three-fold increase between 2004 and 2014 (Okebukola and Fonteyne, 2014). However, several challenges are experienced by these agencies, including a lack of human and institutional capacity to implement their mandate (Shabani *et al.*, 2014) at as many as 60% of the agencies (Shabani *et al.*, 2014). In this context, such entities as the African Quality Assurance Network (AfriQAN) play an important role (AfriQAN, 2013). AfriQAN was established by the Association of African Universities (AAU) to assist institutions concerned with quality assurance in higher education in Africa. AAU works in collaboration with similar organizations in other parts of the world.

There are also regional responses to the quality-assurance challenges in Africa. SADC, for example, has developed its own qualifications framework (SADCQF), which will promote mobility over time and be a regional reference point for comparing qualifications obtained in SADC countries. A Technical Committee on Certification and Accreditation, which is responsible for implementing SADCQF, is also responsible for identifying opportunities to build the capacity of national agencies (Samuels and Jaftha, 2017). At the highest level, heads of African states and governments have expressed their commitment to establishing harmonization and quality-assurance mechanisms. The first Ten-Year Implementation Plan of Agenda 2063 seeks to establish an African accreditation agency to develop and monitor educational quality standards across the continent (HAQAA Initiative, 2017), along with an Africa-wide quality-assurance framework. One of the goals of the framework is to promote the relevance and applicability of skills across African contexts.

Consequently, the African Union Commission initiated the development of a Pan-African Quality Assurance and Accreditation Framework to enhance quality and promote compatible methodologies for doing so (Okebukola and Fonteyne, 2014). One of the necessary instruments for the appropriate functioning of the Framework is the development of standards and guidelines for quality assurance. Addressing this need, the African Standards and Guidelines for Quality Assurance in Higher Education (ASG-QA) have recently been developed by a technical working group of African experts, with support from European partners and through extensive consultations with higher-education institutions and quality-assurance agencies, as well as ministries in Africa and regional bodies. ASG-QA is globally benchmarked and underpinned by existing guidelines in different African countries and regions (HAQAA Initiative, 2017). These are positive developments at the strategic level, but they still need to be implemented on a widespread scale before they will have any tangible impact on the quality and transferability of skills. Several challenges need to be overcome to achieve widespread adoption, including securing political will, sourcing funding to support adoption and implementation and integrating or aligning continental guidelines with regional and national frameworks.

Capacities for Institutional Transformation

Tertiary education institutions in Africa have proved resistant to change. In many cases it is clear what must be done, but understanding is limited on how to eliminate persistent barriers to implementing change. At minimum, change will require a transformation of leadership, the elimination of corrupt and fraudulent practices and significant investment in the holistic capacities of staff to teach, conduct research and engage non-academic communities (see Chs. 6 and 9, this volume). Rapidly transforming institutions require competent, responsive, ethical leaders to navigate them through complex change processes. Not all institutions in Africa have the autonomy to appoint their institution's top leadership and governance structures, and in cases where institutions are autonomous, leaders are often appointed based on their academic track records rather than their managerial skills. This leaves institutions at a disadvantage, having leaders who are not equipped to deal with the complexity of challenges to be contended with. Leadership in institutions should be provided with the opportunity to develop the skills they need to lead, in terms not only of academic direction, but also of effective management. Furthermore, they must be held accountable for their performance. Without a culture of accountability, the impetus for change will be low (see Ch. 10, this volume).

Universities' incentives are typically based on the number of graduates produced rather than their quality, which perpetuates the focus on quantity above the transformation of practices or the employability of graduates. Adopting new methodologies for teaching and integrating development skills into curricula increases the burden on tertiary education institutions, where teaching staff are expected to manage increasingly heavy workloads across teaching, research and community engagement. Without appropriate incentives and capacity strengthening for teaching staff, the transformation of teaching and learning practices may not receive the level of attention needed. To help address this, a pan-African collaboration could be considered to establish a centre for excellence in teaching and learning focused on developing research-led materials and developing training modules for teaching staff.

An Enabling Macro-Environment

Tertiary education functions within broader national and regional macro-environments, directly influenced by governance frameworks and policy. Governance systems in Africa remain underdeveloped, and perceptions of corruption remain high (Transparency International, 2019). These factors undermine economic development, which has multiple ramifications for tertiary education systems. The cost of corruption is high: in 2002, the African Union estimated annual losses of \$150 billion to corruption across the continent. This starkly contrasts the approximately \$50 billion received from overseas development aid (OECD, 2018b). Researchers estimate that strengthening governance and reducing corruption could lead to significant dividends, increasing the region's growth in per capita GDP by about 1–2 percentage points (Hammadi *et al.*, 2019). While the processes needed to achieve this would take considerable time and effort to institute, they should not go unattended. Positive developments include such initiatives as the African Union's African Peer Review Mechanism for monitoring governance performance among member states. Participation is voluntary, however, so not all members have joined, and some implementation challenges have been experienced (Turianskyi, 2019). While efforts to improve governance and bolster accountability will have positive impacts in the educational system; left unattended, failing governance and corruption will seriously undermine the capacity of tertiary education to systematically transform and fulfil its intended function in the development agenda.

4.5 Critical Reflection and Implications for TAE in Africa

It is clear that there are many shared challenges facing global tertiary education, tertiary education in Africa and TAE; some contextual challenges also make TAE unique. In terms of the global trends discussed in this chapter, all five have major implications for agrifood systems in Africa (as is discussed in Ch. 2, this volume) and for TAE. In summary, Africa's growing population will require increased food production and, given the projected demographic patterns (under existing policies), a large proportion of new entrants in the labour market will either work in microenterprises or agriculture. As a sector, agriculture will be heavily influenced by climate change; yet, at the same time, food systems contribute significantly to greenhouse gas emissions and must therefore help seek solutions to reducing global warming. Increasing urbanization has implications for food security at the national level, and, in particular, the urban poor are at increased risk of food insecurity (and associated negative health outcomes). Increased urban populations create an increased demand for food, and although urban populations tend to have better access to food, this does not necessarily translate into access to food of nutritious or safe quality. While 4IR technologies have the potential to positively transform food systems in terms of their sustainability, inclusivity, efficiency and health impacts, the capacity of African countries to adapt, adjust and adopt the changes of the 4IR is a concern.

In this context, it is clear that TAE remains of critical relevance to Africa. The Agricultural Education and Skills Improvement Framework (AESIF) 2015–2025 outlines a plan 'to transform Africa's agricultural education and training in order to leverage the quality and quantity of skilled workforce required for the advancement of agricultural growth and the achievement of targets set in the CAADP Malabo Declaration on accelerated agricultural growth and transformation for shared prosperity and improved livelihoods' (NEPAD, 2015, p. 6). AESIF identifies three priority areas for this transformation, including (i) governance and management policies, (ii) teaching and learning and (iii) the mobilization of public and private partnerships (NEPAD, 2015). The priority areas of teaching and learning, as well as public and private partnerships, are clearly of fundamental importance to the revitalization of TAE. In terms of teaching and learning, AESIF identifies the need to reform and update curricula, improve pedagogies (including an increased use of relevant technologies) and focus on the articulation between agricultural education and training (i.e., technical and higher education) and promote appropriate quality-assurance frameworks (NEPAD, 2015). These three areas of action align with the trends in global higher education discussed in this chapter.

Enrolments in TAE are not growing as rapidly as enrolments in other disciplines, despite the clear need for graduates to work in the agrifood sector and the potential of the sector to contribute to Africa's development agenda (as is explored in Ch. 5, this volume). An emphasis on employment opportunities for Africa's youth and increasing the participation of women in agrifood-related education and employment are areas of alignment between AESIF and the broader African tertiary education context (NEPAD, 2015).

In conclusion, while there may be specific challenges and opportunities for TAE in Africa, several commonalities also exist within the broader context of tertiary education in Africa. Effectively revitalizing TAE will require cognizance of these commonalties, as well as attention to the aspects which make African TAE unique. The critical question, however, is not whether the time is right for the transformation of TAE, but what precisely should be done, by whom, with what resources and through which partnerships and collaborative agreements. The complexity of the challenges facing higher education and TAE discussed in this chapter illustrate the reality that there are no quick fixes or 'blanket solutions'. Systematic, context-specific knowledge on what works is limited, although several cases document innovative solutions implemented with some success. Moving forward it will be important for tertiary education stakeholders - and in particular TAE - to adopt a reflexive learning approach. By documenting and sharing their experiences on academic and practitioner platforms, stakeholders across the continent could cross-pollinate lessons and experience. Accessible, systematic and up-to-date knowledge on African higher-education systems and their transformation could play an important role in empowering and positioning the system to transform, perhaps coordinated and curated through units and centres dedicated to the study of the topic.

Endnotes

[1] The medium-variant projections assume a decrease in fertility in countries where large families are still prevalent, and a slight increase in families where women have fewer than two children.
[2] This term refers to the equipment and technologies needed to connect to the Internet and exchange data with other devices and systems.
[3] Africa's eight flagship universities are the University of Botswana, University of Cape Town, University of Dar es Salaam, Eduardo Mondlane University, University of Ghana, Makerere University, University of Mauritius, and University of Nairobi.

References

360 Research Reports (2021) Global mobile learning market size, status and forecast 2021–2027 (January 18). Available at: https://www.360researchreports.com/global-mobile-learning-market-17203765 (accessed 11 June 2021).

AAI (Africa-America Institute) (2015) *State of Education in Africa Report 2015. A Report Card on the Progress, Opportunities and Challenges Confronting the African Education Sector*. Available at: http://www.aaionline.org/wp-content/uploads/2015/09/AAI-SOE-report-2015-final.pdf (accessed 18 March 2021).

ACBF (African Capacity Building Foundation) (2017) *Africa Capacity Report 2017: Building Capacity in Science, Technology and Innovation for Africa's Transformation*. Available at: https://elibrary.acbfpact.org/acbf/collect/acbf/index/assoc/HASH01ad/e44e7241/b749d69a/1a6c.dir/ACR2017%20English.pdf (accessed 18 March 2021).

Adams Becker, S. Cummins, M. Davis, A. Freeman, A. Hall Giesinger, C. *et al*. (2017) *NMC Horizon Report: 2017 Higher education edition*. The New Media Consortium, Austin. Available at: https://www.unmc.edu/elearning/_documents/NMC_HorizonReport_2017.pdf (accessed 18 March 2021).

ADF (African Development Fund) (2016) *Feed Africa: Strategy for Agricultural Transformation in Africa 2016–2025*. African Development Bank Group, Abidjan. Available at: https://www.afdb.org/fileadmin/uploads/afdb/Documents/Generic-Documents/Feed_Africa-_Strategy_for_Agricultural_Transformation_in_Africa_2016-2025.pdf (accessed 18 March 2021).

AfDB (African Development Bank) (2016) *Feed Africa. Strategy for Agricultural Transformation in Africa 2016–2025*. ADB, Abidjan. Available at: https://www.afdb.org/fileadmin/uploads/afdb/Documents/Policy-Documents/Feed_Africa-Strategy-En.pdf (accessed 18 March 2021).

AfDB (2017) *Agro-Millionaires in the Making*. Enable Youth Program. ADB, Abidjan. Available at: https://www.afdb.org/fileadmin/uploads/afdb/Documents/Generic-Documents/Enable_Youth_brochure.pdf (accessed 18 March 2021).

African Union (2018) *The Revised Migration Policy Framework for Africa and Plan of Action (2018–2027)*. AU, Addis Ababa. Available at: https://au.int/sites/default/files/newsevents/workingdocuments/32718-wd-english_revised_au_migration_policy_framework_for_africa.pdf (accessed 18 March 2021).

AfriQAN (African Quality Assurance Network) (2013) *AfriQAN Constitution*. Available at: https://afriqan.aau.org/about/constitution/ (accessed 18 March 2021).

AGRA (Alliance for a Green Revolution in Africa) (2016) *Africa Agriculture Status Report 2016: Progress towards Agricultural Transformation in Africa*. Issue Paper 4. Nairobi.

Almpanis, T. (2015) Staff Development and Institutional Support for Technology Enhanced Learning in UK Universities. *Electronic Journal of e-Learning* 13(5), 366–375. Available at: https://files.eric.ed.gov/fulltext/EJ1084240.pdf (accessed 18 March 2021).

Altbach, P.G. Reisberg, L. and Rumbley, L.E. (2010) Trends in global higher education: Tracking an academic revolution. In Altbach, P.G. (ed.). *Global Perspectives on Higher Education Vol. 22*. Sense Publishers, Leiden. Available at: https://www.cep.edu.rs/public/Altbach,_Reisberg,_Rumbley_Tracking_an_Academic_Revolution,_UNESCO_2009.pdf (accessed 18 March 2021).

ANIE (African Network for Internationalisation of Education) (2018) *Building PhD Capacity in Sub-Saharan Africa*. British Council and DAAD. Available at: https://www.britishcouncil.org/sites/default/files/h233_07_synthesis_report_final_web.pdf (accessed 18 March 2021).

Arouri, M. Ben Youssef, A. Nguyen-Viet, C. and Soucat, A. (2014) *Effects of Urbanization on Economic Growth and Human Capital Formation in Africa*. PGDA Working Paper 119. Program on the Global Demography of Aging at Harvard University (PGDA). Available at: https://halshs.archives-ouvertes.fr/halshs-01068271 (accessed 18 March 2021).

Ashbolt, N.J. (2004) Microbial contamination of drinking water and disease outcomes in developing regions. *Toxicology* 198: 229–238.

Balán, J. (2015) The crisis of the public mission in higher education. *International Higher Education* 80, 4–5.

Baller, S. Di Battista, A. Dutta, S. and Lanvin, B. (2016) The Networked Readiness Index 2016. In Baller, S., Dutta, S., and Lanvin, B. (eds). *The Global Information Technology Report 2016. Innovating in the Digital Economy*, pp. 3–38. World Economic Forum, Geneva. Available at: http://www3.weforum.org/docs/GITR2016/WEF_GITR_Full_Report.pdf (accessed 18 March 2021).

Battersby, J. (2012) Beyond the food desert: Finding ways to speak about urban food security in South Africa. *Geografiska Annaler: Series B, Human Geography* 94(2), 141–159.

Bertinelli, L. and Zou, B. (2008) Does urbanization foster human capital accumulation? *Journal of Developing Areas* 41(2), 171–184.

Bloom, D.E. Canning, D. and Chan, K. (2015) Higher education and poverty in sub-Saharan Africa. *International Higher Education* 45, 6–7. Available at: https://doi.org/10.6017/ihe.2006.45.7924 (accessed 18 March 2021).

Boer, W. (2015) Why entrepreneurship is key to Africa's development. *World Economic Forum – Agenda*, 9 February 2015. https://www.weforum.org/agenda/2015/02/why-entrepreneurship-is-key-to-africas-development/

Bunting, I. Cloete, N. and Van Schalkwyk, F. (2014) *Data on eight flagship universities in Africa 2009–2011. Harvard Dataverse*, V2. Available at: https://doi.org/10.7910/DVN/25427 (accessed 18 March 2021).

Business Tech (2017) How much you could earn based on your level of education in South Africa. *Business Tech*, 9 January. https://businesstech.co.za/news/finance/149531/how-much-you-could-earn-based-on-your-level-of-education-in-south-africa/ (accessed 18 March 2021).

Calderon, A. (2015) Student mobility continues to boost speed of change. *EAIE News*, 10 September. Available at: https://www.eaie.org/blog/student-mobility-continues-to-boost-speed-of-change.html (accessed 18 March 2021).

CHE (Council on Higher Education) (2014) *Distance Higher Education Programmes in a Digital Era: Good Practice Guide*. CHE, Pretoria. Available at: https://che.absol.co.za/media_and_publications/frameworks-criteria/distance-higher-education-programmes-digital-era-good (accessed 27 September 2021).

Christoffels, A. (2018) Science in Africa: Homegrown solutions and talent must come first. *The Conversation*, 24 May. Available at: https://theconversation.com/science-in-africa-homegrown-solutions-and-talent-must-come-first-97025 (accessed 18 March 2021).

Cilliers, J. (2018) *Made in Africa: Manufacturing and the Fourth Industrial Revolution*. Africa in the World Report 8. Institute for Security Studies, Pretoria. Available at: https://issafrica.s3.amazonaws.com/site/uploads/aitwr-8.pdf (accessed 18 March 2021).

Cincotta, R., Engelman, R. and Anastasion, S.(2003). *The security demographic*. Washington, DC, Population Services International.

Cline, W.R. (2008) Global warming and agriculture. *Finance and Development* 45(1), 23–27. Available at: https://www.imf.org/external/pubs/ft/fandd/2008/03/pdf/cline.pdf (accessed 18 March 2021).

Darvas, P. Gao, S. Shen, Y. and Bawany, B. (2017) *Sharing Higher Education's Promise Beyond the Few in Sub-Saharan Africa*. World Bank, Washington, DC. Available at: https://openknowledge.worldbank.org/handle/10986/27617 (accessed 18 March 2021).

Development Initiatives (2018) *2018 Global Nutrition Report: Shining a light to spur action on nutrition*. Development Initiatives, Bristol, UK. Available at: https://globalnutritionreport.org/documents/352/2018_Global_Nutrition_Report.pdf (accessed 18 March 2021).

Dodman, D. Leck, H. Rusca, M. and Colenbrander, S. (2017) African urbanisation and urbanism: Implications for risk accumulation and reduction. *International Journal of Disaster Risk Reduction* 26, 7–15. Available at: https://doi.org/10.1016/j.ijdrr.2017.06.029 (accessed 18 March 2021).

Dyer, G. and Andrews, J. (2011) *Higher Education's Role in Adapting to a Changing Climate*. Second Nature, Boston, Massachusetts, USA. Available at: https://secondnature.org/wp-content/uploads/Higher_Education_Role_Adapting_Changing_Climate.pdf (accessed 18 March 2021).

Eicker, C.K. and Stroebel, A. (2014) Building African capacity in agriculture, food and nutrition. In: Swanepoel, F.J.C., Ofir, Z. and Stroebel, A. (eds.). *Towards Impact and Resilience: Transformative Change in and through Agricultural Education and Training in sub-Saharan Africa*. Cambridge Scholars Publishing. Available at: https://www.cambridgescholars.com/product/978-1-4438-5680-5 (accessed 18 March 2021).

FAO (Food and Agriculture Organization of the United Nations) (2010). *"Climate-smart"Agriculture: Policies, Practices and Financing for Food Security, Adaptation and Mitigation*. FAO, Rome.

Girmaw, A.A. (2016) Higher education 'massification' and challenges to the professoriate: Do academics' conceptions of quality matter? *Quality in Higher Education* 22(3), 260–276. Available at: https://doi.org/10.1080/13538322.2016.1266230 (accessed 18 March 2021).

Goldstone, J. (2002) Population and security. *Journal of International Affairs* 56(1), 3–21.

Goldstone, J. Marshall, M. and Root, H. (2014) Demographic growth in dangerous places. *International Area Studies Review* 17: 120–133.

Hagenmeier, C.C.A. Lansink, A. and Vukor-Quarshie, G.N.K. (2017) Internationalisation and African intellectual *metissage*: Capacity-enhancement through higher education in Africa. *South African Journal of Higher Education* 31(1), 81–103. Available at: http://dx.doi.org/10.20853/31-1-821 (accessed 18 March 2021).

Hammadi, A. Mills, M. Sobrhino, N. Thakoor, V. and Velleso, R. (2019) *A Governance Dividend for Sub-Saharan Africa?* IMF Working Paper 19/1. Available at: http://www.imf.org/en/Publications/WP/Issues/2019/01/11/A-Governance-Dividend-for-Sub-Saharan-Africa-45871 (accessed 11 June 2021).

HAQAA (Initiative Harmonisation of African Higher Education Quality Assurance and Accreditation Initiative) (2017) *African standards and guidelines for quality assurance in higher education (ASG-QA)*. HAQAA Initiative. Available at: http://hdl.handle.net/2445/126939 (accessed 18 March 2021).

Harvey, R. (2017) The 'fourth industrial revolution': Potential and risks for Africa. *The Conversation*, 30 March. Available at: https://theconversation.com/the-fourth-industrial-revolution-potential-and-risks-for-africa-75313 (accessed 18 March 2021).

Hayward, F.M. and Ncayiyana, D.J. (2014) Confronting the challenges of graduate education in sub-Saharan Africa and prospects for the future. *International Journal of Higher Education* 1(1), 173–216. Available at: https://doi.org/10.6017/ijahe.v1i1.5647 (accessed 18 March 2021).

Hewitson, B.C. (2010) Climate change scenario development in sub-Saharan Africa. In Kotecha, P. (ed.). *Climate Change, Adaptation and Higher Education: Securing Our Future*. SARUA Leadership Dialogue Series 2(4), 46–67. Southern African Regional Universities Association, Johannesburg.

Honorati, M. and Johansson de Silva, S. (2016) *Expanding job opportunities in Ghana*. World Bank, Washington, DC. Available at: http://documents.worldbank.org/curated/en/237991477039577804/pdf/109384-PUB-PUBLIC-PUBDATE10-20-16.pdf (accessed 18 March 2021).

ICEF (International Consultants for Education and Fairs) (2015) African summit calls for major expansion of higher education. *ICEF Monitor*, 16 March. Available at: http://monitor.icef.com/2015/03/african-summit-calls-for-major-expansion-of-higher-education (accessed 18 March 2021).

IFAD (International Fund for Agricultural Development) (2011) *Rural Poverty Report 2011*. IFAD, Rome.

IHME (Institute for Health Metrics and Evaluation) (2013) *The global burden of disease: Generating evidence, guiding policy. Sub-Saharan Africa regional edition*. IHME, Seattle, Washington, USA. Available at: http://www.healthdata.org/sites/default/files/files/data_for_download/2013/WorldBank_SubSaharanAfrica/IHME_GBD_WorldBank_SubSaharanAfrica_FullReport.pdf (accessed 18 March 2021).

ILO (International Labour Office) (2017a) *Global Employment Trends for Youth 2017. Paths to a Better Working Future*. ILO, Geneva. Available at: https://www.ilo.org/wcmsp5/groups/public/---dgreports/---dcomm/---publ/documents/publication/wcms_598669.pdf (accessed 18 March 2021).

ILO (2017b) *Youth and Employment in North Africa: A Regional Overview*. A Report Prepared for the Conference on Youth and Employment in North Africa. Available at: https://www.ilo.org/wcmsp5/groups/public/---africa/---ro-addis_ababa/documents/meetingdocument/wcms_577306.pdf (accessed 18 March 2021).

IPCC (Intergovernmental Panel on Climate Change) (2014) Summary for policymakers. In Field, C.B., Barros, V.R., Dokken, D.J., Mach, K.J., Mastrandrea, M.D. *et al.* (eds). *Climate Change 2014: Impacts, Adaptation, and Vulnerability. Part A: Global and Sectoral Aspects. Contribution of Working Group II to the Fifth Assessment Report of the Intergovernmental Panel on Climate Change*, pp. 1–32. Cambridge University Press, Cambridge, UK. Available at: https://www.ipcc.ch/site/assets/uploads/2018/02/ar5_wgII_spm_en.pdf (accessed 18 March 2021).

IPCC (2019) *Climate Change and Land*. An IPCC Special Report on Climate Change, Desertification, Land Degradation, Sustainable Land Management, Food Security, and Greenhouse Gas Fluxes in Terrestrial Ecosystems. Available at: https://www.ipcc.ch/srccl/ (accessed 26 March 2020).

Jogee, R. Callaghan, N.C. and Callaghan, C.W. (2018) Student acculturation in the context of "feesmustfall". *South African Journal of Higher Education.* 32(2), 122–142. Available at: http://dx.doi.org/10.20853/32-2-1645 (accessed 18 March 2021).

Knight, J. (2003) Updated Definition of Internationalization. *International Higher Education* 33, 2–3. Available at: https://doi.org/10.6017/ihe.2003.33.7391 (accessed 18 March 2021).

Langa, P. Wangenge-Ouma, G. Jungblut, J. and Cloete, N. (2016) South Africa and the illusion of free higher education. *University World News*, 26 February. Available at: https://www.universityworldnews.com/post.php?story=20160223145336908 (accessed 18 March 2021).

Leavy, J. and Hossain, N. (2014) *Who Wants To Farm? Youth Aspirations, Opportunities and Rising Food Prices*. IDS Working Paper Vol. 2014 No. 439. Institute of Development Studies, Brighton, UK. Available at: https://opendocs.ids.ac.uk/opendocs/bitstream/handle/123456789/3550/Wp439r.pdf?sequence=4 (accessed 18 March 2021).

Lee, R. (2016) *Macroeconomics, Aging and Growth*. NBER Working Paper 22310. National Bureau of Economic Research, Cambridge, Massachusetts, USA. Available at: https://doi.org/10.3386/w22310 (accessed 18 March 2021).

Levy, D.C. (2010) The global growth of private higher education. *ASHE Higher Education Report* 36(3), 121–133.

Marinoni, G. van't Land, H. and Trine Jensen, T. (2020) *The Impact of COVID-19 on Higher Education Around the World: Global Survey Report*. Available at https://www.iau-aiu.net/IMG/pdf/iau_covid19_and_he_survey_report_final_may_2020.pdf (accessed 11 June 2020).

Martin-Barbero, S. (2020) *COVID-19 has accelerated the digital transformation of higher education. World Economic Forum* (21 July). Available at: https://www.weforum.org/agenda/2020/07/covid-19-digital-transformation-higher-education (accessed 11 June 2021).

Mentz-Coetzee, M., Stroebel, A. and Swanepoel, F.J.C.(2020) The significance of doctoral education and training for development in sub-Saharan Africa. In: *Education and Development: Outcomes for Equality and Governance in Africa* Ndulo, M. and Assié-Lumumba, N. (eds). Palgrave McMillan, London.

Mo Ibrahim Foundation (2015) *2015 Facts and Figures: African Urban Dynamics*. Available at: http://static.moibrahimfoundation.org/u/2015/11/19115202/2015-Facts-Figures-African-Urban-Dynamics.pdf (accessed18 March 2021).

Mohamedbhai, G. (2014) Massification in higher education institutions in Africa: Causes, consequences and responses. *International Journal of Higher Education* 1(1), 59–83. Available at: https://doi.org/10.6017/ijahe.v1i1.5644 (accessed 18 March 2021).

Mourshed, M. Farrell, D. and Barton, D. (2013) *Education to Employment: Designing a System that Works*. McKinsey & Company. Available at: https://www.mckinsey.com/industries/social-sector/our-insights/education-to-employment-designing-a-system-that-works (accessed 18 March 2021).

NBS (National Bureau of Statistics) (2018) *Labor Force Statistics Vol. 1: Unemployment and Underemployment Report*. NBS, Abuja. Available at: https://africacheck.org/wp-content/uploads/2018/03/q1-q3_2017_unemployment_report_VOLUME_1-1.pdf (accessed 18 March 2021).

NEPAD (New Partnership for Africa's Development) (2015) *Agricultural Education and Skills Improvement Framework (AESIF) 2015–2025*. NEPAD, Johannesburg. Available at: https://www.nepad.org/file-download/download/public/15864 (accessed 18 March 2021).

OECD (Organisation for Economic Co-Operation and Development) (2015) *Education at a Glance 2015: OECD Indicators*. Available at: https://www.oecd-ilibrary.org/education/education-at-a-glance-2015_eag-2015-en (accessed 18 March 2021).

OECD (2017) *Education at a Glance 2017: OECD Indicators*. Available at: https://www.oecd-ilibrary.org/education-at-a-glance-2017_5jfrn2shpfxt.pdf?itemId=%2Fcontent%2Fpublication%2Feag-2017-en&mimeType=pdf (accessed 18 March 2021).

OECD (2018a) *Education at a Glance 2018: OECD indicators*. Available at: https://doi.org/10.1787/eag-2018-en (accessed 18 March 2021).

OECD (2018b). *Development Aid at a Glance: Statistics by Region: Africa 2018*. Available at: https://www.oecd.org/dac/financing-sustainable-development/development-finance-data/Africa-Development-Aid-at-a-Glance-2018.pdf (accessed 11 June 2021).

Okebukola, P.A. and Fonteyne, B. (2014) *Developing a Pan-African Quality Assurance and Accreditation Framework*. Draft report for expert and AUC/EC approval. Harmonisation of African Higher Education Quality Assurance and Accreditation Initiative. Available at: https://haqaa.aau.org/wp-content/uploads/2016/04/Developing-A-Pan-African-Quality-Assurance-And-Accreditation-Framework_Draft-2014.pdf (accessed 18 March 2021).

Oketch, M. (2016) Financing higher education in sub-Saharan Africa: Some reflections and implications for sustainable development. *Higher Education* 72(4), 525–539. Available at: https://doi.org/10.1007/s10734-016-0044-6 (accessed 18 March 2021).

Paterson, M. (2021) COVID-19 and higher education: Damage done, lessons learnt. *University World News* (11 February). Available at https://www.universityworldnews.com/post.php?story=20210211103422612 (accessed 11 June 2021).

Pillay, P. (2010a) *Linking Higher Education and Economic Development. Implications for Africa from Three Successful Systems*. Centre for Higher Education Transformation, Wynberg.

Pillay, P. (2010b) Good practices, possible lessons and remaining challenges. In Pillay, P. (ed.). *Higher Education Financing in East and Southern Africa*. Centre for Higher Education Transformation, Somerset West.

Pillay, P. (2011) *Higher education and economic development: Literature review*. Centre for Higher Education Transformation, Wynberg. Available at: https://www.eldis.org/document/A74389 (accessed 18 March 2021).

Randell, H. and Gray, C. (2016) Climate variability and educational attainment: Evidence from rural Ethiopia. *Global Environmental Change* 41, 111–123. Available at: https://doi.org/10.1016/j.gloenvcha.2016.09.006 (accessed 18 March 2021).

Saint, W. (2015) Tertiary education and economic growth in sub-Saharan Africa: The World Bank report. *International Higher Education* 54, 14–15. Available at: https://ejournals.bc.edu/ojs/index.php/ihe/article/view/8417/7551 (accessed 18 March 2021).

Samuels, J. and Jaftha, C. (2017) *Building trust for better movement across SADC: An update on the SADC Qualifications Framework (SADCQF)*. South African Qualification Authority, Pretoria. Available at: http://www.dhet.gov.za/SiteAssets/Latest%20News/2018/An%20update%20on%20the%20SADC%20Qualification%20Framework.pdf (accessed 18 March 2021).

Schwab, K. (2017) *The Fourth Industrial Revolution*. World Economic Forum, Geneva.

Serdeczny, O. Adams, S. Baarsch, F. Coumou, D. Robinson, *et al*. (2016) Climate change impacts in sub-Saharan Africa: From physical changes to their social repercussions. *Regional Environmental Change* 17, 1585–1600. *Available at*: https://doi.org/10.1007/s10113-015-0910-2 (accessed 18 March 2021).

Shabani, J., Okebukola, P., and Oyewole, O. (2014). Quality assurance in Africa: Towards a continental higher education and research space. *International Journal of Higher Education* 1(1), 139–171. Available at: https://ejournals.bc.edu/ojs/index.php/ijahe/article/view/5646/4978 (accessed 18 March 2021).

Swanepoel, F.J.C. (2019) Building Human Capacity and Strengthening Institutions in Low- and Middle-Income Nations to End Global Hunger. *Keynote Address, Ending Global Hunger Colloquium*, 10–11 April, Purdue University, West Lafayette, Indiana, USA.

Szabo, S. (2016) Urbanisation and food insecurity risks: Assessing the role of human development. *Oxford Development Studies* 44(1), 28–48. Available at: https://doi.org/10.1080/13600818.2015.1067292 (accessed 18 March 2021).

Thwala, N. (2017) Private tertiary education is here to stay. *Mail & Guardian*, 17 February. Available at: https://mg.co.za/article/2017-02-17-00-private-tertiary-education-is-here-to-stay (accessed 18 March 2021).

Transparency International (2019) Sub-Saharan Africa: Undemocratic regimes undermine anti-corruption efforts. (19 January). Available at: https://www.transparency.org/news/feature/cpi2018-subsaharan-africa-regional-analysis (accessed 11 June 2021).

TrustAfrica (2015) *Declaration and Action Plan from the 1st African Higher Education Summit on Revitalizing Higher Education for Africa's Future. Executive Summary*. Available at: http://www.trustafrica.org/images/Executive%20SummaryFINAL.pdf (accessed 18 March 2021).

Turianskyi, Y. (2019) *African peer review: Progress is being made, but there are problems. South African Institute of International Affairs, Opinion and Analysis, African Initiatives* (15 March) Available at: https://saiia.org.za/research/african-peer-review-progress-is-being-made-but-there-are-problems (accessed 11 June 2021).

UN DESA (United Nations, Department of Economic and Social Affairs) (2017) *World Population Prospects: Key Findings and Advance Tables*. 2017 Revision Working Paper ESA/P/WP/248. Available at: https://population.un.org/wpp/Publications/Files/WPP2017_KeyFindings.pdf (accessed 18 March 2021).

UNAS (Uganda National Academy of Sciences) (2019) *Owning Our Urban Future: Urban-Rural Linkages for Balanced Regional Development in Africa*. Consensus Study Report. Kampala, Uganda.

United Nations. (2015) Transforming Our World: The 2030 Agenda for Sustainable Development. Available at: https://sustainabledevelopment.un.org/post2015/transformingourworld/publication (accessed 18 March 2021).

United Nations. (2018) *World Urbanization Prospects: The 2018 Revision. Key Facts*. Available at: https://population.un.org/wup/Publications/Files/WUP2018-KeyFacts.pdf (accessed 18 March 2021).

University of Oxford (2015) *International Trends in Higher Education 2015*. Available at: https://www.ox.ac.uk/sites/files/oxford/International%20Trends%20in%20Higher%20Education%202015.pdf (accessed 18 March 2021).

Urdal, H. (2006) A clash of generations. *International Studies Quarterly* 50, 607–629.

Van der Wende, M. (2017) *Opening Up: Higher Education Systems in Global Perspective*. Centre for Global Higher Education Working Paper 22. Centre for Global Higher Education, London. Available at: https://www.researchcghe.org/perch/resources/publications/wp22.pdf (accessed 18 March 2021).

Vermeulen, S.J. Campbell, B.M. and Ingram, J.S.I. (2012) Climate change and food systems. *Annual Review of Environment and Resources* 37, 195–222.

Walker, R. (2015) *Demographic Vulnerability*. The Population Institute, Washington, DC.

Wangenge-Ouma, G. (2012) Tuition fees and the challenge of making higher education a popular commodity in South Africa. *Higher Education* 64(6), 831–844. Available at: https://doi.org/10.1007/s10734-012-9531-6 (accessed 18 March 2021).

WEF (World Economic Forum) (2016) *The Future of Jobs. Employment, Skills and Workforce Strategy for the Fourth Industrial Revolution*. Global Challenge Insight Report. Available at: http://www.hsrc.ac.za/uploads/pageContent/9352/WEF_Future_of_Jobs%20(002).pdf (accessed 18 March 2021).

WEF (2017a) *The Africa Competitiveness Report 2017. Addressing Africa's Demographic Dividend*. Available at: https://www.afdb.org/fileadmin/uploads/afdb/Documents/Publications/Africa_Competitiveness_Report_2017.pdf (accessed 18 March 2021).

WEF (2017b) *The Future of Jobs and Skills in Africa. Preparing the Region for the Fourth Industrial Revolution*. Available at: http://www3.weforum.org/docs/WEF_EGW_FOJ_Africa.pdf (accessed 18 March 2021).

WEF (2019) *Innovation with a Purpose: Improving Traceability in Food Value Chains through Technology Innovations*. System Initiative on Shaping the Future of Food. Available at: http://www3.weforum.org/docs/WEF_Traceability_in_food_value_chains_Digital.pdf (accessed 18 March 2021).

World Bank (2008) *Accelerating Catch-Up: Tertiary Education for Growth in Sub-Saharan Africa*. World Bank, Washington, DC.

World Bank (2010) *Financing Higher Education in Africa*. Available at: https://doi.org/10.1596/978-0-8213-8334-6 (accessed 18 March 2021).

World Bank (2013) *World Development Report 2013: Jobs*. World Bank, Washington, DC.

World Bank (2015) *The Impact of Aging on Economic Growth*. South East Europe Regular Economic Report, No. 8S, Fall 2015. Available at: http://hdl.handle.net/10986/23670 (accessed 18 March 2021).

World Bank (2016) How does an aging population affect a country? *World Bank News, Feature Story*, 1 September. Available at: http://www.worldbank.org/en/news/feature/2016/09/05/uruguay-como-afecta-pais-envejecimiento-poblacion (accessed 18 March 2021).

World Bank (2017) *Higher Education for Development. An Evaluation of the World Bank Group's Support*. Available at: https://ieg.worldbankgroup.org/sites/default/files/Data/Evaluation/files/highereducation.pdf (accessed 18 March 2021).

5 Trends in Tertiary Agricultural Education Capacity in Africa

Nienke Beintema[1]*, John Lynam[2] and Florence Nakayiwa*
[1]*Independent Consultant and former head of ASTI, based in the Netherlands;* [2]*Independent Scholar, based in Kenya;* [3]*Regional Universities Forum for Capacity Building in Agriculture, Uganda*

5.1 Introduction

The availability of tertiary education has risen faster in Africa than in most other regions of the world. Student enrolments have increased from less than half a million students in 1970 to more than 7 million as of 2016, and numbers are expected to triple in the next three decades (Darvas *et al.*, 2017; Calderon, 2018). This growth has driven substantial expansion in the number of tertiary institutions over the past five decades, including a surge in the establishment of private universities. Rather than having a single flagship university, many African countries now have numerous public and private universities (Darvas *et al.*, 2017), and yet the demand for higher education in sub-Saharan Africa (SSA) still outweighs supply. In fact, the region has the world's lowest gross enrolment ratio, and access to higher education remains biased against traditionally-underserved segments of the population (Mohammedbhai, 2014; Darvas *et al.*, 2017).

This chapter provides an assessment of trends in institutional and human-resource capacity in Africa's tertiary education sector and, specifically, in its tertiary agricultural education (TAE) sector. The next two sections focus on the changing institutional landscape since the 1970s, including an analysis of whether more recent TAE expansion mirrors overall growth in the tertiary education sector. Sections 5.4 and 5.5 present an overview of growth in enrolments within Africa's tertiary education sector, along with information on student populations for sample countries and institutions for which TAE data are available. Section 5.6 provides an overview of developments in TAE teaching capacity across these sample countries, both over time and by various demographic indicators. The chapter concludes with an analysis of estimated future demand for PhD graduates in TAE compared with the potential supply, followed by a discussion of implications and recommendations.

5.2 An overview of Africa's Tertiary Education Sector

The establishment of the first African tertiary education institutions dates back to the 19th century, when a number of training colleges were established in South Africa. This led to the creation of Africa's first institution to award degrees, the University of Cape of Good Hope in 1873. But the establishment of the sector across Africa did not really begin until five decades later. In anglophone Africa, seven university colleges were established in the late 1940s and

*Corresponding author: nienke.nld@gmail.com

DOI: 10.1079/9781789246544.0005

1950s as part of the University of London. Their organizational structure, curricula, admission policies and governance were determined by the London-based university. Similarly, the institutions established during the 1950s in francophone Africa mirrored and remained linked with existing universities in France. While the anglophone colonies were expected to fund their own activities, their francophone counterparts were mostly funded by the French government (see Ch. 3, this volume). Only one tertiary education centre was established in the Belgian colonies, in Congo in 1949 (Beintema *et al.*, 1998).

In the initial years following independence, these institutions largely continued to operate as they had done in the colonial period. In addition, the majority of the teaching staff at that time were expatriates of European and other western countries, employed to train African staff to replace them over time (Osiru *et al.*, 2016). New public universities were established in the 1970s, and student enrolments and public funding for tertiary education began to rise. For example, total spending grew from $4 billion in 1970 to $10 billion in 1980 (both in inflation-adjusted 1983 prices). This growth was a vital component of the overall post-independence development plan, driven by the new governments and supported by the donor community (Court, 1980; Moock, 1988; Beintema *et al.*, 1998). Although the number of new tertiary education institutions and enrolments continued to rise through the 1980s and 1990s, funding for these agencies did not grow at the same rate. Many countries lowered their national budgets for tertiary education in response either to structural adjustment programmes (Ch. 3, this volume) or to economic recession (or both). At the same time, donor support to tertiary education was waning in favour of support for primary and secondary education. As a result, average spending per student declined across Africa. A study by the World Bank estimated that public spending per student for a sample of 33 countries fell from $6,800 in 1980 to $981 in 2007/08 (World Bank, 2010). This contrasted with other global regions, where average growth in public resources for tertiary education mirrored growth in enrolments.

In the late 1990s, recognizing higher education's important role in economic development and poverty reduction, the donor community began shifting its attention back to the sector. This led governments in some SSA countries to increase their commitments to tertiary education (Bloom *et al.*, 2014). The number of tertiary institutions continued to rise at a high rate, mainly due to a surge in the establishment of private institutions. During 1990–2014, the estimated number of public institutions across Africa grew from 100 to 500, with private institutions rising from only 30 in 1990 to more than 1,000 in 2014 - more than a 30-fold increase (Darvas *et al.*, 2017). In Kenya, for example, the number of tertiary education institutions doubled from 33 to 73 during 2012–2018, while in Nigeria close to 300 new private universities were awaiting accreditation as of 2018 (Trines, 2018). Many of the new private institutions operate on a for-profit basis and only offer diploma- and certificate-level training, but a small share are non-profit and often faith-based. A number of foreign providers of higher education offer satellite - or similarly styled - campus(es) in Africa (Mohamedbhai, 2014). The number of tertiary education institutions providing distance-learning and regional partnerships also increased in recent years (Bloom *et al.*, 2014; Moock 2016)[1].

Lack of adequate resources continues to be a major concern in tertiary education institutions today. Public education spending in Africa continues to focus on primary and secondary schooling, which reflects the demographic composition of the population (see Ch. 3, this volume). The majority of the 28 African counties for which data are available spent about 20% or less of their education budgets on tertiary education in the mid-2010s (Table 5.1). Exceptions were Gabon (38%) and Ethiopia (48%). Spending per enrolled student varied widely across countries, and variability increased with education level (i.e. BSc-, MSc- or PhD-degree programmes). It does appear, however, that spending levels rebounded from the aforementioned 2007/08 low (World Bank, 2010), at least in current (i.e. non-inflation-adjusted) prices. Not surprisingly, the costs of tertiary education are substantially higher than those of primary and secondary education. Higher tuition rates would have improved the quality of the programmes over the past couple of decades, but such rates would be beyond the means of most students based on income limitations in most African countries.

Table 5.1. Government expenditure on education in selected African countries, mid-2010s

	Share of total by education level (%)				Expenditure per student (current US$)		
Country	Primary	Secondary	Tertiary	Not specified	Primary	Secondary	Tertiary
Burkina Faso	58	28	14	—	78	96	503
Côte d'Ivoire	51	35	14	0.1	259	355	1,237
Ethiopia	29	23	48	—	49	119	1,665
Ghana	29	37	18	16	141	295	879
Kenya	38	47	13	2	na	na	na
Mozambique	49	31	14	6	83	261	708
Rwanda	34	35	21	—	39	195	784
Senegal	32	17	28	24	150	155	1,541
South Africa	40	32	15	12	1,140	1,383	2,938
Uganda	59	25	16	—	na	na	na
Tanzania	55	23	21	—	99	200	1,853
Zimbabwe	50	33	17	—	224	417	2,140
Sample average	41	33	18	8	242	507	1,878
Sample range	[24–68]	[23–65]	[4–48]	[0–24]	[32–1,140]	[50–1,383]	[341–5,379]

Source: Constructed by authors from UIS (2019).
Note: Data are for the latest year available; na = not available. Samples include 29 countries for total spending and 22 countries for spending per student. Primary includes pre-primary education; secondary includes post-secondary, non-tertiary education (see UIS, 2012 for definitions of qualification levels).

Moreover, governments lacked the resources to be able to offer comprehensive student loans. Facing these realities, (i) many universities became overcrowded, (ii) facilities were poorly maintained and (iii) salaries for teachers - which were already low - were further eroded over time. This in turn led to high staff turnover, low staff motivation and the need for teachers to supplement their incomes with alternative employment (Osiru *et al.*, 2016; Blair and Jordan, 1994).

5.3 Africa's Tertiary Agricultural Education System

Many of today's universities, colleges and other tertiary education institutions focus on a wide array of sciences, often to the exclusion of agriculture and related disciplines. This is especially true for private institutions, only a handful of which offer agricultural training - and if they do, it is often only at the diploma or certificate level[2]. In fact, as of 2016, only 178 tertiary education institutions covering 37 SSA countries were found to offer BSc-, MSc- or PhD-degree programmes in agricultural sciences. Of these, more than a quarter were established during 2000–2009 (Figs 5.1 and 5.2). Only 12% of degree-conferring TAE institutions were private in 2016, which is much lower than the share for the tertiary education sector overall. This is not surprising, however, considering the costs involved in providing agricultural science education are substantially higher than the comparable costs, for example, of business, accounting and other science disciplines (Osiru *et al.*, 2016). A further 33 institutions not included in the sample provided degree training in disciplines such as natural resources and environmental and biological sciences[3]. Also of note, one-third of the sample TAE institutions were located in Nigeria. Kenya and South Africa represent 9% and 7% of the sample's TAE institutions, respectively[4]. Most other SSA countries only have one to three TAE institutions offering degree-level agricultural science programmes.

Various TAE institutions have more than one faculty, school or other type of unit providing training. It is common, for example, for many larger universities to have a faculty of agricultural sciences and a faculty of veterinary sciences. For this reason, the 178 sample institutions

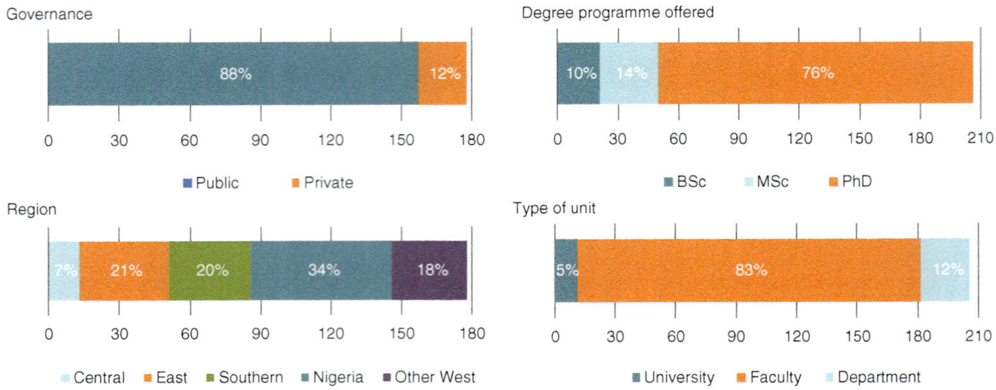

Fig. 5.1. Overview of tertiary agricultural education institutions, 2016
Sources: Constructed by authors from ASTI and RUFORUM (2018), complemented by Internet searches.
Notes: TAE = tertiary agricultural education. 'TAE institutions' include 178 universities, colleges and training institutes; TAE units include 206 agricultural universities (or equivalent), faculties of agriculture (or equivalent) and departments with agricultural studies (or equivalent) which fall under a faculty within a broader field of science. Agriculture includes animal, veterinary, fishery and forestry sciences. Biological, environmental, food and natural resource sciences are excluded. Only TAE institutions offering at least BSc-level training and with identifiable units offering graduate-level agricultural training are included, which resulted in a sample of the following 37 countries (figures in parentheses indicate the number of institutions with TAE units per country): Central Africa includes Cameroon (2), Central African Rep. (1), Chad (1), DR Congo (7), Rep. of Congo (1), Gabon (1); East Africa includes Burundi (2), Ethiopia (7), Kenya (16), Rwanda (2), Tanzania (2) and Uganda (9); Southern Africa includes Botswana (1), Eswatini (1), Lesotho (1), Madagascar (2), Malawi (1), Mauritius (1), Mozambique (5), Namibia (2), South Africa (13), Zambia (1) and Zimbabwe (7); and West Africa includes Benin (3), Burkina Faso (3), Cabo Verde (1), Côte d'Ivoire (1), The Gambia (1), Ghana (8), Guinea (1), Liberia (2), Mali (2), Mauritania (1), Niger (4), Nigeria (59), Senegal (3), Sierra Leone (1) and Togo (1).

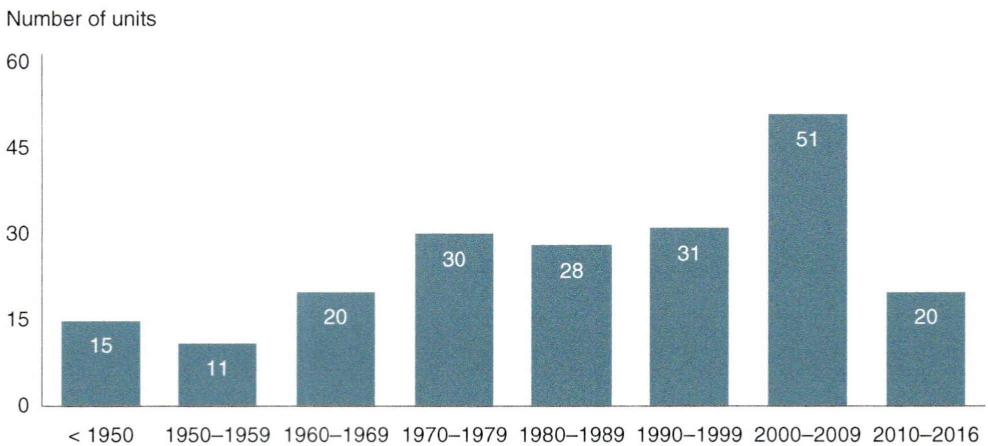

Fig. 5.2. Number of TAE units by their establishment date, pre-1950s–2016
Sources: Constructed by authors from ASTI and RUFORUM (2018), complemented by Internet searches.
Note: TAE = tertiary agricultural education.

actually comprise 206 different units. These range from fully-fledged agricultural universities (e.g. Sokoine University in Tanzania) to departments within faculties (e.g. the Department of Agriculture within the Faculty of Applied Sciences at Methodist University in Ghana). The majority of units, however, are faculties, schools or an equivalent unit (83%) offering agricultural sciences, often across a broad range of disciplines. A small number of units specialize in focused areas - such as veterinary medicine, forestry or fisheries - and many of the broad-based agricultural units include such areas as natural resources and environmental sciences.

About two-thirds of the 206 units were established prior to 2000 (Fig. 5.2). Notably, 51 units were established during 2000–2010, including 10 agricultural science units within newly-established private universities, and 14 new public universities in Nigeria. Between 2010 and 2016, the expansion appeared to have tapered off to levels similar to those of the last four decades of the 20th century. Surprisingly, 76% of the units offer PhD-degree programmes - at least double the share observed in 1996 (Beintema *et al.*, 1998). Although this share is high, many TAE units, especially the smaller ones, often only have a few PhD students enrolled. This share is lower among the newer units: 60% of units established after 2000 provided PhD-level training, compared with more than 80% of units established in the 20th century. This is a substantial increase from the mid-1990s, when only one in three TAE units offered doctoral training (Beintema *et al.*, 1998). Most of these units follow the anglophone system, whereby PhD degrees are based on research only. As previously indicated, growth in the number of TAE units was considerably lower than growth in the overall number of tertiary education institutions (Table 5.2). As of 2016, the number of TAE units had doubled since 1990, whereas the number of tertiary education institutions had grown tenfold during that time. Even just focusing on the public sector, overall growth in the number of tertiary education institutions was four times higher than growth in the number of TAE units.

5.4 Growth in Tertiary Education Enrolments in Africa

The view that education (including tertiary education) is an important factor in achieving national economic growth and that it should be available to all has led to the substantial increase in the number of African tertiary education institutions (as discussed in Section 5.2). Another important phenomenon is what is known as the 'massification' of tertiary education in terms of enrolments. The term was first used in the context of the tertiary education sector by Trow (1973) to classify the transformation of tertiary education institutions from exclusive access by the few, to egalitarian access by the many. Trow (1973) identified three transition phases using the gross enrolment ratio (GER), which measures the total number of students enrolled in tertiary education as a share of those aged 20–24 years. A GER of 15% or lower indicates an 'elite' system, a GER of 16–50% indicates a 'massive' system and a GER of more than 50% indicates a 'universal' system[5]. Following these classifications, and using data

Table 5.2. Growth in tertiary education institutions versus TAE institutions, 1990 and 2016

Educational category	1990	2016	Absolute growth
Tertiary education institutions			
Total no. of institutions	130	1,500	1,053%
No. of public institutions	100	500	400%
Tertiary agricultural education institutions			
Total no. of units	104	206	98%
No. of public units	100	184	84%

Sources: Constructed by authors from Darvas *et al.* (2017) and ASTI and RUFORUM (2018), complemented by Internet searches.
Note: TAE = tertiary agricultural education.

from the Institute for Statistics of the United Nations Educational, Scientific and Cultural Organization (UNESCO), Calderon (2018) found that, as of 2018, only 7 of a sample of 41 SSA countries had massive systems (17%), and none had universal systems[6]. This starkly contrasts with Asian and Latin American countries, where 30% had universal systems as of 2018. It should be noted that, although Trow (1973) used the term massification in the context of education systems, it is more commonly used to describe significant growth in student enrolment numbers at institutions over time (Mohamedbhai, 2014).

The total number of students enrolled across all fields of study in tertiary education institutions in SSA grew substantially over time, from less than 200,000 in 1970 to over 7 million as of 2016 (Fig. 5.3). Growth accelerated from the beginning of the 21st century, both across countries and universities. Mohamedbhai (2014), for example, found that student enrolments at universities in Burkina Faso, Ghana, Kenya, Mozambique, Senegal and Zimbabwe grew by four to eight times during 1986–2006. Calderon (2018) projected that student enrolments across all fields of study in African tertiary education institutions will reach 14 million by 2025 and 22 million by 2040, resulting in a substantial increase in the already very high annual growth rates, especially during 2025–2040. These estimates are based on projected high population growth in most African countries, coupled with the expectation that countries will both increase their rates of completion of secondary education, and strengthen their tertiary education systems (Calderon 2018).

During 1990–2016, tertiary student enrolments across all fields of study increased 37-fold, whereas the number of tertiary education institutions increased only tenfold. As a result, the number of students per institution increased from around 1,500 in 1990 to almost 5,000 as of 2016 (Darvas *et al.*, 2017; Calderon, 2018). Despite this large increase, SSA is still well behind other regions of the world in terms of the share of its student-aged population with access to higher education. In 2016, SSA's GER was 9% compared with 48% for Latin America and the Caribbean and 34% for Asia and the Pacific (Table 5.3). Participation was substantially higher in North America and Europe (77%) compared with the developing world. Rates across African countries varied, but only 7 of 41 sample countries had GER shares of 15% or higher as of 2016. Despite the projected growth in total enrolments by 2040, SSA's GER is expected to remain the lowest in the world (Calderon 2018).

Unsurprisingly, in most countries, the majority of the tertiary education students in a mid-2010s sample of 21 countries were enrolled in BSc-degree programmes (Table 5.4). The share of students following MSc-degree training ranged widely, from 3–5% in Burundi, Cabo Verde, Ethiopia, Lesotho, Mozambique and Togo, to more than 20% in Côte d'Ivoire, Mali and Niger. Far fewer students were enrolled in PhD programmes (5% of total enrolments or less). Universities in some countries (in particular, South Africa) attract students from other African countries. Furthermore,

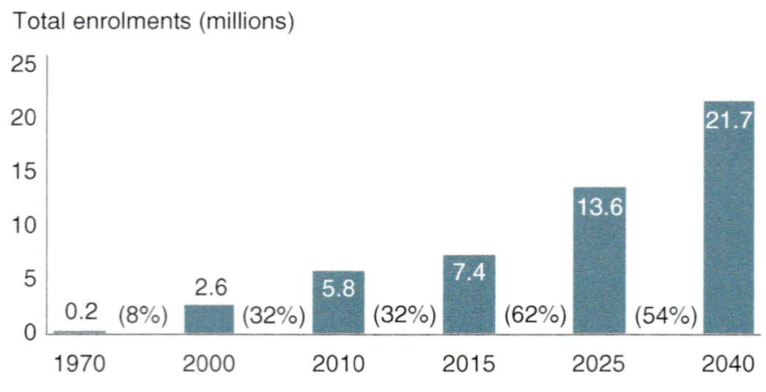

Fig. 5.3. Projected growth in tertiary education enrolments, 1970–2040
Sources: Constructed by authors from Bloom *et al.* (2014) for 1970 and Calderon (2018) for all other years.
Notes: Data include enrolments across all fields of study at degree, diploma and certificate (or equivalent) levels. Average annual growth rates are shown in parentheses.

Table 5.3. Global tertiary education enrolments by region, 2000 and 2016

Region	Total enrolments (million)		Gross enrolment (ratio)	
	2000	2016	2000	2016
Sub-Saharan Africa	2.6	7.4	4.4%	8.5%
Asia–Pacific	37.5	113.1	12.4%	34.2%
Latin America and Caribbean	11.5	26.2	22.6%	48.4%
Arab States and Central Asia	6.6	12.8	19.2%	30.8%
North America and Europe	41.8	56.4	53.2%	77.0%
Global total	99.9	215.9	19.0%	36.8%
Sub-Saharan African share of global total	2.6%	3.4%	—	—

Source: Constructed by authors from Calderon (2018).
Note: Data include enrolments across all fields of study, as well as diploma and certificate (or equivalent) levels as preparation for BSc-degree training.

Table 5.4. Share of enrolled students by degree level in selected African countries, mid-2010s

Country	BSc level	MSc level	PhD level
Benin	86.9[a]	11.4	1.6
Burundi	95.3	3.5	1.2
Cabo Verde	96.8	3.0	0.2
Côte d'Ivoire	71.2	23.8	5.0
Ethiopia	94.6	5.1	0.3
Ghana	89.9	9.3	0.8
Guinea	87.6	10.2	2.2
Kenya	87.7	9.6	2.7
Lesotho	96.5	3.4	0.1
Mali	77.5	22.0	0.4
Mauritania	81.4	18.0	0.6
Mauritius	87.7	11.3	1.0
Mozambique	95.5[a]	4.2	0.3
Niger	64.5	32.1	3.4
Rwanda	93.5	6.3	0.2
Senegal	77.8	18.3	3.9
Seychelles	84.7	15.0	0.3
South Africa	89.2	7.9	2.9
Tanzania	92.9	6.6	0.5
Togo	93.8	4.4	1.8
Zimbabwe	86.7	13.0	0.3

Source: Constructed by authors from UIS (2019).
Note: Data include students enrolled in all fields of study.
[a] BSc-level data for Benin and Mozambique include students enrolled in certificate and diploma (or equivalent-level) programmes.

a sizeable number of African students pursue training at tertiary education institutions around the world. The number of students enrolled in institutions outside Africa grew substantially, from around 30,000 in the early 1970s to just over 400,000 in the mid-2010s (Table 5.5). While most African students in the 1970s were enrolled in American, French, and British universities, in recent years African students are more widely dispersed across the globe.

A critical concern is that many of the students never return home after they finish their studies. Although reliable data are difficult to obtain, a 2019 article in *The Economist* claimed that 1 in 9 African graduates live in Organisation for Economic Co-operation and Development

Table 5.5. African students enrolled in tertiary institutions outside Africa, early 1970s to mid-2010s

Country/ region	Number of students			Share of total (%)		
	Early 1970s	Early 1990s	Mid-2010s	Early 1970s	Early 1990s	Mid-2010s
United States	7,838	17,123	40,271	25	19	10
France	5,182	25,924	100,506	17	29	25
United Kingdom	5,092	7,377	33,433	17	8	8
Canada	1,281	4,199	18,557	4	5	5
Belgium	1,571	5,043	5,470	5	6	1
Other	9,421	28,968	202,722	31	33	51
Other Europe	na	na	52,683	na	na	13
Arab States	na	na	62,622	na	na	16
Asia	na	na	46,956	na	na	12
Total	30,385	88,634	400,960	100	100	100

Source: Constructed by authors from Beintema *et al.* (1998) and UIS (2019).
Notes: Data include students enrolled across all fields of study, as well as students from North African countries. A few host countries are excluded due to lack of data; na = not available.

countries, compared with 1 in 13 Latin American graduates and 1 in 30 Asian graduates (*The Economist*, 2019). Growth in student enrolments over time has generated both positive and negative effects. One key negative effect is graduate unemployment (see Ch. 7, this volume), which has been attributed to the poor quality and lack of relevance of qualifications. As many of the chapters in this volume discuss, a mismatch exists between the training tertiary education institutions provide and the skills required in the workplace. In a sample of 29 countries for the mid-2010s, the share of the unemployed workforce with tertiary qualifications (including diplomas and certificates) ranged from 5% in Ethiopia, Ghana, Tanzania and Uganda to 15% or higher in Burundi, Rwanda, Senegal and Mali (authors' calculations based on World Bank, 2019).

5.5 Overview of Student Populations in Africa's Tertiary Agricultural Institutions

Data on enrolments in African TAE institutions were available through UNESCO and the Association of Commonwealth Universities until the early 2000s, when both organizations stopped publishing such information. Analysing these two datasets, Beintema *et al.* (1998) found that, for a sample of 19 countries, only 5% of TAE students were undertaking MSc and PhD degrees in the early 1990s (excluding those enrolled in diploma and certificate, or equivalent, training). The authors also found that, from the early 1980s to the early 1990s, growth in the number of enrolments across all fields of study grew faster than the corresponding growth in enrolments in agricultural sciences. Osiru *et al.* (2016) argued that, based on qualitative evidence, African universities had also experienced substantial growth in agricultural science enrolments in more recent years, but it was unclear whether this growth was as strong as growth across all fields of science[7]. Furthermore, in most countries only a small proportion of students were enrolled in agricultural sciences as of the mid-2010s, ranging from 0.2% in the small island states Cabo Verde and Seychelles, to more than 5% in Kenya, Eswatini, DR Congo, Ethiopia, Eritrea and Sierra Leone. But shares for the majority of the 29 countries for which data were available only ranged between 1% and 3% (UIS, 2019). These low shares are partly explained by students' limited interest in obtaining agricultural degrees, combined with the fact that science faculties are relatively more expensive to establish and maintain (Osiru *et al.*, 2016).

To obtain more evidence, the ASTI programme of the International Food Policy Research Institute and the Regional Universities Forum for Capacity Building in Agriculture

(RUFORUM) partnered in a pilot study on TAE units in SSA, which was incorporated into ASTI's existing data collection activities[8]. This effort was hindered by the lack of central databases at most of the surveyed TAE institutions, which meant that the information had to be collected from the individual departments and other units, and was often unusable (due to incompleteness or other issues of quality). Ultimately, this meant that the resulting data were incomplete, and the goal of the study was thwarted. Consequently, the discussion presented in this section is limited to data for the 2015/16 student populations of key universities in 15 East and Southern African countries.

Despite the increased focus on expanding and diversifying graduate training programmes to meet the demand for agricultural researchers and other specialists, the majority of students pursuing agricultural science degrees were enrolled in BSc-degree (or equivalent) programmes. Growth within this category of students had overloaded the capacity of teachers and facilities, such as classrooms, laboratories and library space (Osiru *et al.*, 2016). The composition of enrolments across the sample universities was quite varied (Table 5.6 and Fig. 5.4a). At South Africa's University of Pretoria and University of Stellenbosch, 19% and 42% of agricultural science students were undertaking graduate (i.e. MSc and PhD) degrees, respectively, and many were enrolled in doctoral programmes. Haramaya University and Hawassa University in Ethiopia, Makerere University in Uganda and the University of Zimbabwe recorded similar shares of students in graduate programmes, but the majority were pursuing MSc degrees. Many of the universities in smaller African countries only established their graduate degree programmes in recent years and, hence, had very small numbers of students enrolled, both overall and per programme. For example, at Botswana University of Agriculture and Natural Resources, Egerton University, the University of Namibia and the University of Rwanda, only 2 to 3 PhD students were enrolled per programme; numbers were substantially higher at Haramaya University (19) and the University of Zimbabwe (17).

Most PhD programmes were not well developed as of 2015/16 (Osiru *et al.*, 2016). Furthermore, many universities were following the European model, whereby PhD students only prepare a dissertation based on their research rather than undertaking coursework (Osiru *et al.*, 2016). Institutional expansion should have generated institutional differentiation, but this has been limited because most universities mimic the flagship model of academic programmes and practices. This can partly be explained by isomorphism (i.e. structural correspondence) and the sector's need to establish legitimacy (DiMaggio and Powell, 1983; Boxenbaum and Jonsson, 2017). While some universities can clearly be designated as teaching universities, others need recognition as highly-productive, research-intensive institutions (Cloete *et al.*, 2018). The TAE sector would benefit greatly from differentiating its value chain into institutions intended to develop the capacity of agricultural practitioners, field agents and agribusiness entrepreneurs, as distinct from those designed to absorb a higher concentration of graduates as preparation for careers in research institutes and universities. Shifting the often negative perception of agricultural technical and vocational education and training (ATVET) would support this differentiation (see Ch. 13, this volume).

It also makes financial sense to group MSc and PhD programmes at national and regional levels rather than spreading resources across individual programmes at multiple schools and faculties. This is particularly so given the wide range of agricultural science disciplines requiring coverage. RUFORUM, for example, has already implemented these regionally-based programmes, and they should be scaled out across Africa (Osiru *et al.*, 2016). Fragmentation of the degree programmes provided by TAE units applies to both the number of agricultural science programmes and the disciplines offered (Fig. 5.5). While a majority of the students were enrolled in crop and animal science programmes, the range of other programmes was substantial and included natural resources, environmental sciences, social sciences and - occasionally - non-agricultural areas, such as tourism.

Rates of female participation in graduate agricultural science programmes vary greatly across agencies (Fig. 5.4b). In 2015/16, more women than men were enrolled at the University of Nairobi, National University of Lesotho and University of Namibia; in contrast, women constituted less than 10% of graduate students at Haramaya University.[9] The shares of female students across degrees also vary widely, but this

Table 5.6. Number of agricultural enrolments, graduates and programmes at selected East and Southern African universities, 2015/16

University and unit name	Introduction of PhD programme	No. of enrolments			No. of graduates			No. of programmes		
		BSc level	MSc level	PhD level	BSc level	MSc level	PhD level	BSc level	MSc level	PhD level
Botswana University of Agriculture and Natural Resources (BUANR)	2011	947	46	6	186	8	—	10	4	2
University of Eswatini, Faculty of Agriculture (UNISWA-FA)	2012	992	35	19	175	22	—	10	6	—
Haramaya University, College of Agriculture and Environmental Sciences (HU-CAES), Ethiopia	2005	1,531	233	252	na	na	na	9	13	13
Hawassa University, College of Agriculture and Wondo Genet College of Forestry and Natural Resources (HwU-CA&WCFNR), Ethiopia	2010	3,067	768	76	na	na	na	11	28	8
Egerton University, Faculty of Agriculture (EU-FA), Kenya	1996	573	59	36	515	35	15	8	16	13
University of Nairobi, Faculty of Agriculture and Faculty of Veterinary Medicine (UoN-FA&FVM), Kenya	1974	na	324	124	na	45	12	15	35	17
National University of Lesotho, Faculty of Agriculture (NUL-FA)	—	488	9	—	na	na	na	6	na	—
University of Mauritius, Faculty of Agriculture (UoM-FA)	na	423	34	20	115	8	8	20	2	na
University of Namibia, Faculty of Agriculture and Natural Resources (UNAM-FANR)	na	475	16	9	68	7	3	3	3	4
University of Rwanda, College of Agriculture and Veterinary Medicine (UR-CAVN)	2014	3,197	58	2	na	na	na	11	2	1

University of KwaZulu-Natal, School of Agriculture, Earth and Environmental Sciences (UKZN-SAEES), South Africa	na	na	123	120	na	43	23	na	na	na
University of Pretoria, Faculty of Natural and Agricultural Sciences and Faculty of Veterinary Sciences (UPFNAS&FVS), South Africa	na	1,360	81	231	361	89	25	na	na	na
University of Stellenbosch, Faculty of Agricultural Sciences (US-FAS), South Africa	na	362	147	113	142	71	10	na	na	na
Makerere University, College of Agricultural and Environmental Sciences, and College of Veterinary Medicine, Animal Resources, and Biosecurity (MAK-CAES&COVAB), Uganda	na	1,200	342	74	532	157	23	27	25	9
University of Zimbabwe, Faculty of Agriculture and Faculty of Veterinary Sciences (UZ-FA&FVS)	na	486	42	33	na	na	na	7	8	2

Source: Constructed by authors from ASTI and RUFORUM (2018).

Notes: na = not available. Not all universities offer PhD-degree programmes, which is the case at the University of Mauritius.

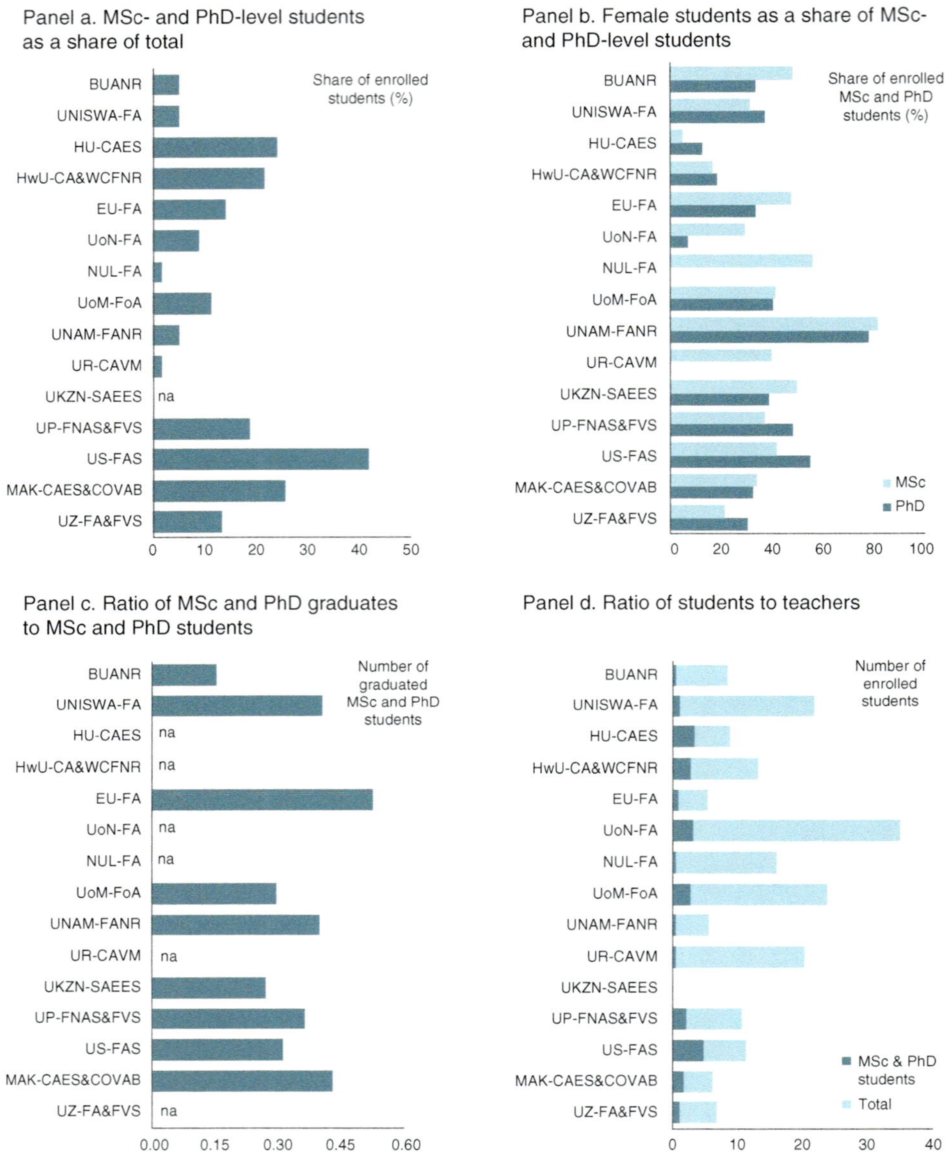

Fig. 5.4. Overview of TAE students enrolled in selected East and Southern African universities, 2015/16
Source: Constructed by authors from ASTI and RUFORUM (2018).
Notes: See Table 5.6 for key to institute abbreviations. TAE = tertiary agricultural education; teachers = deans, department heads, (assistant) professors and (senior) lecturers. Data for the University of Nairobi are only for the Faculty of Agriculture.

is partly a function of small student populations, especially at the MSc- and PhD-degree levels. The ratio of BSc-level students to teachers (i.e. deans, professors, lecturers etc.) is an indicator of the quality of training offered, with lower ratios being preferable (Fig. 5.4d); in the case of MSc and PhD students, however, low ratios are more an indicator of the high cost of these programmes than their quality. Most of the selected universities reported ratios of three or fewer MSc and PhD students per teacher (Fig. 5.4c). Exceptions were the University of Stellenbosch and

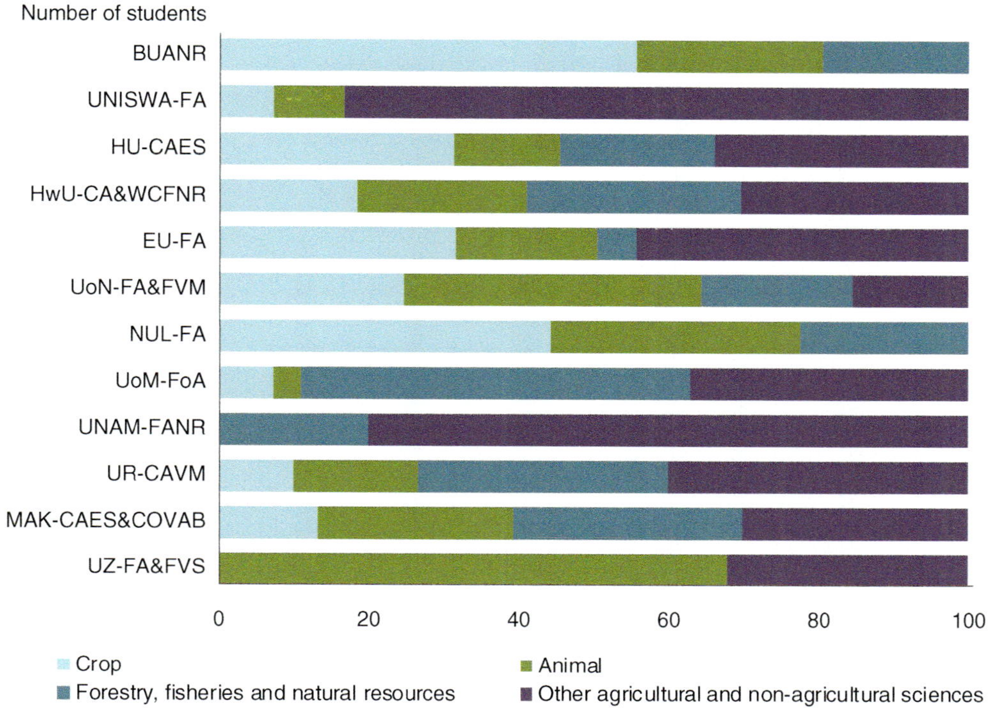

Fig. 5.5. Graduate agricultural enrolments at selected East and Southern African universities by major discipline, 2015/16
Source: Constructed by authors from ASTI and RUFORUM (2018).
Notes: See Table 5.6 for key to institute abbreviations. Data include student enrolments in MSc- and PhD-level programmes in agricultural sciences. Other agricultural and non-agricultural sciences include food and nutrition science, and socioeconomics.

University of KwaZulu-Natal, with ratios of five and six students, respectively. A number of universities had very high student–teacher ratios in their BSc-level agricultural science programmes.

5.6 Overview of Teaching Capacity in Africa's Tertiary Agricultural Institutions

The significant growth in the number of tertiary education institutions and accompanying influx of students led to a substantial increase in the number of teachers - defined as deans, department heads, (assistant) professors, and (senior) lecturers - although qualitative evidence suggests that this growth was substantially slower (Cloete *et al.*, 2018). The absolute number of teachers in SSA's TAE sector grew more than six-fold during 1981–2016 and totalled close to 14,000 as of 2016 (Fig. 5.6). In most SSA countries, this growth was driven by the expansion of existing universities, the creation of new units within established universities and the establishment of new universities. In some countries, teaching capacity rose at far higher rates during this 35-year period. In Rwanda and Uganda, for example, the number of agricultural science teachers increased 12-fold during 1981–2016, and in Mozambique numbers increased 20-fold - the highest level among the sample countries. The average number of teachers per institution rose from 23 in 1981 to close to 67 in 2016. Despite this overall increase, the teaching capacity of many faculties remained small, and it is very unlikely that growth in the number of teachers kept pace with growth in the number of students.

Absolute numbers of teachers varied considerably across the 39 countries for which data were available. The 60 Nigerian TAE units

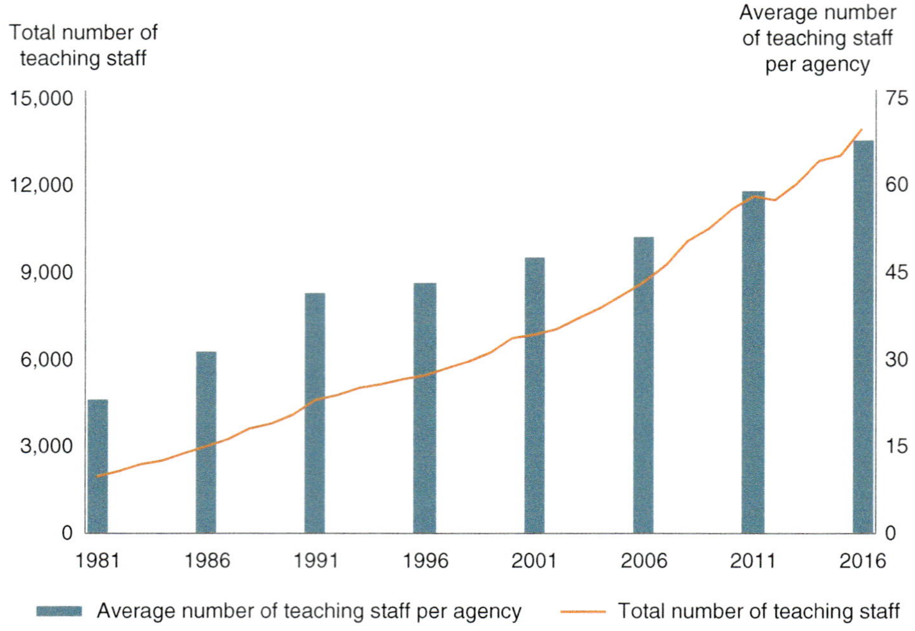

Fig. 5.6. Growth in the number of TAE teachers, 1981–2016

Sources: Constructed by authors from ASTI (2019), complemented by ASTI and RUFORUM (2018) and various secondary sources.

Notes: TAE = tertiary agricultural education; teachers = deans, department heads, (assistant) professors and (senior) lecturers employed in degree-conferring TAE units. Data were estimated for Angola, Comoros, Equatorial Guinea, Sao Tome and Principe, and Seychelles based on trends in the other 39 countries; data were estimated for some years for other countries due to data gaps. For Nigeria, the country with the highest number of teachers, data for 2009–2016 were estimated based (i) on data for 2008 and (ii) growth in the number of researchers employed at Nigeria's government agricultural research agencies during this period.

identified as offering degree training in agricultural sciences employed an estimated 3,900 teachers in 2016 (Fig. 5.7)[10]. The Ethiopian and Kenyan TAE sectors each employed more than 1,000 teachers that year, while DR Congo, Ghana, South Africa, Tanzania and Uganda each employed between 500 and 1,000 teachers. In contrast, 12 of the 39 sample countries employed fewer than 100 teachers in 2016. No tertiary education institutions in Guinea-Bissau offered degrees in agricultural sciences. Vast differences in the size of countries and populations across Africa make it useful to look at the number of teachers per capita (Fig. 5.8). Of the 39 sample countries, 11 employed fewer than 10 teachers of agricultural science per million people. In contrast, Botswana, Eswatini and Namibia recorded particularly high ratios. All three of these countries have relatively small populations, and their universities were established in the 1980s or early 1990s.

On average (excluding Nigeria), half of all teachers in the 39-country sample held PhD degrees and 38% were trained to the MSc-degree level (Fig. 5.9). The total number of teachers with BSc degrees declined somewhat in 2012 and remained fairly stagnant thereafter. Unsurprisingly, the share of PhD-qualified teachers differed across countries. Of teachers employed at TAE institutions in Benin, Burkina Faso, Côte d'Ivoire, Madagascar, Mauritania and Niger, 80% or more held PhD degrees (Fig. 5.10). In contrast, less than 20% of the teachers at TAE institutes in Cabo Verde, Ethiopia, Guinea, Liberia, and The Gambia were PhD-qualified.

In many SSA countries, the share of PhD-qualified TAE teachers in their 50s and 60s is problematically high, and many are nearing retirement age (Fig. 5.11). As of 2016, for a 33-country sample (excluding Nigeria and a few other smaller countries), an average of nearly half of all teachers with PhD degrees were over 50 years old. This is

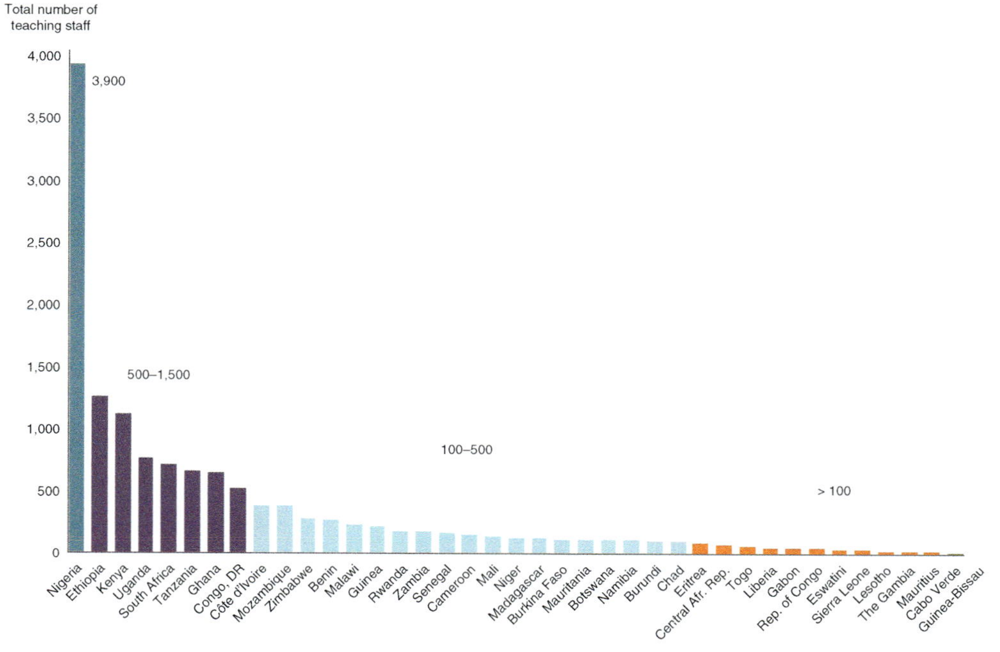

Fig. 5.7. Number of TAE teachers by country, 2016
Source: Constructed by authors from ASTI (2019), complemented by ASTI and RUFORUM (2018) and various secondary sources.
Note: TAE = tertiary agricultural education; teachers = deans, department heads, (assistant) professors and (senior) lecturers employed in degree-conferring TAE units. Data for Burkina Faso and Malawi are for 2014; data for Eritrea, Liberia and Sierra Leone are for 2011; data for Nigeria are for 2008. Guinea-Bissau has no tertiary agricultural education institutions.

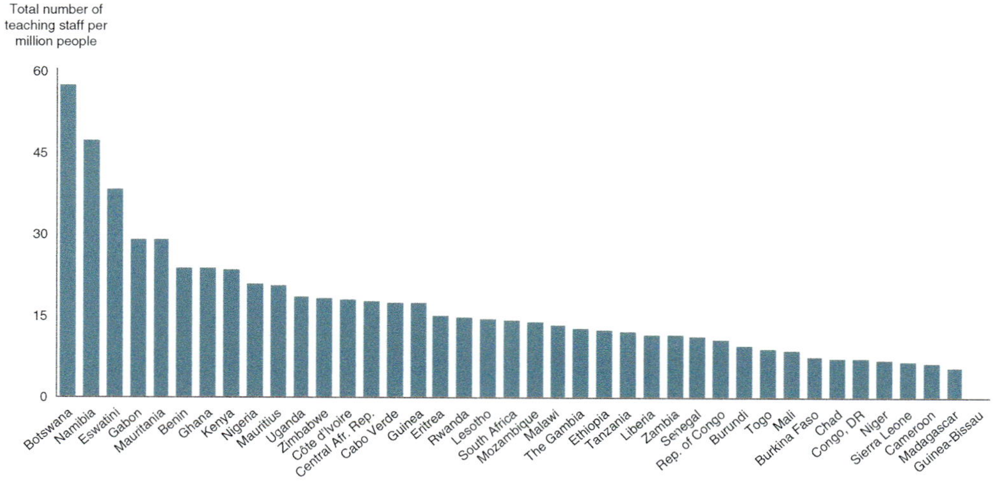

Fig. 5.8. Number of TAE teachers per million people by country, 2016
Source: Constructed by authors from ASTI (2019) and FAO (2018), complemented by ASTI and RUFORUM (2018) and various secondary sources.
Note: TAE = tertiary agricultural education; teachers = deans, department heads, (assistant) professors and (senior) lecturers employed in degree-conferring TAE units. Data for Burkina Faso and Malawi are for 2014; data for Eritrea, Liberia and Sierra Leone are for 2011; data for Nigeria are for 2008. Guinea-Bissau has no TAE institutions.

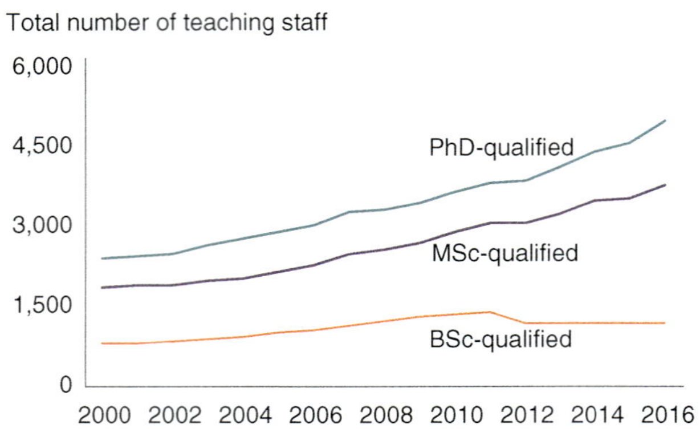

Fig. 5.9. Number of TAE teachers by qualification level, 2000–2016
Source: Constructed by authors from ASTI (2019), complemented by ASTI and RUFORUM (2018) and various secondary sources.
Notes: TAE = tertiary agricultural education; teachers = deans, department heads, (assistant) professors and (senior) lecturers employed in degree-conferring TAE units. Data are for a 38-country sample and exclude Nigeria and a few countries with very small tertiary agricultural education teaching capacity.

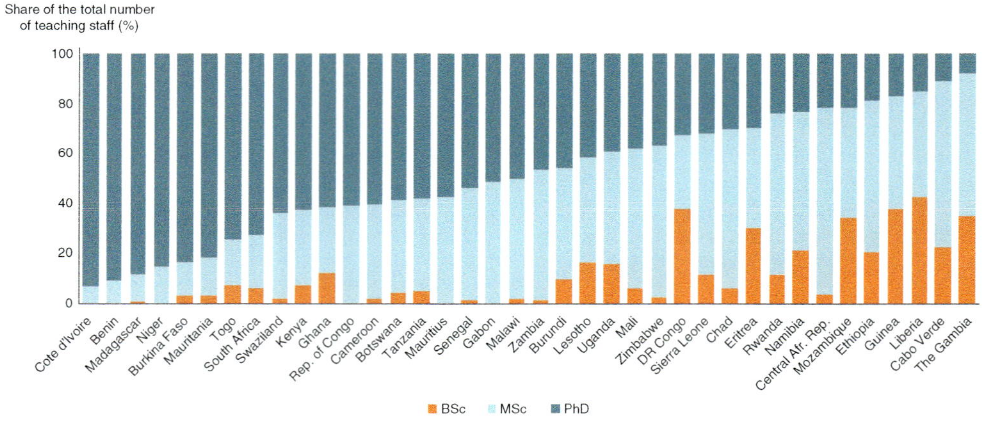

Fig. 5.10. Share of TAE teachers by qualification level and country, 2016
Source: Constructed by authors from ASTI (2019), complemented by ASTI and RUFORUM (2018) and various secondary sources.
Notes: TAE = tertiary agricultural education; teachers = deans, department heads, (assistant) professors and (senior) lecturers employed in degree-conferring TAE units. Data are for a 38-country sample and exclude Nigeria and a few countries with very small tertiary education sectors in terms of teaching capacity.

particularly serious in several countries with small TAE teaching capacities, such as Eritrea, Eswatini, Chad, Ghana, Guinea, Liberia, Mali and Mauritius. For example, both of the PhD-qualified teachers employed by the School of Agriculture and Environment Science at the University of The Gambia were over 50 years old in 2016. In contrast, a number of countries - such as Cabo Verde, Cameroon, Mauritania, Rwanda and Zambia - employ higher levels of younger teachers, many of whom need further mentoring and training.

Female participation in TAE provision has grown over time. Women constituted an average of 18% of the teachers in a 36-country sample of TAE units in 2008, and this share had risen to 24% in 2016 (ASTI, 2019). In general, TAE systems in Southern Africa employ relatively more women than do the systems of West and East Africa. In most systems, female teachers are, on average, less-qualified and younger than their male counterparts.

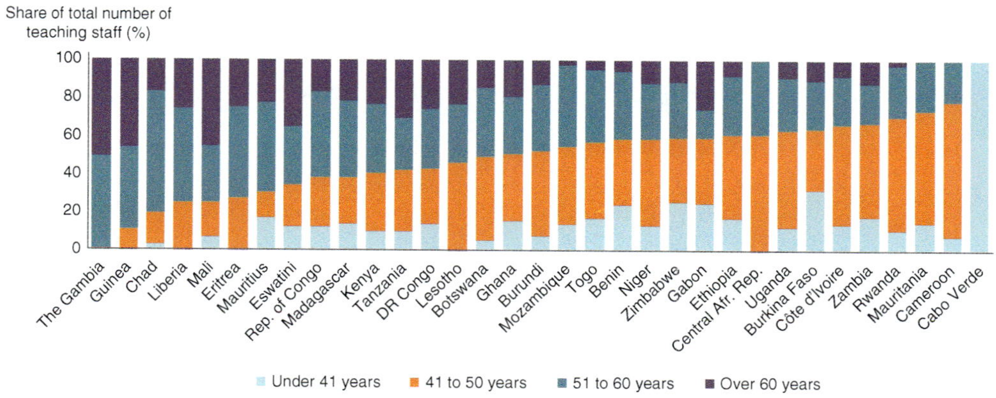

Fig. 5.11. Share of PhD-qualified TAE teachers by age bracket and country, 2016
Source: Constructed by authors from ASTI (2019), complemented by ASTI and RUFORUM (2018) and various secondary sources.
Notes: TAE = tertiary agricultural education; teachers = deans, department heads, (assistant) professors and (senior) lecturers employed in degree-conferring TAE units. Data for Burkina Faso are for 2014; data for Eritrea and Liberia are for 2011. Data are for a 33-country sample and exclude Nigeria and a few countries with very small tertiary education sectors in terms of teaching capacity.

5.7 Training the Next Generation of Africa's Agricultural Researchers and Teachers

Many SSA countries face serious challenges in maintaining a critical mass of teaching and research capacity within their TAE and agricultural research sectors. Significant numbers of senior staff in government-based agricultural research institutes, and especially those with PhD degrees, were approaching retirement age as of 2016 (Beintema, 2020). This cohort was largely trained with donor support at universities in the North in the 1980s and early 1990s. Their retirement puts many countries in jeopardy of losing their most experienced and well-qualified agricultural teachers and researchers, who are needed to train agricultural science students, lead agricultural research programmes and provide workplace training and mentoring for their junior colleagues. This trend - combined with existing capacity gaps and high shares of more recently-recruited junior staff - has left many countries even more vulnerable (Beintema and Stads, 2017).

Projected future demand for TAE and agricultural research staff has two components. Firstly, additional PhD-qualified staff are needed to fill existing gaps in capacity, which are estimated to represent 3% growth, annually, over actual 2016 levels. Secondly, additional staff with doctoral degrees are needed to replace those who will be retiring by 2030. On this basis, in Kenya, for example, projected demand in 2030 is estimated to be 485 PhD-qualified staff to fill existing gaps and 439 PhD-qualified staff to replace those due to retire by 2030. For small countries with small TAE and agricultural research systems, these numbers are far lower, but their existing staff capacity and supply of new PhD graduates are also substantially lower.

Lack of national data on agricultural science graduates makes it difficult to determine whether the supply of PhD graduates will be sufficient to meet future demand for senior agricultural research and teaching staff. A partial assessment was possible, however, from a small sample of universities (see Table 5.7). At Makerere, Uganda's main TAE university, for example, 27 students graduated with PhD degrees in 2015/16. Taking this figure as an approximate yearly average, and assuming all students seek employment in the TAE and agricultural research sectors - which is highly unlikely - the university is projected to produce 378 new PhD-qualified teachers/researchers, enough to meet the projected demand by 2030 based on existing staffing levels. This does not, however, take account of the expected increased demand stemming from (i) population

Table 5.7. Projected demand for PhD-qualified staff in agricultural research and tertiary agricultural education in selected East and Southern African countries, 2016 and 2030

	Actual, 2016		Projected, 2017–2030		
Country	Existing number of PhD-qualified staff	Share of PhD-qualified staff over 50 years old	Number of PhD-qualified staff needed to reach 3% growth per annum	Number of PhD-qualified staff needed to replace retirees	Total number of PhD-qualified staff needed in 2030
Botswana	77	45	40	37	76
Eswatini	32	67	16	21	37
Ethiopia	405	38	208	154	362
Kenya	946	61	485	439	924
Lesotho	15	30	8	7	15
Mauritius	23	55	12	12	24
Uganda	410	38	210	108	318
Zimbabwe	115	38	59	45	104

Sources: Constructed by authors based on ASTI (2019).

growth, (ii) the implementation of national development strategies and (iii) expansion of the coverage needed across an increasing number of disciplines and issues, including natural resources, socioeconomics and climate change. This tight market for PhD graduates of agricultural sciences will draw attention to the quality of PhD programmes. Limited class size for PhD training at national TAE institutions suggests the need for differentiation across TAE institutions and the use of regional approaches. Care must be taken now to ensure that a competitive market does not further fragment and erode the quality of TAE provision in Africa. Instead, effective differentiation of institutions, programmes and skills is needed to (i) maximize resource use, (ii) improve the quality and diversity of TAE programmes offered, which will in turn (iii) improve the quality and diversity of the research undertaken and (iv) provide a critical mass of well-trained, innovative agricultural scientists capable of successfully solving the diverse challenges facing Africa's agricultural sector.

5.8 Implications and Recommendations

Despite significant growth in tertiary education enrolments in agricultural and other field of sciences, Africa still has the lowest gross enrolment ratio in the world, female participation levels and research outputs remain low, and the number of graduates of agriculture and related sciences is inadequate - not only to meet demand, but also to drive the necessary innovation to solve development challenges. Several institutions run outdated curricula and their pedagogic capacities are deficient. Importantly, a mismatch exists between the knowledge provided by the tertiary education institutions and the skills required in the workplace.

As the number of institutions and units continues to expand across Africa's tertiary education sector, the necessary development of human-resource capacity has fallen behind in terms of (i) graduate numbers, (ii) the quality of training and (iii) an appropriate level of differentiation to meet workplace demand but avoid inefficient and resource-depleting competition and duplication, as distinct from value-adding complementarity. In effort to establish legitimacy, many tertiary education institutions have mimicked the structure of traditional flagship universities, offering a full spectrum of academic programmes, including business studies, arts and humanities, when their original focus is science and technology. Movement towards vertical and horizontal specialization and integration is needed to differentiate the roles and purposes of TVET, undergraduate education and graduate education. For example, ATVET institutions have a distinct mandate to prepare agricultural and agribusiness employees, entrepreneurs and extension agents for the workplace and provide a vital link between TAE and the community. And while some universities have a clear teaching and learning mandate, others should be recognized as research-intensive institutions focused on generating new knowledge and innovations.

Alignment with overarching regional and national development strategies has emerged as a key area of concern for Africa's tertiary education systems, in which agricultural and food system development often plays a pivotal role. Most countries have articulated their aspirations in strategic plans which prioritize science and technology and depend on a significant increase in the competencies of graduates. Nevertheless, inadequate investments in tertiary education institutions belie these aspirations. Critical challenges remain in terms of quality, private-sector linkages and necessary learning and research infrastructure, including laboratories, equipment and communications technologies. Partnerships with universities in the North have compensated for these weaknesses to some degree, as has the mobility of students to acquire graduate degrees and experience outside Africa. This, however, has weakened demand for local interactions: UIS (2015) conceded minimal levels of collaboration among African researchers and universities compared with their collaborations with American and European researchers and universities.

The expansion of the TAE sector - in terms of the number of institutions and the number of PhD programmes offered - has not been accompanied by strong development of teaching capacity. This trend has been ongoing since independence, but the situation has become more serious over time. The current characteristics of Africa's TAE sector have in many ways been defined by the structural adjustment programmes of the 1990s, which liberalized the economy and emphasized primary and secondary education. Budgetary restraint and limits on deficit spending removed the subsidy to students in the higher levels of education and significantly reduced government financing to higher education. Consequently, in strengthening the capacity of TAE institutions, a period of consolidation and differentiation is called for, accompanied by more targeted national priorities in the allocation of higher-education budgets. Moreover, market demand for greater differentiation in the skillsets of PhD graduates suggests the need to further differentiate PhD programmes and enhance their quality, a re-occurring theme throughout this book.

Endnotes

[1] Despite technological barriers (the so-called digital divide), online education is on the rise in SSA, not only in terms of the number of students enrolled, but also in terms of the number of online courses being offered in South Africa and other developing and industrialized countries. With future population growth, some (e.g. Bill Gates) argue that distance learning may be one solution to meet rising future demand for education (Trines, 2018).

[2] Agricultural technical and vocational education and training (ATVET) institutions are mainly excluded from this chapter due to lack of available data (see Ch. 13, this volume, for a discussion of ATVET institutions).

[3] Note that a substantial share of the 178 higher education institutions which provide training in agricultural science also provide training in these related sciences.

[4] The regional total excludes Sudan, which is classified as a North African country. As of 2014, 14 Sudanese universities, at minimum, offered at least bachelor's-level training in agricultural sciences (ASTI, 2019).

[5] The United Kingdom, United States and many other countries with mass or universal tertiary-education systems still have a sub-set of elite universities which are not accessible to all (Mohamedbai, 2014).

[6] UIS uses 18–23 years of age for its GER calculations, whereas Trow (1973) uses 20–24 years of age.

[7] Malawi's Bunda College of Agriculture, for example, experienced an increase in its annual enrolments, from 350 in 2006 to 919 in 2016 (Kekeli Kobla *et al.*, 2018).

[8] See ASTI's website (www.asti.cgiar.org) for more information about the programme's data-collection procedures and methodologies.

[9] This 10% share reflects the overall low rate of female participation in (agricultural) sciences. For example, in 2016, 10% of the total number of Ethiopian researchers was female (Beintema and Haregewoin, 2018).

[10] The latest year for which ASTI's capacity data are available for Nigeria's TAE sector is 2008, at which time the country's teaching staff accounted for 28% of the sample total; estimated 2016 data are based on the assumption that Nigeria's share remained constant between 2008 and 2016.

References

ASTI and RUFORUM (2018) Unpublished survey data. International Food Policy Research Institute, Washington, DC, and Regional Universities Forum for Capacity Building in Agriculture, Kampala, Uganda.

ASTI (2019) ASTI Database. International Food Policy Research Institute, Washington, DC. Available at: https://www.asti.cgiar.org/data/overview (accessed 24 March 2021).

Beintema, N.M. Pardey, P.G. and Roseboom, J. (1998) *Educating Agricultural Researchers: A Review of the Role of African Universities*. EPTD Discussion Paper 36, International Food Policy Research Institute, Washington, DC. Available at: https://www.ifpri.org/publication/educating-agricultural-researchers (accessed 24 March 2021).

Beintema, N.M. and Stads, G.J. (2017) *A Comprehensive Overview of Investments and Human Resource Capacity in African Agricultural Research*. An ASTI Synthesis Report. International Food Policy Research Institute, Washington, DC. Available at: https://www.asti.cgiar.org/sites/default/files/pdf/SSA-Synthesis-Report-2017.pdf (accessed 24 March 2021).

Beintema, N.M. and Haregewoin, T. (2018) *Ethiopia: Agricultural R&D Indicators Factsheet Update*. International Food Policy Research Institute, Washington, DC. Available at: https://www.asti.cgiar.org/sites/default/files/pdf/Ethiopia-Factsheet-2018.pdf (accessed 19 April 2021).

Beintema, N. (2020) *ASTI in Retrospect 03: High Shares of Senior Researchers Aged Over 50*. International Food Policy Research Institute, Washington, DC. Available at: https://www.asti.cgiar.org/sites/default/files/pdf/asti20/ASTIat20-Age.pdf (accessed 19 April 2021).

Blair, R. and Jordan, J. (1994) *Staff Loss and Retention at Selected African Universities: A Synthesis Report*. AFTHR Technical Note 18 (September). World Bank, Washington, DC. Available at: http://documents1.worldbank.org/curated/en/217191468145171957/pdf/multi-page.pdf (accessed 24 March 2021).

Bloom, D.E., Canning, D. Chan, K. and Luca, D.L. (2014) Higher education and economic growth in Africa. *International Journal of African Higher Education* 1(1), 22–57.

Boxenbaum, E., and Jonsson, S. (2017) Isomorphism, diffusion and decoupling: Concept evolution and theoretical challenges. Chapter 3 in Greenwood, R. Oliver, C. Lawrence, T.B. and Meyer, R.E. (eds.) *The SAGE Handbook of Organisational Institutionalism*. SAGE Publications, London.

Calderon, A.J. (2018) *Massification of Higher Education Revisited*. RMIT University Melbourne. Available at: http://cdn02.pucp.education/academico/2018/08/23165810/na_mass_revis_230818.pdf (accessed 24 March 2021).

Cloete, N., Bunting, I. and van Schalkwyk, F. (2018) *Research Universities in Africa*. African Minds, Cape Town.

Court, D. (1980) The development ideal in higher education: The experience of Kenya and Tanzania. *Higher Education* 9, 657–680. Available at: https://link.springer.com/article/10.1007/BF02259973 (accessed 19 April 2021).

Darvas, P. Gao, S. Shen, Y. and Bawany, B. (2017) *Sharing Higher Education's Promise beyond the Few in Sub Saharan Africa*. Directions in Development, World Bank, Washington, DC. Available at: https://openknowledge.worldbank.org/handle/10986/27617 (accessed 24 March 2021).

DiMaggio, P.J. and Powell, W.W. (1983) The iron cage revisited: Institutional isomorphism and collective rationality in organisational field. *American Sociological Review* 48(2), 147–160. Available at: https://www.jstor.org/stable/2095101?seq=1 (accessed 24 March, 2021).

FAO (Food and Agriculture Organization of the United Nations) (2018) FAOSTAT annual population database. Available at: http://www.fao.org/faostat/en/#data/OA (accessed 24 March, 2021).

Kekili Kobla, A.O. Mirriam, B. and Wang, L-.H. (2018) The impact of massification on higher education in Africa: A case study of Bunda College of Agriculture in Malawi and University of Cape Coast in

Ghana. *European Journal of Economic and Business* 3(1), 24–28. Available at: http://lawarence-press.com/ojs/index.php/EJEB/article/view/505 (accessed 24 March, 2021).

Mohamedbhai, G. (2014) Massification in higher education institutions in Africa: Causes, consequences and responses. *International Journal of African Higher Education* 1(1), 59–83. Available at: https://doi.org/10.6017/ijahe.v1i1.5644 (accessed 24 March, 2021).

Moock, P.R. (1988) *Education in Sub-Saharan Africa: Policies for Adjustment, Revitalization, and Expansion*. World Bank, Washington, DC.

Moock, J.L. (2016) Network innovations: Building the next generation of agricultural scientists in Africa. Chapter 10 in Lynam, J. Beintema, N.M. Roseboom, J. and Badiane, O. (eds.) *Agricultural Research in Africa: Investing in Future Harvests*. International Food Policy Research Institute, Washington, DC. Available at: https://www.ifpri.org/publication/agricultural-research-africa-investing-future-harvests (accessed 24 March 2021).

Osiru, M. Nampala, P. and Ekwamu, A. (2016) African faculties of agriculture within an expanding university sector. Chapter 9 in Lynam, J. Beintema, N.M. Roseboom, J. and Badiane, O. (eds.) *Agricultural Research in Africa: Investing in Future Harvests*. International Food Policy Research Institute, Washington, DC. Available at: https://www.ifpri.org/publication/agricultural-research-africa-investing-future-harvests (accessed 24 March 2021).

The Economist. (2019) A booming population is putting strain on Africa's universities: Tertiary education in Africa. *The Economist* (10 August). Available at: https://www.economist.com/middle-east-and-africa/2019/08/10/a-booming-population-is-putting-strain-on-africas-universities (accessed 24 March 2021).

Trines, S. (2018) Educating the masses: The rise of online education in Sub-Saharan Africa and South Asia. *World Education News + Reviews*. August 14. Available at: https://wenr.wes.org/2018/08/educating-the-masses-the-rise-of-online-education (accessed 24 March 2021).

Trow, M. (1973) *Problems in the Transition from Elite to Mass Higher Education*. Carnegie Commission on Higher Education, Berkeley, California, USA. Available at: https://eric.ed.gov/?id=ED091983 (accessed 24 March 2021).

UIS (UNESCO Institute for Statistics) (2012) *International Standard Classification of Education: ISCED 2011*. UNESCO Institute for Statistics, Montreal, Canada. Available at: http://uis.unesco.org/sites/default/files/documents/international-standard-classification-of-education-isced-2011-en.pdf (accessed 19 April 2021).

UIS (2015) *UNESCO Science Report: Towards 2030*. Executive Summary. UNESCO Institute for Statistics, Montreal, Canada. Available at: http://uis.unesco.org/sites/default/files/documents/unesco-science-report-towards-2030-ex-sum-en.pdf (accessed 24 March 2021).

UIS (2019) Education stats database. UNESCO Institute for Statistics, Montreal, Canada. Available at: http://data.uis.unesco.org/ (accessed 24 March 2021)

World Bank (2010) *Financing Higher Education in Africa*. World Bank, Washington, DC. Available at: https://openknowledge.worldbank.org/handle/10986/2448 (accessed 24 March 2021).

World Bank (2019) *World Development Indicators*. World Bank, Washington, DC. Available at: https://databank.worldbank.org/reports.aspx?source=World-Development-Indicators (accessed 24 March 2021).

Part III

Pathways of Transformation

6 Leadership and Change Management to Transform Tertiary Agricultural Education Institutions

Mabel Imbuga[1]*, Daniel N. Sila[1] and John Wesonga[1]
[1]*Jomo Kenyatta University of Agriculture and Technology, Kenya*

6.1 Introduction

Against a backdrop of labour-market demand for creative and innovative graduates, underfunding and graduate unemployment (Pitso, 2019), African tertiary agricultural education (TAE) institutions seek to attain world-class standards and be globally competitive. To achieve these goals, effective leadership is essential (Aguirre and Martinez, 2002). Broadly speaking, in the context of the TAE sector, leadership encompasses the structures, practices and people supporting the attainment of defined organizational and societal goals. It entails inspiring people to identify with these goals, and changing the culture in which organizations operate (Aguirre and Martinez, 2002). Eddy *et al.* (1997) calls for change in leadership practices for higher education in the 21st Century in order to raise standards and status, and improve the overall learning environment. Maxel (2013) argues that East African universities have stagnated in terms of their leadership hierarchies, rules and procedures, formality, standardization, specialization, low integration levels and weak networking mechanisms.

This chapter explores the concepts of leadership and change management within the context of transforming Africa's TAE sector, and more specifically its universities. The next two sections focus on existing leadership practices in African TAE institutions and a vision for their future transformation. Sections 6.4 and 6.5 more specifically address the role of leadership in effecting this transformation, and best practices for such leadership. Section 6.6 analyses the challenges associated with successfully managing organizational change. Section 6.7 provides a brief conclusion.

6.2 Current Leadership Practices

A key role of leadership in TAE institutions is to devise and communicate clear visions and goals through the formulation of vision and mission statements, mottos and slogans, value propositions (i.e. an innovation, service or feature intended to make a company or product attractive to customers) and core values among other elements often included in strategic plans. Indeed, most higher education institutions (HEIs) have formulated and publicized their strategic intentions of attaining global excellence. This enables 'core stakeholders to have shared common values, philosophy, beliefs and mind programming' (Maxel, 2013, p. 158). This also contributes to resolving conflicts between TAE management and employee unions which cripple the operation

*Corresponding author: mimbuga@yahoo.com

 Transforming Tertiary Agricultural Education in Africa (eds D. Kraybill, J. Lynam and Adipala Ekwamu)
DOI: 10.1079/9781789246544.0006

of TAE institutions in most parts of Africa. Guided by shared values, philosophies and beliefs, institutions are equipped to collectively develop strategic knowledge and direction. Kenya's Jomo Kenyatta University of Agriculture and Technology (JKUAT), with its well-defined and communicated strategic intentions, has focused on generating strategic knowledge in the fields of agriculture, engineering, and information and communications technology (ICT). Other African universities, such as Makerere University (Uganda), the University of Nairobi (Kenya), and Egerton University (Kenya), among others, have also crafted strategic niches. As a result of this leadership, these universities have succeeded in attracting funding to host specialized institutions which raises their status, impact and reach. JKUAT, for example, hosts the Pan African Institute for Basic Sciences, Innovation and Technology. The University of Ibadan, which seeks to be 'a world-class institution for academic excellence geared towards meeting societal needs' (University of Ibadan, 2021, Vision), hosts the Pan African University of Life and and Earth Sciences Institute.

Many TAE institutions place a premium on external rankings of excellence, which often pit them against more established institutions in the Global North. While such rankings can provide pressure for institutions to improve, creating a culture of internal evaluation and audit would arguably be more worthwhile. Internal evaluation would focus more on assessing organizational structures and contexts, and devising measures to promote an internal culture of excellence. Indeed, literature indicates that the capacity of institutions to manage and improve their own processes promotes improvement-oriented and sustainable approaches to quality assurance (Botha *et al.*, 2008). External rankings are criticized for their partial coverage and imprecise ranking criteria (Enserink, 2007; Marginson and van der Wende, 2007).

Many institutions have established internal structures and policies to coordinate academic quality assurance, such as JKUAT's Directorate of Academic Quality Assurance, and Academic Quality Assurance policy. In addition, JKUAT pursues a culture of excellence through certification to such standards as ISO 9001 (which provides a framework for systematically improving an organization) and ISO 14001 (which is an international standard for environmental management systems). Accreditation of programmes by professional bodies - such as the Engineers Board of Kenya, Kenya Veterinary Board, Kenya Nutritionists and Dieticians Institute and the Commonwealth Association of Architects - fosters a positive organizational culture. In the context of agricultural science, accreditation is currently limited to such programmes as veterinary medicine, food science and nutrition, but it should be developed for other programmes.

Leadership is needed to overtly promote high-impact teaching, research and innovation. Africa faces a myriad of challenges which can be addressed by well-trained graduates. Teaching should be geared towards producing well-grounded graduates who can solve challenges, such as low agricultural productivity. Innovative methodologies integrating practical skills, such as on-farm internships, should be promoted (Yusuf *et al.*, 2019). Pockets of expertise exist across diverse fields of agriculture, but efforts to strengthen the capacity for teaching and research through training at MSc and PhD levels are limited in both number and size (see Ch. 5, this volume). Current efforts by the African Union Commission and World Bank to establish pan-African universities and African centres of excellence in higher education have formed an initial nucleus which governments ought to upscale and sustain. This is largely donor-driven, and sustainability is still questionable due to limited commitment by African governments. Laboratories in most TAE institutions continue to be dilapidated, poorly equipped and managed, overcrowded and lacking a clear research focus. Where equipment is available, it is poorly maintained due to lack of technicians with the necessary repair and maintenance skills. This has led to limited innovation in the field of agriculture and, hence, the persistence of perennial food and nutrition insecurity and safety concerns.

TAE leadership should ensure the alignment of their institutional agendas with national, regional and continental priorities. Agendas should accommodate such priorities as the Sustainable Development Goals (SDGs), Agenda 2063 and STISA–2024, among others (Aarts *et al.*, 2020). Such alignment would ensure institutions' relevance, along with support from the wider community. As an example, Mount

Kenya University established its Graduate Enterprise Academy in 2013 in response to the 'Big Four Agenda' of the Kenyan government (Wanderi and Makandi, 2019). Public universities in Kenya factor programmes and activities which support or promote government initiatives into their annual performance targets and strategic plans. The relevance of higher education could also be enhanced by prioritizing emerging issues and trends. For instance, in the wake of rampant cases of malnutrition - often referred to as 'hidden hunger' - institutions may heed the call of Fan *et al.* (2019) to integrate programmes and initiatives which promote nutrition. Similarly, prioritizing climate change would be timely. In this regard, JKUAT established nutrition programmes and, subsequently, a Department of Human Nutrition Sciences. A centre on climate change adaptation was also established. Strathmore University (Kenya) and the University of Nairobi established programmes on climate change as an urgent response to emerging issues. Arguably, however, it would be more efficient to mainstream these cross-cutting issues within institutions instead of introducing less sustainable standalone programmes, as was argued by Aguirre and Martinez (2002) regarding mainstreaming diversity in US universities.

Most TAE institutions in Africa have largely depended on government funding for their survival, and such funding has declined over time (see Ch. 5, this volume). According to ASTI (2019), for a sample of 32 African countries during 2007–2016, agricultural research spending increased in 13 countries but declined in 19. When expressed as a share of gross domestic product (GDP), research funding only increased in 7 countries, whereas it declined in 25. Alternative funding mechanisms are therefore needed, along with a clear strategic direction of how such funds can be generated and used. The role of university councils in mobilizing resources and providing the necessary strategic direction to transform TAE institutions must be well conceived and articulated. Councils appear to be more interested in external audits than in creating and cultivating a culture of internal evaluation and auditing. TAE institutions should leverage apparent growth in the willingness of African governments to provide them with greater oversight and support. For example, Kenya is signatory to the African Union's protocol which obliges member states to commit 1% of their GDP to support scientific research. The National Research Fund, established under the 2013 Science and Technology Innovation Act, facilitates research to advance science, technology and innovation. Kenya's current allocation for research amounts to 0.5 percent of the country's GDP, a negligible amount compared with the country's research needs and potential. However, this commitment is a step in the right direction since it is supposed to be reviewed upwards, but such efforts will be futile if TAE leadership remains bureaucratic, and organizational structures and policies remain static and unresponsive to change.

6.3 A Vision for Africa's Tertiary Agricultural Education Institutions

African TAE institutions aspire to having world-class standards and being globally competitive. These aspirations are evident in their visions. For instance, the University of Cape Town (South Africa) seeks, among other things, to 'influence the global higher education landscape' (University of Cape Town, 2016, About UCT). Stellenbosch University (South Africa) seeks to be 'leading research-intensive university, globally recognised as excellent, inclusive and innovative' (Stellenbosch University, no date, Our vision). In East Africa, the University of Dar es Salaam (Tanzania) has a vision of becoming 'a leading Centre of intellectual wealth spearheading the quest for sustainable and inclusive development' (University of Dar es Salaam, 2021, Mission & vision), while the University of Nairobi commits to being 'a world-class university committed to scholarly excellence' (University of Nairobi, 2020, UON vision, mission & core values). Although no consensus exists regarding the meaning of 'world-class university', the concept may offer inspiration and strategic direction for institutions to pursue and improve their performance. Salmi (2009) presents the attributes of world-class universities as a high concentration of talent, abundant financial resources and good governance (see Fig. 6.1).

Talent consists of highly-motivated students, teaching staff and researchers capable of producing highly-competitive graduates and

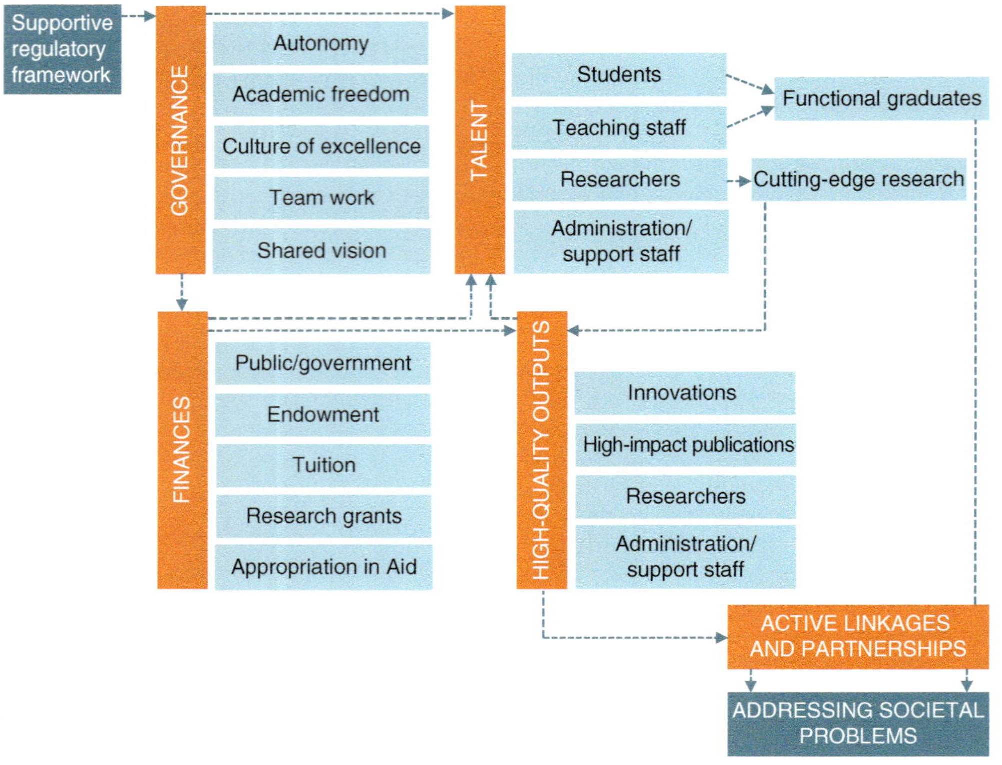

Fig. 6.1. Schematic representation of attributes of a world-class university
Source: Adapted by authors from Salmi (2009, Fig. 1.2).
Note: Appropriation in aid refers to funds generated by an institution that the institution is authorized by government to use directly without remitting to government first.

innovative research outputs. University leadership should institute decisive measures to attract this calibre of staff and students, and to motivate and support them in undertaking leading-edge teaching and research. In addition, staff mobility and collaboration are critical for incubating new ideas and ensuring a vibrant culture. All of this rests on attracting financial resources through government and other sources to support day-to-day operations, research programmes and capital investments. Funding can be internally generated through (i) student tuition fees, (ii) the provision of contract-based research for the public and private sectors and (iii) securing gifts and endowments. Diversifying funding sources enhances institutions' financial sustainability and enables them to develop the resources needed to implement programmes and mandates effectively. JKUAT, for example, generates about 70% of its own operating budget, and the Government of Kenya contributes the remaining 30%. A good reputation, as reflected in university rankings, can significantly support institutions in attracting financial resources from a variety of sources. Monks and Ehrenberg (1999) report a positive correlation between a university's ranking and the size of its funding endowment.

Governance is defined to include institutions' overall regulatory framework, their internal environment and their degree of academic and managerial autonomy. Good governance will encourage a strategic vision, innovation and flexibility, and enable institutions to make decisions and manage resources without being encumbered by bureaucracy (Liu *et al.*, 2016). Improvements in governance may be achieved through training to enhance the capacity of governing bodies to effectively fulfil their roles. This has been reported in the health sector, where one study observed that well-informed boards act more effectively in developing policies to oversee the work of healthcare providers and managers, and in determining how finances will be allocated (Parand *et al.*, 2014). To achieve world-class status, African TAE institutions need to formulate appropriate strategies and policies, and establish strategic partnerships and a strong human resource base. In view of the negative impacts of rapidly growing student enrolments (Salmi, 2009), institutions should guard against increasing student admissions to unmanageable numbers. Due to reduced government support, most universities have focused on increasing the enrolment of fee-paying students, a strategy which may be unsustainable. To bridge the gap in government funding, some universities in Kenya have established regional campuses to make programmes more accessible and cost-effective for students (e.g by reducing accommodation costs). Another approach employed by universities in Kenya and Tanzania is to establish income-generating enterprises on campus, such as shopping malls, university hospitals, hotels, factories, food outlets and (in one case) a tailoring unit. An alternative approach - practised by some private universities in Kenya - would be to focus on offering high-quality academic programmes at premium tuition rates.

TAE institutions should be wary of endogamy, the phenomenon of relying on their own undergraduates either to continue into graduate programmes or join the institutions' teaching staff. Special attention should be given to the competencies of graduates to ensure their readiness for the workplace when they graduate, a common theme throughout this volume. This entails ensuring that graduates are employable or able to become self-employed. Adopting a learner-centred approach emphasizing learning outcomes and competencies, as promoted by Kenya's Commission for University Education, is a sound course of action. Donors, such as the Netherlands Initiative for Capacity Development in Higher Education (NICHE)[1] have supported several African TAE institutes to embrace competence-based curricula, align their training with the demand of the labour market and increase their accessibility to women and other marginalized groups. The learning environment and methods should connect students to real-world contexts and enable them to develop skills in problem-solving, critical thinking and systems-thinking (Knobloch *et al.*, 2019). Such approaches as (i) the 'flipped classroom' model, (ii) problem-based, experiential learning and (iii) critical-thinking and decision-making (McNamara *et al.*, 2019) have great potential for building students' practical skills and competencies required and should be promoted. The inclusion of innovative ICT use to deliver cutting-edge teaching would also be prudent. Open, distance and e-learning options already exist in many African universities; however, their application in agricultural training is still limited. Nevertheless, it should be possible to apply ICT-enabled learning options through simulation models and virtual farms. Use of multimedia should provide opportunities for sharing practical experiences with learners. It is imperative that TAE institutions strike a balance in focusing on teaching quality and research quality in order to navigate these challenging times.

6.4 Leadership in the Context of Organizational Change

Change can be positive or negative, exciting or demoralizing, vital or unnecessary, easy or difficult - and often all of the above. The role of leaders in effecting change is critical. In order to generate alignment with commitment to the process of change, and achieve success, leaders must translate the vision for change, help direct the transition process and positively influence change agents. Studies show that between 50% and 70% of efforts to effect planned change fail. To gain the desired results from a new direction, system or initiative, organizations need the benefit of leadership and change management. Leaders who bring about change often possess the attributes described below:

1. *Charismatic influence and inspirational motivation.* Leaders must be able to inspire the people they lead through trust, honesty and integrity. Those being led must respect their leaders in order to be willingly guided. Acting as an example is important, as is acting beyond self-interest. Charismatic leaders inspire confidence in those they lead, are uncompromising in the pursuit of their mandated goals and are effective in delegating responsibilities to ensure goals are achieved (Sparks, 2014). The primary goal of transformational leaders should be to elicit the full support of all involved towards achieving targeted goals. Open, honest and inspiring communication fosters a unifying organizational culture and a sense of belonging and commitment (Cimino *et al.*, 2005).

2. *Intellectual stimulation.* Leaders who encourage rationality in the face of the kinds of emotional challenges which arise in the educational setting create a more positive environment with less conflict and more growth. Innovation and hard work are encouraged through intellectual pursuits, such as critical thinking and problem-solving. Keeping students intellectually engaged in the classroom keeps them on track, and the same can be said for keeping teachers intellectually engaged and on track (Ogola *et al.*, 2017).

3. *Individualized consideration.* Forging relationships is essential, and this requires two-way communication. Followers who feel heard and valued will be much more invested in the process of change, and interested in pursuing larger, organizational goals. This individualized process allows leaders to know the strengths of the people within their organization, so as to support and guide them in achieving success. Administrators who seek connection with their staff in small, regular ways are able to build trust and a sense of shared purpose, which encourages a productive overall relationship (Church and Rotolo, 2016).

4. *Persuasion and resilience.* Change and influence are inextricably linked. Influence is not only about gaining compliance, but also about gaining the commitment necessary to drive change. It is also about mapping the critical change agents who must be brought on board, and defining what 'buy-in' from each stakeholder will look like to produce a successful outcome (Perloff, 2008; Sparks and Areni, 2008). Resilience, on the other hand, helps people handle the pressure, uncertainty and setbacks associated with the process of change. Managers need to build their own reserves and resilience to support their mental and physical health. They can also guide others in facing change in healthy and sustainable ways.

5. *Vision.* A manager leading change must be visionary. Vision provides guidance to an organization by articulating its goals and direction. It serves as 'a signpost pointing the way for all who need to understand what the organization is and where it intends to go' (Nanus, 1992, p. 12–15). By providing a conceptual picture, an organization's vision not only describes its direction or goal, but also the means of accomplishing it. For educational leaders who implement change in their schools, vision is 'a hunger to see improvement' (Pejza, 1985, p. 12). Nanus (1992, p. 24–31) maintains that the 'right vision': (i) attracts commitment and energizes people, (ii) creates meaning in workers' lives, (iii) establishes a standard of excellence, (iv) bridges the present to the future and (v) transcends the status quo.

6. *Ethical leadership.* The term ethical leadership refers to the principle that true leaders are concerned with 'doing the right thing' and not merely 'doing things right'. Ethical leaders are not afraid to do what they truly believe to be right - even if it is unpopular, unprofitable or inconvenient. Studies have found ethical treatment to be among the most appreciated qualities of leaders by their employees. The advantages of such leadership go beyond success in decision-making processes. The characteristics of ethical leadership include justice, respect for others, honesty, team building, acting humanely and leading by example, among others (Hegarty and Moccia, 2018).

7. *Recruiting talented people.* Transformational leaders recruit, develop and retain great people in their teams and organizations. Institutions only generate results commensurate with the calibre of the people they employ. Effective managers are creative 'talent scouts' who take an active role in team building. They perpetually seek team members, whether established experts or individuals with the potential – who will become major contributors to their teams (Church and Silzer, 2014)

6.5 Best Practices for Transformative Leadership

Numerous TAE institutions - located in Europe, the Americas, Asia and within Africa itself - provide valuable lessons for African universities. The management and operation of such universities will not necessarily provide a blueprint for African institutions, but a number of best practices can be tailored to fit African circumstances and provide motivation and direction for transformative change. Some of these best practices are briefly discussed below.

Establishing Working Policies

Establishing effective policies and procedures is a key pillars for transformative TAE leaderships. The first step in developing policy is ensuring that the organization's departmental leaders understand the need for the policy and are actively involved. Working procedures should be developed to support the implementation process. Policies and related procedures should be logical and available for reference to staff at all levels. Wageningen University (the Netherlands), for instance, operates under Europe's Charter of European Agricultural Education and Training, which provides an operating framework. The charter is also supported by European policies for agriculture and rural areas. An important characteristic of the charter is administrative autonomy, which enables individual institutions to determine regionally-pertinent topics and projects, and make best use of available resources. Each country has the flexibility to determine the legal and budgetary conditions it needs to create or adapt in order to achieve the goals outlined.

National policies can play a major role in promoting excellence in universities. Several examples of such policies are reviewed by Liu *et al.* (2016). Numerous countries have instituted initiatives to create world-class universities, including Australia (Group of Eight), China (Project 211 and Project 985), Germany (TU9), Sweden and the United Kingdom (Russell Group). In Latin America, the University of Costa Rica (UCR) ranks first among Central American and the Caribbean universities, and 19th among those of Latin America (QS World University Rankings, 2019). UCR maintains international alliances with (i) the German Academic Exchange Service, (ii) the governments of France, Japan, Mexico, Spain and Taiwan (Republic of China), (iii) the European Union, (iv) the Organization of American States and (v) other entities, such as the Organization for Tropical Studies, which is a non-profit consortium of about 50 universities, colleges and research institutions worldwide.

In Kenya, JKUAT has well-documented governance and operating structures. These include functional policies for strategic direction and operations manuals and guidelines which strengthen policy implementation. JKUAT's strategic plan (2018–2022) aligns with the national government's medium-term plan (2018–2022). Policies address mechanisms to achieve the university's core mandates, which - among others - include training, research and innovation for development. Documentation is available within every department or unit for reference and is also available on the university's website (www.jkuat.ac.ke). JKUAT also has clear guidelines and user manuals for each of its divisions, e.g. human resources, financial management, transport, training procedures, waste management, etc. Three months before the beginning of each financial year, units draft budgets for their planned activities for approval by the Finance Department. Implementation of the strategic plan is monitored through performance contracts with the government based on clear indicators and targets which reflect each units goals. Performance contracts are also linked to staff performance appraisals to capture their contributions to the university's strategic plan. Systematic staff development protocols are guided by a staff training policy, implemented through a human resource manual aligned with departmental needs. Staff development is coordinated by a training committee comprising representatives of the respective departments.

Building a Vision and a University Ranking

Responsible and dedicated future planning is critical (i) to people's fundamental quality of life, (ii) to the sustainable management of the natural base and (iii) to people's capacity to navigate a diverse and ever-changing, competitive global environment. World-class universities anticipate emerging concerns and opportunities, and marshal their intellectual resources to address them. JKUAT's leadership recognizes the power of teams to deliver its vision to be 'a university of global excellence in training, research and innovation for development' (JKUAT, 2021, University vision). From that perspective, JKUAT has aligned its operations with national priorities embedded in the Kenyan government's socio-economic blueprint, Vision 2030. This vision provided guidance for the government's first three medium-term plans and, in particular, the role of science, technology and innovation in development. JKUAT takes responsibility for its future by assessing the implications of its actions in the short and long term. Most of JKUAT's activities are defined within the strategic plan and annual workplans.

Over the past decade, many universities have embraced the 'knowledge economy' in their visions. In this context, global university rankings of universities in comparisons are becoming increasingly important to students, employers, governments and peers. Consequently, many tools for measuring the quality of higher education have been developed for such purposes as peer review, accreditation, benchmarking and for checking newly-introduced degree programmes. Although rankings have gained worldwide importance in the past decade, the practice dates back to to the beginning of the 20th century with the English publication *Where We Get our Best Men* (UNESCO, 2014), which ranked universities by their numbers of successful alumni. The release of the Academic Ranking of World Universities (ARWU) by Shanghai Jiao Tong University (China) has become a yardstick for university performance. Many ranking systems currently exist based on a variety of indicators, propelling universities to improve their performance in order to become more competitive. When used well, university rankings can create a culture of success by exploiting the universities' key mandates of training, research and outreach for societal development.

Wageningen University (The Netherlands) is a leading agricultural university globally. It is ranked (i) first among agricultural universities worldwide according to the National Taiwan University, (ii) first among agricultural and forestry universities according to QS World University Rankings and (iii) first in the agricultural life sciences by ARWU rank (2019). Wageningen University's mission is 'improving the quality of life and contributing to international development, (WUR, no date, Vision and mission). In combining research, education and the creation of value, it has succeeded in attracting European Union and international funding for research, as well as high international rankings and citation indexes. Some of Wageningen's key success factors include: (i) aligning its strategic plan with global issues - e.g. the United Nations' Sustainable Development Goals (SDGs), (ii) fast-tracking decision-making by giving research programmes autonomy, (iii) conducting regular, collective reviews of the university's branding, (iv) maintaining strong linkages with relevant government ministries and alumni, (v) making resource mobilization the responsibility of all staff members, (vi) providing a strong enabling environment in terms of policies, infrastructure, etc. and (vii) integrating an international approach to its activities to foster diversity and inclusion. These values have made Wageningen University a global leader in agricultural sciences and a university worth emulating.

Developing Relevant Academic Programmes

Universities are increasingly being challenged with training graduates with the competencies needed by society and industry (see Ch. 7, this volume). To address perceived gaps in this area, many industries are now providing targeted training to their recruited graduates prior to assigning them to their eventual positions. Such training is both time consuming and expensive, not only in terms of operating costs and loss of staff productivity, but also because practice-oriented educational programmes must constantly be updated.

In Kenya, massification of higher education led to closure of many technical and vocational education and training institutions, leading many to assert that graduates lack the methodological and personal competencies needed for problem-solving, cannot handle complex issues independently and, hence, are not ready to enter the workforce when they graduate (see Chs. 7 and 13, this volume). In contrast, in Germany, most highly-skilled workers have undergone a dual system of vocational education and training comprising classroom study at specialized trade schools, and supervised work experience on the job. This unique programme allows students to obtain theoretical knowledge for their occupation of choice, followed by industry training to gain practical, hands-on knowledge and experience. So-called soft skills (e.g. interpersonal skills) are also acquired during the training. This value proposition ultimately ensures that academic programmes match the needs of industry. Integrating flexible learning modules, credit transfer systems and student/staff mobility, combined with harmonizing the scheduling of academic semesters, has yielded excellent results. This can only be achieved, however, through effective leadership and commitment to inculcating learning and research with practical skills.

Undertaking Cutting-Edge Research

In technologically advanced parts of the world, research outputs have contributed immensely to growth and development. The presence of research universities is now widely viewed as a necessary condition for bringing about innovation-based economic development. Stimulation of a research culture needs strong support from university management in an environment where senior scientists mentor their younger colleagues (see Ch. 5, this volume). Research must be relevant to industry and the community, and include both basic and applied components. At JKUAT, research is largely informed by the SDGs, Agenda 2063 (see Ch. 1, this volume), and the national development agenda, represented in the Kenyan government's Vision 2030 and elaborated in its medium-term plans. The creation of Kenya's Big Four Agenda - food and nutrition security, manufacturing, universal health coverage and affordable housing - has further crystallized JKUAT's priority areas of research focus for the 2018–2023 period. The 'research chair' concept is a further enhancement, whereby one of JKUAT's senior scientists receives government funding to spearhead research in relevant areas of interest (see Ch. 8, this volume). JKUAT's leadership intends to position the university to become Kenya's and the region's best research university with the accompanying global reputation and rankings in its core areas. Currently, JKUAT has a research chair in manufacturing, through which the university spearheads related research and innovation working with the relevant industry and government bodies. The priority focus of this research chair is to strengthen the coconut value chain, from production to creating diverse value-added products. The only other research chair in Kenya focuses on Health and is based at Moi University.

Research contributes to the development of innovative raw materials and services which can be commercialized by industry. Companies which have incorporated innovative research in their strategies are ensuring themselves against competition. In this spirit, many of the world's universities are (i) commercializing intellectual property, (ii) building research parks to test and showcase their pilot products and (iii) establishing incubation centres to nurture start-up companies and mentor nascent entrepreneurs. These approaches are intended to cross-fertilize creative and critical thinking, leading to innovation. The importance of science technology parks and incubation centres in development is evident in their worldwide presence (e.g. in China, Europe, Japan and the United States), as well as in newly-industrialized countries, such as Brazil, India, Malaysia,Russia, Singapore, South Africa and South Korea. A culture of innovation will only be realized with the conviction, commitment and determination of governments and university leadership.

Establishing Linkages with Industry

The hallmark of an educational institution is to provide graduates with a high-quality education. A key factor in the United States' technological power is its long tradition of close ties and frequent collaborations with industry, combined with its network of superior universities. A similar trend is unfolding in India. In forging close relationships with industry, the University of Agricultural Sciences (Bangalore, India) has made remarkable strides in providing high-quality education, meaningful research and effective dissemination of its resulting technologies through extension services. This has spurred the transformation of the agricultural sector in the Indian state of Karnataka. The university provides human resources, technical knowhow and private-sector consultancies to foster growth in agro-processing, including fish processing and poultry processing, meat and hide processing, sericulture (silk farming), floriculture, vegetable processing and - to a certain extent - dairy processing.

To bridge the gap that often arises between academia and industry, JKUAT has organized a monthly 'industry-linkage' workshop since 2018, which brings together industry stakeholders and academicians to deliberate on what needs to be done to enhance collaboration. Alumni are also valued as contributors to strengthening academia–industry relationships. These collaborative interactions help academicians to understand the practical realities of working in industry, and members of both sectors can formalize their intentions and commitments by entering into memoranda of understanding. In some cases, contracts with specific terms of reference are appropriate. Intellectual property rights and non-disclosure agreements should clearly stipulate agreements to avoid litigation and loss of trust. For such collaborations to thrive, standard bureaucratic procedures, common in African universities, need to be averted in favour of prompt decision-making. Some Kenyan industries have breached their non-disclosure agreements and commercialized universities' intellectual property, causing a major setback and generating mistrust. Additional dialogue and periodic contract review might reduce or solve such problems.

In the United States, many universities act as hubs for innovation, whereby scientists from various disciplines share their new technologies, procedures and systems with industry stakeholders. Productive linkages provide a foundation for a strong innovation system, which attracts huge demand from industry. These linkages also ensure that researchers focus on providing technical solutions and innovations, which - once adopted by industry - make them competitive in local and international markets. This leads to increased competitiveness, diversification of economic activities and market development. In addition, partnerships between industry and academia can play a major role in the successful recruitment and retention of students and ultimately in developing the local workforce and economy. Industry has now become an important source of funding for academic research through new product development, prototyping, laboratory testing and consultancies. Building this unique relationship between universities and industry requires government support and strong university leadership. Knowledge economies are created when governments provide an enabling environment within which academia and industry can flourish. In such an environment, private industry commercializes university innovations, creating a win–win situation for all.

In Africa, the linkage between universities and industry is still weak. This is largely influenced by differences in the visions held by the two sectors. Universities are often considered 'ivory towers' whose researchers have little or no connection to industry - which is an overly pessimistic view. The prevailing entrepreneurial spirit at JKUAT has enabled the university to seize opportunities and quickly adapt to changing circumstances, and this has yielded positive results. From developing online educational programmes and services, to extending laboratory research into the commercial domain, JKUAT has adopted a culture of collaboration and co-operation with industry. A key factor in this success is the adoption of the Japanese concept of 'Monozukuri' (the Japanese term for manufacturing), which entails commitment to ongoing relationship building and continuous development of innovations from conceptualization, to prototyping, to commercial testing and ultimately piloting. Feedback from users is considered critical prior to commercialization. JKUAT prioritizes listening to (industry and community) end users to foster a culture of research that delivers its intended impacts. To this end, JKUAT has adopted the industrial park concept to incubate and develop new businesses beyond their initial concept phases. JKUAT has also established JKUAT-NISSIN Co. Ltd., a private–public partnership to produce noodles.

Changing Paradigms and Priorities

An excellent institution is one that facilitates growth among its students and faculty. Universities have often been accused of being slow to change; however, recent worldwide changes suggest otherwise. In Kenya, the massification of higher education created tremendous growth and demand for higher education nationally. Universities which failed to offer attractive programmes also failed to attract students and, consequently, lost out on the opportunities sectoral expansion provided. Most public universities grew four-fold in student numbers within a decade (see Ch. 5, this volume). Competition for students was strong, and universities actively sought to attract every available student. Many self-sponsored students enrolled, which was a relief for university management. On the downside, however, this created excessive pressure on available resources for teaching and research. Growth in student numbers was not matched by growth in infrastructure and teaching staff, ultimately compromising the quality of training in some cases. This illustrates the reality that, to be effective, change must be measured and well thought out.

Developing talent is an important goal in tertiary education, which may be lost when universities grow too fast. A good example of a university that has integrated the model of talent development as a means of improving degree-based training is the University of California, Davis (United States). UC-Davis focuses on academic excellence centred on developing students' talents and abilities. The university's course, 'Best Practices in Talent Development and Management', offers participants practical tips, tools and frameworks for managing, developing and engaging talent effectively. Real-world case studies highlight best practices for aligning the development of talent with strategic human-resource processes to drive changes in institutional culture, prepare talented students through a period of orientation or 'on-boarding', and develop and integrate leadership skills, including mentoring and succession and career planning for high performance and engagement (UC-Davis Continuing and Professional Education, no date). The talent-development approach to higher education excellence emphasizes intellectual and personal development of students as a fundamental component of the institution's purpose. This approach has become a practice at UC-Davis, making it another TAE institution worthy of emulating.

American universities evaluate university professors on the basis of their teaching, service and research. One aspect of this tradition is that professors are expected to devote time and effort to the intellectual and moral development of their students. Teaching, therefore, is strongly emphasized. Service to the community, including informing the community, is also viewed as a core aspect of a university's mission. A third goal is advancing knowledge through research. A new paradigm adopted in some US universities, however, focuses on four components: discovery, integration, application and teaching. This approach, known as the Boyer model (Long, 2014) expands the definition of scholarly activity to include not only published research but also teaching, service and the application of knowledge to real-world problems. University faculty members who adopt the Boyer model are rewarded with tenure and promotion - for example, for engaging in practical activities such as prototyping and commercializing innovations. The model seeks to eliminate artificial barriers between academic disciplines by encouraging faculty members to think, teach, research and write in ways which transcend their area of expertise. A balance between faculty members' own published research and unpublished research conducted with students is envisaged as a means of focusing on solving societal problems. Faculty members are encouraged to tailor their (teaching, research and dissemination) activities to create change in the community by improving livelihoods (Boyer, 1990). Priority-setting based on the Boyer model has created a new wave of innovation which has strengthened the role of US universities in economic, technological and social development (Boyd, 2013).

In Kenya and, more specifically, at JKUAT, the idea of building an inclusive community which (i) celebrates differences, (ii) brings people together from a wide range of backgrounds and perspectives and (iii) recognizes that excellence cannot be achieved without diversity is considered critical for success. JKUAT management is endeavouring to create a university community which values and energizes all its members. This is achieved through developing clear staff appraisal systems and policies to guide personnel training and development. Faculty members must demonstrate their succession plans, by matching younger scientists with seasoned professors, and reviewing these relationships annually. Practical learning is preferred to theoretical sessions, and high-performing students gain an opportunity to enter into employment as junior staff. Entrepreneurship is a mandatory course for all finalists, whose talents and innovations undergo incubation. Finalists in JKUAT's Faculty of Agriculture undertake one-year internships in Israel. This innovative approach to teaching made JKUAT the most preferred Kenyan university for first-year students in science, technology, engineering and mathematics programmes in 2018, 2019 and 2020.

Utilizing Information and Communications Technology

Embracing the power of technology to ensure the highest quality training, research and outreach - and indeed to change the way the world works - is increasingly becoming important. Leaders in most universities and research institutions need to spearhead the development and use of ICT, for example, to (i) respond to the changing student demand for courses, (ii) pioneer the application of technology in the field of development, (iii) increase connectivity with the community through wireless Internet services, (iv) invent new technologies and devices to fuel economic change and (v) exploit the power of technology to enhance the way communities interact and function. ICT is fundamental to economic development and the attainment of a national competitive advantage. This is exemplified by the wide access and usage of internet and mobile phone technology in many parts of Africa, and particularly in Kenya and Rwanda. Mobile phones have led to a new way of doing business in East Africa and led to the creation of new business empires. Leaders in the higher-education sector should spearhead the development of ICT policies, e.g. to support infrastructure development, e-learning strategies (e.g. blending learning models with open- and distance-learning models) and information policies.

As an example, European ICT policies for educational systems fall into three general categories summarized (i) increase, (ii) protect and (iii) change (Fig. 6.2). A further nine related sub-categories involve access, production, quality, cooperation, culture, economy, systems, power relations and view of knowledge. The *increase* category comprises policy goals, such as increasing (i) internet access either to all or to special groups, (ii) the number of computers in schools, (iii) the quality of education, (iv) the production of digital content, and (v) partnerships between schools and the private sector. The *protect* category encompasses measures to actively protect and preserve (i) language, (ii) historical and cultural heritage and (iii) national employment opportunities and the economy. The *change* category comprises the change necessary to achieve leaders' visions for the education system. Some universities want to use ICT to gain a competitive advantage. In other cases, strong universities with recognized international names prefer not to disclose details of their vision and strategies. They seek well-known, high-status international partners and target elite students nationally and

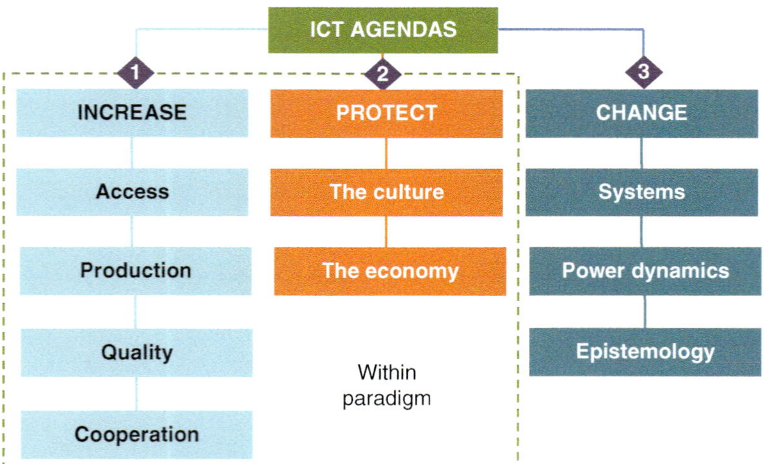

Fig. 6.2. European information and communications technology policies for educational systems Source: Adapted by authors from Hansson and Holmberg (2003, p. 89).

globally. Achieving these goals will depend on how ICT is integrated into university systems and how university leadership positions itself to exploiting available knowledge.

Information technology is increasingly playing a critical role in the transformation of teaching, research and dissemination in TAE institutes. The adoption of e-learning models is growing globally, and non-traditional (ICT-based) approaches to data collection, analysis and storage are evolving. If used wisely and artfully, ICT can aid the acceleration of innovation and change. Dynamic institutional change depends on institutes' having a solid core of ICT infrastructure, including (i) wired and wireless networks enabling universal connectivity, (ii) tools to enable collaboration through seamless and robust communications and (iii) social networking platforms to support expanded communication, collaboration and dissemination across multiple communities.

In Kenya, the government is increasingly supporting digital learning in primary and secondary schools. In 2016/2017, JKUAT helped the Kenyan government to deliver more than 329,000 laptops to individual students in 21 of Kenya's 47 counties through the Digital Learning Programme. This was made possible through the establishment of a Computer Assembly Centre at JKUAT's School of Computing and Information Technology (located on the main campus). To aid teachers, the university provided 14,064 laptops, 7,032 projectors, and 7,032 content access points (i.e. devices which provide a local wireless network) to primary schools. This was done in collaboration with POSITIVO, a technology company from Brazil, which once again illustrates the critical role of relationships with the private sector. JKUAT has become the reference centre for ICT innovations in Kenya, working directly with government and numerous development partners.

Building National, Regional and Global Partnerships

To extend the quality, breadth and impact of university programmes, leaders of African universities would be wise to take advantage of the power of partnerships. The greatest academic institutions of the 21st century will be those willing and able to partner with other stakeholders to achieve what no one institution or organization could achieve on its own. This calls for regional, pan-African and global partnerships. A good example of a regional network is the Association of Southeast Asian Institutions of Higher Learning, a non-governmental organization founded in Bangkok in 1956. The association aims to assist the development of its members to achieve distinction in teaching, research and public service (ASAIHL, 2016). In Africa, the Regional Universities Forum for Capacity Building in Agriculture (RUFORUM), which is located at Makerere University in Uganda, was established in 2004 as a consortium of 32 universities in Eastern, Central and Southern Africa. Its membership has grown to include 126 universities across all regions of Africa. RUFORUM's mandate is to oversee graduate training and networks of specialization in the field of agricultural sciences. It also assists in brokering partnerships between African institutions and global partners and has helped strengthen the African Union initiative of creating pan-African universities.

The vice chancellor's office at JKUAT has placed a high priority on developing partnerships. JKUAT is a RUFORUM member and host of the pan-African University Institute for Basic Sciences, Technology and Innovation. To date, over 500 students from 42 African countries have been admitted, of which three cohorts totalling 237 graduate students have graduated. With RUFORUM funding, JKUAT has launched an MSc programme on research methods and PhD training in food science and nutrition, drawing students from across Africa. JKUAT seeks to create a global network with universities in Southeast Asia (Japan, Singapore, Thailand and Vietnam) and Europe (Belgium, Germany, the Netherlands, Sweden and the United Kingdom, among others).

6.6 Change Management

Transforming TAE institutions entails substantial organizational change. Most people prefer the status quo and will, therefore, resist the changes required to transform institutions unless the introduction and control of the change is well-managed (Bruckman, 2008). Bringing about the required transformation will therefore require that the change process is managed effectively. Resistance to change can be addressed through various actions, including effective planning and implementation, and engagement with stakeholders. Resistance often arises due to lack of understanding of the planned change and its purpose and impacts. Improved communication helps stakeholders understand the intentions underlying the changes and reduce associated fears (Henry, 1997; Tyler *et al.*, 2013). Tyler *et al.* (2013) suggest that the critical factors in overcoming resistance to change in the context of nursing homes were communication, education and training. For JKUAT, the introduction of performance contracting created anxiety and resistance from staff led by their unions. With effective communication, however, the staff eventually embraced the process.

Self (2007) describes five important points for leaders to address, including (i) the need for change, (ii) the appropriate change, (iii) the ability to adopt the change, (iv) the support required to make the change and (v) the benefits of the change to individuals and organizations. Having clarity on these issues and communicating them improves stakeholders' perception of the change and reassures them that the change is being managed effectively. Frequent and open communication helps to alleviate fear of unknown outcomes resulting from the change and bolsters support for the change process. Among the things that should be communicated are clear policies and instructions so that staff are guided in their participation in planned changes. If, despite clear and open communication, resistance remains, it is critical that leaders understand its sources and reasons in order for appropriate action to be taken (Aladwani, 2001; Self and Schraeder, 2009).

Key stakeholders should be able to accept and have ownership of the change being made. A study by Lines (2006) found that, as stakeholder involvement increases, resistance decreases. Stakeholders should be actively involved to ensure the sustainability of the change process, and its outcomes. Leaders may initiate the change process, but beyond that stakeholders should be empowered and facilitated to move the process forward. Consultation is necessary to obtain stakeholders' buy-in in order to minimize resistance. This should include unions, which are instrumental in influencing staff and other stakeholders. Resistance often stems from fear of loss of jobs, power, control or influence. To alleviate such fears, capacity building will be vital to ensure staff have the skills and training needed to comply with changes and adapt to new roles, which will reduce potential tension and resistance to the planned changes (Gallant, 2000). Gallant (2000) emphasises the need to accompany the introduction of changes with infrastructure support and strategic planning - e.g. new learning technologies. Support for change also includes adequate resources and an improved working environment in order to reduce work pressure and any negativity leading to resistance (Gallant 2000). Rewards and sanctions may also be used (Miller, 2001). Staff who support the change process may be rewarded and motivated in various ways. Where staff actively engage in subversion, sanctions may be considered. According to the Vice Chancellor of University of Cape Coast, Ghana, excellent leadership requires gratitude, honesty and teamwork (D. Dabire Kuupole, Cape Coast, Ghana, 2019, personal communication).

6.7 Conclusion

Modern universities should actively adapt their training, research and innovation to meet the changing needs of the economy and society. This requires astute and visionary leadership. A culture of innovation should be fostered, and young talent recruited and supported. Linkages with industry and communities are critical to ensure that universities remain relevant. To successfully develop the TAE sector, governments, academia and the private sector must work together with focus and commitment. Leadership must provide clear strategic direction and nurture foundational relationships and partnerships if they are to build transformative institutions as a foundation for knowledge economies.

Endnotes

[1] African universities supported by NICHE include Jimma University (Ethiopia), JKUAT, the University of Eldoret (Kenya), and South Eastern Kenya University, among others.

References

Aarts, H. Greijn, H. Mohamedbhai, G. and Jowi, J.O. (2020) The SDGs and African higher education. In Ramutsindela M., and Mickler D. (eds) *Africa and the Sustainable Development Goals. Sustainable Development Goals Series*. Springer International Publishing. Available at: https://doi.org/10.1007/978-3-030-14857-7 (accessed 14 April 2021).

Aguirre, A. and Martinez, R. (2002) Leadership practices and diversity in higher education: Transitional and transformational frameworks. *Journal of Leadership Studies* 8(3), 53–62.

Aladwani, A. (2001) Change management strategies for successful ERP implementation. *Business Process Management Journal* 7(3), 266–275. Available at: https://doi.org/10.1108/14637150110392764 (accessed 14 April 2021).

ASAIHL (Association of Southeast Asian Institutions of Higher Learning) Union of International Associations. Available at: https://uia.org/s/or/en/1100006729 (accessed 12 April 2021).

ASTI (2019) ASTI database. International Food Policy Research Institute, Washington, DC. Available at: https://www.asti.cgiar.org (accessed 12 April 2021).

Botha, J. Favish, J. and Stephenson, S. (2008) Institutional audits: A comparison of the experiences of three south African universities. *Journal of Quality in Higher Education* 14(1), 29–53.

Boyd, W.E. (2013) Does Boyer's Integrated Scholarships Model work on the Ground? An adaption of Boyer's model for scholarly professionald Development. *International Journal for the Scholarship of Teaching and Learning* 7(2), Article 25. Available at: https://doi.org/10.20429/ijsotl.2013.070225 (accessed 14 April 2021).

Boyer, E. (1990) *Scholarship Reconsidered: Priorities of the Professoriate*. Wiley and Sons, New York.

Bruckman, J.C. (2008) Overcoming resistance to change: Causal factors, interventions, and critical values. *Psychologist-Manager Journal* 11(2), 211–219. Available at: https://doi.org/10.1080/10887150802371708 (accessed 14 April 2021).

Church, A.H. and Rotolo, C.T. (2016) Lifting the veil: What happens when you are transparent with people about their future potential? *People & Strategy* 39(4), 28–32.

Church, A.H. and Silzer, R.F. (2014) Going behind the corporate curtain with a BluePrint for Leadership Potential: An integrated framework for identifying high-potential talent. *People & Strategy* 36(4), 51–58.

Cimino, J. Buswick, T. and Seifter, H. (2005) The story factor: Inspiration, influence and persuasion through the arts of storytelling. *Journal of Business Strategy* 26(5), 78.

Eddy, J.P. *et al.* (1997) 21st century leadership practices needed for higher education. *Education* 117(5841), 1026–1028.

Enserink, M. (2007) Who ranks the university rankers? *Science* 317, 1026–1028. Available at: https://doi.org/10.1126/science.317.5841.1026 (accessed 14 April 2021).

Fan, S. Yosef, S. and Pandya-Lorch, R.I. (2019) *Agriculture for Improved Nutrition: Seizing the Momentum*. CABI, Wallingford, United Kingdom. Available at: http://dx.doi.org/10.1079/9781786399311.0000 (accessed 14 April 2021).

Gallant, G.M. (2000) Professional Development for Web-Based Teaching: Overcoming Innocence and Resistance. *New Directions for Adult and Continuing Education* 2000(88), 69–78. Available at: https://doi.org/10.1002/ace.8807 (accessed 14 April 2021).

Hansson, H. and Holmberg, C. (2003) A European and Swedish perspective on ICT – policies and strategies in education. National and institutional pathways: Crossings, blind alleys and uphill slopes. *Journal of Systemics, Cybernetics and Informatics* 2(6), 86–91. Available at: http://www.iiisci.org/Journal/CV$/sci/pdfs/P243705.pdf (accessed 30 May 2021).

Hegarty, N. and Moccia, S. (2018) Components of ethical leadership and their importance in sustaining Organizations Over the Long Term. *Journal of Values-Based Leadership*: 11 (1), Article 7.

Henry, P.K. (1997) Overcoming resistance to organizational change. *Journal of the American Dietetic Association* 97(10), S145–S147. Available at: https://doi.org/10.1016/S0002-8223(97)00751-7 (accessed 14 April 2021).

JKUAT (Jomo Kenyatta University of Agriculture and Technology) (2021) Vision, mission & objectives. Available at: http://www.jkuat.ac.ke/mission-vision-objectives/ (accessed 31 May 2021).

Knobloch, N.A. Charoenmuang, M. Cooperstone, J.L. Patil, B.S. (2019) Developing interdisciplinary thinking in a food and nutritional security, hunger, and sustainability graduate course. *Journal of Agricultural Education and Extension* 26(1), 113–127. Available at: https://doi.org/10.1080/1389224X.2019.1690014 (accessed 14 April 2021).

Lines, R. (2006) Influence of participation in strategic change: Resistance, organizational commitment and change goal achievement. *Journal of Change Management* 4(3) 193–215. Available at: https://doi.org/10.1080/1469701042000221696 (accessed 14 April 2021).

Liu, Z. Kipchumba, S.K. and Liu, L. (2016) Paths for world-class universities in agricultural science. *Higher Education* 71, 97–118. Available at: https://doi.org/10.1007/s10734-015-9891-9 (accessed 14 April 2021).

Long, B. (2014) The tenure process in higher education: An application of the Boyer model of the professoriate to create a coherent record of scholarly productivity. *International Journal of Humanities and Social Science* 4(9), 43–50.

Marginson, S. and van der Wende, M. (2007) To rank or to be ranked: The impact of global rankings in higher education. *Journal of Studies in International Education* 11(3/4), 306–329.

Maxel, O.J.M. (2013) Strategic organization for competitiveness in East African universities. *European Journal of Business and Management* 5(20), 151–160.

McNamara, J. Sweetman, S. Connors, P. Lofgren, I. and Greene, G. (2019) Using Interactive Nutrition modules to increase critical thinking skills in college courses. *Journal of Nutrition Education and Behavior* 52(4), 343–350. Available at: https://doi.org/10.1016/j.jneb.2019.06.007 (accessed 14 April 2021).

Miller D. (2001) Successful change leaders: What makes them? What do they do that is different? *Journal of Change Management* 2 (4), 359–368. Available at: https://doi.org/10.1080/714042515 (accessed 14 April 2021).

Monks, J. and Ehrenberg, R. (1999) The Impact of U.S. News & World Report College Rankings on Admissions Outcomes and Pricing Policies at Selective Private Institutions. NBER Working Paper Series 7227 [electronic version]. National Bureau of Economic Research, Cambridge, Massachusetts, USA. Available at: https://doi.org/10.3386/w7227 (accessed 14 April 2021).

Nanus, B. (1992) *Visionary Leadership: Creating a Compelling Sense of Direction for Your Organization*. Jossey-Bass, San Francisco.

Ogola, M. Sikalieh, D. and Linge, T. (2017) The influence of intellectual stimulation leadership behaviour on employee performance in SMEs in Kenya. *International Journal of Business and Social Science* 8(3), 89–100. Available at: https://www.ijbssnet.com/journals/Vol_8_No_3_March_2017/9.pdf (accessed 12 April 2021).

Parand, A. Dopson, S. Renz, A. and Vincent, C. (2014) The role of hospital managers in quality and patient safety: A systematic review. *BMJ Open* 4: e005055. Available at: https://doi.org/10.1136/bmjopen-2014-005055 (accessed 14 April 2021).

Pejza, J.P. (1985) The Catholic school principal: A different kind of leader. Paper presented at the Annual Meeting of the National Catholic Educational Association, St. Louis, Missouri, USA. Available at: https://files.eric.ed.gov/fulltext/ED256053.pdf (accessed 12 April 2021).

Perloff, R.M. (2008) *The Dynamics of Persuasion: Communication and Attitudes in the 21st Century*. Lawrence Erlbaum Associates, New York.

Pitso, T. (2019) Invigorating innovation and entrepreneurship: Insights from selected South African and Scandinavian universities. *Southern African Journal of Entrepreneurship and Small Business Management* 11(1), a187. Available at: https://doi.org/10.4102/sajesbm.v11i1.187 (accessed 14 April 2021).

QS World University Rankings. (2019) QS Latin America University Rankings 2019. Available at: https://www.topuniversities.com/university-rankings/latin-american-university-rankings/2019 (accessed 12 April 2021).

Salmi, J. (2009) *The Challenge of Establishing World-class Universities. Directions in Development: Human Development*. World Bank, Washington, DC. Available at: https://doi.org/10.1596/978-0-8213-7865-6 (accessed 14 April 2021).

Self, D. (2007) Organizational change: Overcoming resistance by creating readiness. *Development and Learning in Organizations* 21(5), 11–13. Available at: https://doi.org/10.1108/14777280710779427 (accessed 14 April 2021).

Self, D. and Schraeder, M. (2009) Enhancing the success of organizational change: Matching readiness strategies with sources of resistance. *Leadership & Organization Development Journal* 30(2), 167–182. Available at: https://doi.org/10.1108/01437730910935765 (accessed 14 April 2021).

Sparks, G. (2014) Charismatic leadership: Findings of exploratory investigations of the techniques of influence. *Journal of Behavioral Studies in Business* 7, 1–11.

Sparks, J.R. and Areni, C.S. (2008) Style versus substance. Multiple roles of language power in persuasion. *Journal of Applied Social Psychology* 38(1), 37–60.

Stellenbosch University. Our vision. Available at: http://www.sun.ac.za/english/about-us/strategic-documents#Vision (accessed 31 May 2021).

Tyler, D.A. Lepore, M. Shield, R.R. Looze, J. and Miller, S.C. (2013) Overcoming resistance to culture change: nursing home administrators' use of education, training, and communication. *Gerontology & Geriatrics Education* 35(4), 321–336. Available at: https://doi.org/10.1080/02701960.2013.837049 (accessed 14 April 2021).

UNESCO (United Nations Educational, Scientific and Cultural Organization) (2014) *Higher Education in Asia: Expanding Out, Expanding Up: The Rise of Graduate Education and University Research*. UNESCO Institute for Statistics, Montreal. Available at: http://uis.unesco.org/sites/default/files/documents/higher-education-in-asia-expanding-out-expanding-up-2014-en.pdf (accessed 12 April 2021).

UC-Davis Continuing and Professional Education (no date) Best practices in talent development and management. Available at: https://cpe.ucdavis.edu/course/best-practices-talent-development-and-management (accessed 15 April 2021).

University of Cape Town (2016) About UCT: Mission: Vision. Available at: https://www.uct.ac.za/main/about/mission (accessed 12 April 2021).

University of Dar es Salaam (2021). Vision & mission. Available at: https://www.udsm.ac.tz/web/index.php/mission-and-vision (accessed 31 May 2021).

University of Ibadan (2021) Vision. Available at: https://www.ui.edu.ng/content/vision-and-mission (accessed 12 April 2021).

University of Nairobi (2020) UON vision, mission & core values. Available at: https://a-qualityassurance.uonbi.ac.ke/node/2504 (accessed 27 September 2021).

WUR (Wageningen University and Research) (no date) Vision and mission. Available at: https://www.wur.nl/en/Research-Results/Chair-groups/Social-Sciences/KnowledgeTechnology-and-Innovation-Group/Vision-and-Mission.htm (accessed 15 April 2021).

Wanderi, P.M. and Makandi, L. (2019) Kenyan universities contribution towards the President's "Big Four" Agenda: The case of Mount Kenya University's Graduate Enterprise Academy. In Universities, Entrepreneurship and Enterprise Development in Africa: Conference Proceedings 2018. Available at: https://ideas.repec.org/h/sau/ueedcc/07202-216.html (accessed 12 April 2021).

Yusuf, S.F.G. Popoola, O.O. and Olutegbe, N.S. (2019) Perceptions of Undergraduate Students to Practical Agriculture Experience at the University of Fort Hare, South Africa. *Journal of Agricultural Extension* 23(1), 210–222.

7 Transformative Curricula and Teaching Practices to Meet Labour Market Needs in Tertiary Agricultural Education in Africa

Keba Hulela[1]*, Joseph Mukuni[2], Might Kojo Abreh[3], Joseph Amooti Kasozi[4] and David Kraybill[5]

[1]Botswana University of Agriculture and Natural Resources; [2]Virginia Tech, United States; [3]University of Cape Coast, Ghana; [4]School of Education, Botswana Open University; [5]Ohio State University, United States

7.1 Introduction

Many observers have noted the large gap between the needs of today's African economy and the training being offered by higher education institutions (HEIs), which is based on traditional approaches (McCowan, 2014). The vast majority of Africa's university students pursue advanced studies with the ultimate goal of acquiring employment, but many graduates remain unemployed long after graduation. At the same time, employers have become increasingly vocal about the inadequacy of the skills and knowledge imparted by HEIs (Griesel and Parker, 2009). Employability has become a major concern in higher education worldwide, and universities around the globe are re-examining their curricula and pedagogy with the aim of helping graduates succeed in gaining employment, creating self-employment and generally adapting with resilience to ever-changing and unpredictable economic and social conditions (Pegg *et al.*, 2012). In cutting-edge institutions of higher learning, traditional curricula and didactics, which primarily transfer theoretical knowledge through lectures, are being replaced with more active approaches to learning (Freeman *et al.*, 2014).

This chapter describes and analyzes labour market conditions and policy, as well as programme options for improving the economic and social relevance of agricultural training in African HEIs. The intended audience is teachers and administrators of tertiary agricultural education (TAE) who have little or no training in curriculum development or pedagogy. The aim is to present practical steps for reforming curricula and pedagogical approaches to enable TAE institutions to meet the needs of communities more effectively and to address the demands of dynamic labour, information and technology markets. The discussion also describes how tertiary education teachers and administrators can develop the vision, methods and institutional culture required to prepare students for employability and lifelong learning.

7.2 Labour Market Conditions in Sub-Saharan Africa

Labour markets in most countries of sub-Saharan Africa (SSA) are characterized by a mismatch between the demand and supply of workers,

*Corresponding author: khulela@buan.ac.bw

 Transforming Tertiary Agricultural Education in Africa (eds D. Kraybill, J. Lynam and Adipala Ekwamu)
DOI: 10.1079/9781789246544.0007

resulting in widespread unemployment and underemployment. One of the major causes of low labour demand is low labour productivity, resulting from inadequate levels of capital. Financial, physical and human capital are all necessary to labour productivity. When financial capital is difficult to access, businesses are unable to acquire the physical capital to make workers more productive. Low levels of human capital mean that workers, managers and entrepreneurs lack the skills needed to enhance labour productivity.

The vast majority of labourers in SSA work in the informal economy, an environment characterized by minimal physical capital, low labour productivity and little or no taxes paid by businesses. In the informal economy, workers labour within their own households or work for others with no day-to-day guarantee of employment. The work is typically strenuous manual labour, often performed in dangerous settings, and wages are low. On the supply side, an enormous population of workers in SSA are employed in the informal economy, yet they aspire to move into the formal economy where capital is far more abundant. Working conditions in the formal economy are generally safer and less strenuous than in the informal economy; workers have greater assurance of continuous employment; and wages are typically higher.

A relatively-high percentage of workers in SSA fall into the youth demographic (generally considered 15–25 years of age), a phenomenon driven by a large drop in infant mortality rates without a parallel drop in fertility rates over the past two decades. Many youth have not completed primary school, or, if they have, their literacy and numeracy skills are sub-standard because of poor schooling quality (Garcia and Fares, 2008). Low levels of human and financial capital in their households entrap them within the informal economy. Lack of financial, physical and human capital in SSA also dampens labour demand by preventing the economy from diversifying into value-added activities, such as processing of raw materials, manufacturing and advanced services (e.g. technology-based services). Instead, production activities remain confined to agriculture, forestry, mining, manufacturing of a narrow range of locally-consumed products and low-value services. Without industrial and agribusiness development, formal-sector employment remains low.

7.3 The Skills Gap

Employers in the private and public sectors often complain that university graduates are not prepared for the workforce (Egeru *et al.*, 2016). They point to the lack of practical technical skills and fundamental 'soft' skills, including critical thinking, oral and written communication, time management and the ability to set and monitor goals, collaborate with others and resolve conflicts. Employers, community leaders and other outside stakeholders often summarize these problems by telling educators that university curricula are too theoretical and not practical enough for employment or self-employment. Graduates leave institutions unprepared to find or create jobs, and those graduates who succeed in finding jobs often need to be retrained by their employers in basic skills and knowledge. Those who cannot find formal-sector employment lack the entrepreneurial skills needed to become self-employed or create jobs for others.

The effects of the training gap are especially acute in Africa given the slow growth of formal-sector employment, large and rapidly-growing numbers of university graduates and a high unemployment rate. Agricultural and food systems in Africa are changing quickly (Kabasa *et al.*, 2015). In many countries, agricultural production is still mainly carried out by numerous smallholder producers, though medium- and large-size farms now account for a growing share of production. Technologies of production are also evolving quickly in an effort to increase productivity, and modern food-distribution and processing systems have emerged to feed the rapidly-growing urban population. TAE institutions must stay abreast of the evolving competencies required in agricultural food systems if they are to succeed in training highly-sought-after and valued graduates capable of creating employment opportunities for themselves and others.

HEIs can improve labour productivity and boost their graduates' employment and income potential by revamping their curricula and pedagogy to help students acquire the relevant competencies - defined as skills, knowledge and attitudes - to meet the needs of the labour market. The concept of employability addresses the reality that employment is not guaranteed following graduation, and that even those who find employment may become job-seekers again in the future if their employment is terminated or if they wish to advance to a better job (Forrier and

Sels, 2003). Training for employability focuses not only on employment immediately after graduation, but also on helping graduates find employment throughout their careers in jobs that make full use of their evolving capacity.

The competencies required by employers and entrepreneurs can be identified through information exchange and collaboration between training institutions and the private sector. Collaboration can be established by creating advisory committees consisting of industry representatives and instructional staff, and by appointing industry representatives to the boards of training institutions. Two aims of such collaborations are (i) identifying priority skills for which training and training standards are required and (ii) planning and implementing workplace experience for students, e.g. through internships. Other forms of collaboration include guest lectures by industry experts, jointly-sponsored courses, the development of certification standards and procedures, the provision of training machinery and the mobilization of resources. Through these endeavours, students can become familiar with current developments in their field of study and acquire practical skills and knowledge to prepare them for the workplace when they graduate. A special set of competencies is needed by entrepreneurs. Given the dearth of jobs in many African countries, entrepreneurship is vital for economic and social development. TAE institutions can make valuable contributions in this area. EARTH University in Costa Rica, for example, has developed an undergraduate degree programme to foster entrepreneurship in its students (see Box 7.1).

As part of their degree training, EARTH University's students create and operate their own business enterprise over a three-year period to learn about entrepreneurship through experience. This model of training is highly relevant to the African context, and elements of it are being adapted for use in programmes at Gulu University in Uganda and Egerton University in Kenya (Zaglul, 2016). The entrepreneurship programmes at all three of these universities help students develop the problem-solving, critical and innovative thinking - and flexibility - needed to manage private enterprises. Students develop both the mind set and the skillset to become employable once they graduate. Training in entrepreneurship is one way HEIs can address the paucity of employment opportunities in Africa's formal labour market.

7.4 Transformative Curricula

In many of Africa's TAE institutions, curricula lack relevance to continental, national and local

Box 7.1. EARTH University's academic model

EARTH University's curriculum is based on four formative areas designed to prepare students with the competencies required by today's labour market:

1. **Technical and scientific knowledge:** to ensure that EARTH alumni have the professional competencies to sustainably manage agriculture and natural resources
2. **Personal development, attitudes and values:** to promote self-awareness, empathy, respect, tolerance, teamwork, effective communication and lifelong learning, along with values which promote peace, dialogue and understanding
3. **Ethical entrepreneurship:** to foster an entrepreneurial mind set in students and to enable them to solve problems and generate opportunities proactively
4. **Social and environmental awareness and commitment:** to develop social and environmental responsibility in students and strengthen their capacity to generate positive change

EARTH University promotes a dynamic and participatory teaching and learning process, whereby its professors serve as facilitators of learning. The educational approach is based on two principles:

1. **Student-centred learning.** Students, alone or in groups, explore real challenges and become active participants in generating knowledge, as opposed to being passive receivers of information.
2. **Experiential learning.** Students learn through planned experiences designed to encourage learning; they are given opportunities to build knowledge and develop abilities of practical relevance.

Source: EARTH University (no date).

issues, including environmental sustainability, gender, farmer participation in research and extension, and changing career patterns within the agricultural sector (Wallace and Nilsson, 1997). Relevance can be increased through review and evaluation of curricula and training materials by key stakeholders - including researchers and extension workers, farmers and agribusiness managers - to transform what is taught and how it is taught. Many education scholars broadly define 'curriculum' as the totality of student experiences occurring through the educational process, including but not limited to classroom activities (Kelly, 2009). For the purposes of this chapter, curriculum is defined to encompass the underlying rationale, aims, content, methods, resources, scope, time, sequencing and assessment of the teaching and learning process. Some educators view the syllabus and the curriculum as the same thing, but this is an insufficiently comprehensive view of the curriculum. The 'syllabus' specifies a list of topics, student responsibilities and the resources on which they can draw; the curriculum comprises not only the syllabus, but also information on proposed teaching and learning methods, available instructional resources (such as venues and equipment), recommended sequencing of instructional components and pedagogical guidance on assessment procedures.

Curriculum development and revision occur across international, national, institutional, classroom and individual levels. Beyond simply borrowing material from other parts of the world, the aim of curriculum development and revision is to improve what is taught and learned to ensure the provision of content and learning experiences of relevance to the local context. For a curriculum to remain relevant, it must be revised as new knowledge is acquired and the needs of the labour market change. Filling gaps in the knowledge and relevance of existing curricula is an ongoing process. In addition to updating subject-matter, curriculum revision involves changes to teaching and learning methods in response to educational research on the effectiveness of new methods compared with existing ones. Curriculum development and revision should be guided by an understanding of the learning process. Anderson *et al.* (2001, p. 25) describe six levels of learning, building on earlier research by Bloom on a taxonomy of educational objectives (Bloom, 1956):

1. *Remembering.* The most basic level is remembering, which includes retrieving, recognizing and recalling relevant knowledge. Moving towards greater transformation of students' thought processes.

2. *Understanding.* The second level is understanding, in which students construct meaning by interpreting, exemplifying, classifying, summarizing, inferring, comparing and explaining.

3. *Applying.* The third level is applying, which entails students use of procedures to execute or implement.

4. *Analyzing.* The fourth level is analyzing, in which students break information into parts to explore relationships through organizing and differentiating.

5. *Evaluating.* The fifth level is evaluating, which consists of making judgements based on criteria and standards through checking and critiquing.

6. *Creating.* The sixth and highest level is creating, in which students generate new perspectives, ideas or products.

Each level of learning builds on the foundation of the preceding level, so curricula can only be transformative to the extent that they provide a range of assignments, activities or experiences across all six.

Unfortunately, the curricula implemented in many TAE institutions in Africa do not go much beyond the two lowest levels, and even if they call for the third level, limited financial, human and physical resources often preclude its implementation. As a result, agricultural graduates are ill-equipped to address current challenges in the agricultural and food system - yet this is what many employers and stakeholders now require. Beyond application, the three higher levels of learning are vital to develop students' capacity for independent thinking and action. Students whose learning incorporates these higher levels graduate with the ability to apply their knowledge even as technical, economic and social circumstances change over time. Higher-level learning is a new experience for many TAE students, particularly if their primary and secondary schooling did not encourage independent thinking.

In 1987, the United Nations Bruntland Commission formally defined sustainable development as meeting 'the needs of the present without compromising the ability of future generations to meet their own needs' (Bruntland, 1987, p. 16). To be effective in supporting the sustainable transformation of African agriculture, curricula must be designed and implemented to promote all six levels of learning. Such curricula will address the real-world issues, both locally and nationally, examining strengths, weaknesses, opportunities and threats within Africa's agricultural sector. Such curricula will not only ensure graduates' employability, but also equip them with the competencies they need to develop the innovative and transformative agricultural solutions Africa requires.

7.5 Transformative Teaching Practice

Teaching practice involves both teacher-centred and student-centred approaches. In teacher-centred instruction, the teacher acts as an authority figure possessing expert knowledge, while the student is viewed as a passive vessel into which that knowledge is poured. Lecturing is the dominant - and often only - method of instruction used, and assessment is considered the final step in the learning process. In student-centred instruction, the student is encouraged and empowered to acquire knowledge and skills through active discovery (Weimer, 2002; Wright, 2011). The teacher designs learning activities and sets them in motion, serving more as a guide in the process than a giver of knowledge. The methods of instruction are active rather than passive, emphasizing 'doing' on the part of the student. In the lecture method, the teacher may intersperse dialogue or small group activities to maintain students' attention and involvement. The learning process continues beyond the end of the course through the application of acquired knowledge and skills in real-life settings at work, at home or in the community.

TAE in Africa today remains heavily teacher-centred. If HEIs and programmes are to succeed in filling the skills gap highlighted by employers and other stakeholders, teaching practices must be transformed to give students the opportunity and responsibility to play a more active role. HEIs should provide teacher training on student-centred learning strategies and encourage teachers to supplement lectures with active learning exercises. Such exercises include guided discussion, individual and collaborative projects, analysis of case studies, debates and practical exercises. In addition to traditional learning environments, such as the classroom or laboratory, teachers can make use of the university library, individual or group field visits, guest speakers, hands-on practical work assignments either on or off campus and internships. As teachers and institutions become more confident in the use of student-centred approaches, they can adapt them to fit their local context.

Four principles are a useful starting point for teachers to improve their teaching and enhance their students' learning (Bulger *et al.*, 2002). These principles focus on intended outcomes, clarity, engagement and enthusiasm. Intended outcomes should be identified in the curriculum but, unfortunately, not all curricula are sufficiently developed to do this. However, whether or not the curriculum addresses outcomes, teachers preparing courses, modules or particular class sessions must decide on the most appropriate mode of teaching. Identifying intended outcomes helps teachers decide which material to present and which teaching methods will most likely help students learn the material. In the classroom, the teacher should share the intended outcome with students so they know where they are headed. Bulger *et al.* (2002, Ace 1) pose the following questions to help teachers understand how to incorporate a focus on intended outcomes into their teaching:

1. What were the desired student outcomes for your last class meeting?
2. Were the outcomes directly stated or implied?
3. What did your students actually learn, and how was that learning documented?
4. Did the employed instructional strategies effectively contribute to each student's ability to accomplish the stated outcomes?

Clarity in teaching can be achieved by building on the knowledge students already possess, and by making the content as simple to grasp as possible. Bulger *et al.* (2002, Ace 2) pose practical questions to help teachers understand the concept of clarity:

1. During your last class meeting, what instructional techniques did you employ to provide the

students with a clear explanation of the lesson content?
2. What types of illustrations, demonstrations, heuristics and the like were used to supplement and clarify verbal explanation?
3. Were there any concepts or skills that you were able to incorporate from previous lessons and courses?
4. Did you allocate sufficient time for your students to ask questions so that you could clarify information?

Engagement by students in the learning process is essential for mastery of knowledge (Guthrie *et al.*, 2012). This is done most effectively by instructional methods in which students grapple with what they are learning. Practical, hands-on exercises, appropriate for some topics but not for others, are highly effective for engaging students in the learning process. Other methods include small group tasks inside or outside the classroom, writing assignments, oral presentations and in-class quizzes. Engaging students is easier when the class is small but can also be done when the class is large. Even when the lecture format is used in teaching, student engagement can be achieved by interactive lecturing which engages dialogue rather than one-way communication. The lecturer can ask probing questions about particular topics in the course material, then give students the opportunity to answer or ask questions. Bulger *et al.* (2002, Ace 3) ask the following questions:

1. In your last class, how much time were your students engaged in learning activities other than note taking?
2. On how many occasions during your last class did students have the opportunity to be actively engaged in the learning process?

Enthusiasm on the part of the teacher is an important input into the learning process, according to the literature on effective teaching (Keller *et al.*, 2016). A teacher who is enthusiastic about a topic is motivated to prepare the teaching material carefully and to present it in the most effective way possible. Student enthusiasm for learning is dampened by an evident lack of enthusiasm on the part of the teacher. Bulger *et al.* (2002, Ace 4) ask:

1. Were your students excited about attending your last class?
2. Were you excited about teaching your last class?
3. What have you done to effectively communicate your passion for the subject matter you teach to your students?
4. What strategies do you employ to stay current in your field of study and communicate your excitement about new developments?
5. How have your past teaching, research and service been used to positively affect the teaching-learning environment for your students?

7.6 Methods for Transforming Curricula and Teaching Practices

An important step in transforming curricula and teaching practices to achieve greater relevance for current and emerging economic and social conditions is to gather information about the needs of the labour market. This information is then used to identify knowledge and skills gaps in existing training programmes. Gathering information through stakeholder forums is essential in developing curricula. If financial resources are available, it is also useful to conduct tracer studies and occupational analyses. Many universities conduct curriculum reviews in consultation with stakeholders every two to five years. To obtain useful information, it is important that stakeholders understand they are being brought into the review process to have positive input into the curriculum, not merely to fulfil a ministerial requirement that they be present. In the case of the largest employers of an institution's graduates, it is wise to conduct individual interviews with human resource managers to elicit their views on knowledge and skills gaps. Individual interviews should be conducted on an ongoing basis, not just during the formal curriculum review, so that when it comes time to revise the curriculum, a partnership will already have been established.

Tracer studies are based on interviews with graduates to identify their type of employer, job profile and wage level, the knowledge and skills required for their job, and view of how the educational institution could more effectively prepare its students for employment. Interviewees should include self-employed graduates so that gaps in training focusing on entrepreneurship can also be identified. Tracer studies should not simply be one-time events but be repeated periodically. Understanding knowledge and skills gaps is the most important purpose of graduate tracer studies, but they also serve other purposes. Tracer studies can be used in (i) accrediting academic programmes, (ii) identifying potential quality improvements in academic programmes, facilities and services and (iii) providing information for marketing academic programmes (Schomburg, 2003). In addition, tracer studies can be used as a means of reconnecting universities with their graduates for the purpose of developing or sustaining an alumni association.

Occupational analyses go deeper than tracer studies in identifying workplace knowledge and skills requirements. Such analyses are appropriate when an educational institutional is developing specialized courses or programmes intended to prepare students for high-demand occupations. In face-to-face meetings lasting several hours, trained interviewers gather information about the tasks carried out by the interviewee, who is often a graduate of the educational institutional sponsoring the study (Halbrooks, 2003). To reduce reporting error, interviewers can arrange for interviewees to record tasks in advance on a daily basis. The interview is then structured to gather information for use in subsequent analysis of the skills, knowledge and attitudes required for carrying out each task successfully. 'Developing a Curriculum' (DACUM) is a widely used methodology of occupational analysis (Norton, 2009).

Once tracer-study and occupational interviews are completed, the resulting information can be used to modify course content and teaching/learning methods. The aim is to identify skillsets, a body of knowledge and attitudes which the university will incorporate into its curricula and which students will be expected to master before they graduate. Mastery generally requires active learning, so experiential learning is a necessary supplement to lecture-based teaching methods. For each academic programme - and with stakeholder input, including tracer studies and occupational analyses, if they have been conducted - faculty members develop a profile of the desired graduate. Appropriate course content and teaching strategies are also determined to ensure that students achieve mastery in the competencies identified via the profile. After the new curriculum is in place, teachers should be encouraged to frequently update teaching materials, teaching methods and learning environments in response to changing conditions and the external environment. When the time comes for the next mandated curriculum revision, faculty should expect to make major changes to the entire curriculum in response to feedback from external stakeholders.

7.7 Conclusion and Recommendations

Trends and patterns in African labour markets are dynamic and ever-changing due to rapid population growth and an exponential increase of the youth population. The processes which underpin agricultural and food systems have also changed significantly, especially in terms of production, transportation and communications technologies. Unfortunately, a lag exists between the training offered by many African HEIs and the skills needed in the job market. The goal of curriculum development and revision is to improve the quality of academic instruction by ensuring its relevance to labour market conditions. This is achieved through a shift from teacher-centred to learner-centred curricula and teaching methods. Such approaches enhance graduates' employability and capacity to aquire lifelong learning skills, including self-management and social networking skills, so that they can adapt to changing economic and social circumstances throughout their careers. HEIs should establish and maintain active communication and strong collaboration with outside stakeholders so as to address the current and emerging needs of employers and entrepreneurs.

To nurture a relevant and innovative teaching culture, HEIs should establish formal criteria for teaching quality, and reward innovative teachers through annual prizes and awards. To publicly acknowledge and promote good teaching, prize-winning teachers could be given the opportunity to share their teaching methods and practices with other members of the teaching staff through workshops. Human resource policies should also assign weight to teaching quality as a criterion for tenure and promotion. HEIs should also engage in ongoing assessments of the quality of their curricula, teaching methods and learning environments based on regular feedback from external stakeholders, students and graduates.

To provide institutional support to teaching and learning, it is useful for HEIs to establish a teaching and learning centre offering faculty members training, coaching and follow-up in support of curriculum development. Centres could oversee graduate tracer studies, occupational analyses and curricular and pedagogical revision. They could also plan and coordinate guest lectures, workplace experience programmes and internships, as well as developing case studies, business simulations, innovation incubators and job fairs.

Transforming curricula and teaching methods requires active leadership from vice chancellors, deans and department chairs/heads. For this transformation to become widespread, champions of change within the teaching staff should be identified, empowered and rewarded as a means of influencing their peers to embrace and adopt the new methods. HEIs that have adopted learner-centred education focused on employability have found that their graduates succeed in finding or creating jobs, and that their employers report increasing satisfaction with them. This in turn enhances the institution's reputation, attracts students and - over time - creates a pool of alumni who directly and indirectly support the institution financially, through networking, by providing ongoing inputs into curricula revision and by becoming a tangible example of the efficacy of their alma mater. In this way, TAE institutions can better serve their students by ensuring programme relevance. This will ultimately succeed in contributing to the goal of transforming the agricultural sector across African countries, and better the lives of African people.

References

Anderson, L.W., Krathwohl, D.R., Airasian, P.W., Cruikshank, K.A., Mayer, R.E., *et al.* (2001) *A Taxonomy for Learning, Teaching, and Assessing: A Revision of Bloom's Taxonomy of Educational Objectives*. Abridged edition. Longman, White Plains, New York.

Bloom, B.S. (1956) *Taxonomy of Educational Objectives, Handbook: The Cognitive Domain*. David McKay, New York.

Bruntland, G.H. (1987) *The Bruntland Report. Our Common Future. World Commission on Environment and Development*. Oxford University Press, UK. Available at: https://sustainabledevelopment.un.org/content/documents/5987our-common-future.pdf (accessed 10 April 2021).

Bulger, S.M. Mohr, D.J. and Walls, R.T. (2002) Stack the deck in favor of your students by using the four aces of effective teaching. *Journal for Effective Teaching* 5(2), 1–7. Available at: http://uncw.edu/cte/et/articles/bulger (accessed 10 April 2021).

EARTH University (no date) About EARTH: EARTH Facts. *Academic Model.* Available at: https://www.earth.ac.cr/en/about-earth/earth-facts/ (accessed 10 April 2021).

Egeru, A. Nampala, P. Okalany, E. Owuor, C. Nnam, J. *et al.* (2016) *Taking Pro-Active Actions to Enhancing the Quality of African University Graduates for the Labour Market*. RUFORUM Working Document Series 14(1), 247–256.

Forrier, A. and Sels, L. (2003) The concept employability: A complex mosaic. *International Journal of Human Resources Development and Management* 3(2), 102. https://doi.org/10.1504/IJHRDM.2003.002414.

Freeman, S. Eddy, S.L. McDonough, M. Smith, M. K. Okoroafor, *et al.* (2014) *Active Learning Increases Student Performance in Science, Engineering, and Mathematics. Proceedings of the National Academy of Sciences*, 111(23), 8410–8415. https://doi.org/10.1073/pnas.1319030111.

Garcia, M. and Fares, J. (2008) *Youth in Africa's labor market*. World Bank, Washington, DC.

Griesel, H. and Parker, B. (2009) Graduate Attributes: A Baseline Study on South African Graduates from the Perspective of Employers. Available at: http://www.saqa.org.za/docs/genpubs/2009/graduate_attributes.pdf (accessed 10 April 2021).

Guthrie, J.T. Wigfield, A. and You, W. (2012) *Handbook of Research on Student Engagement*. https://doi.org/10.1007/978-1-4614-2018-7.

Halbrooks, M.C. (2003) DACUM as a model for horticulture curriculum development and revision: A case study. *Horttechnology*, 13(2), 569–576. Available at: https://journals.ashs.org/downloadpdf/journals/horttech/13/3/article-p569.pdf (accessed 10 April 2021).

Kabasa, J.D. Kirsten, J. and Minde, I. (2015) Implications of changing agri-food system structure for agricultural education and training in Sub-Saharan Africa. *Journal of Agribusiness in Developing and Emerging Economies* 5(2), 190–199.

Keller, M.M. Hoy, A.W. Goetz, T. and Frenzel, A.C. (2016) Teacher enthusiasm: Reviewing and redefining a complex construct. *Educational Psychology Review* 28(4), 743–769. https://doi.org/10.1007/s10648-015-9354-y.

Kelly, A.V. (2009) *The Curriculum: Theory and Practice* (4th edition). Sage Publications, London.

McCowan, T. (2014) *Can Higher Education Solve Africa's Job Crisis? Understanding Graduate Employability in Sub-Saharan Africa*. British Council. Available at: https://www.britishcouncil.org/sites/default/files/graduate_employability_in_ssa_final-web.pdf. (accessed 10 April 2021).

Norton, R. E. (2009) *Competency-Based Education via the DACUM and SCID Process: An Overview*. Ohio State University. Available at: https://unevoc.unesco.org/e-forum/CBE_DACUM_SCID%20article(2).pdf (accessed 10 April 2021).

Pegg, A., Waldock, J., Hendy-Isaac, S. and Lawton, R. (2012) *Pedagogy for Employability*. Higher Education Academy, York, UK.

Schomburg, H. (2003) *Handbook for Tracer Studies*. International Centre for Higher Education Research, University of Kassel. Available at: https://www.qtafi.de/images/schomburg/schomburg_2003_tracer_handbook_v2.pdf. (accessed 10 April 2021).

Wallace, I., and Nilsson, E. (1997) *The Role of Agricultural Education and Training in Improving The Performance of Support Services for the Renewable Natural Resources Sector*. Available at: https://cdn.odi.org/media/documents/2919.pdf (accessed 10 April 2021).

Weimer, M. (2002) *Learner Centered Teaching: Five Key Changes to Practice*. Jossey-Bass, San Francisco.

Wright, G.B. (2011) Student-centered learning in higher education. *International Journal of Teaching and Learning in Higher Education* 23. https://doi.org/10.4324/9781315559605.

Zaglul, J.A. (2016) EARTH University educational model: Perspective on agricultural educational models for the twenty-first century. *Frontiers in Life Science* 9(3), 173–176.

8 Transformative Research and Innovation Capacity in Tertiary Agricultural Education in Africa

Clesensio Tizikara[1]*, Paul Nampala[2], David Nielson[3], Nienke Beintema[4], Patrick Okori[5] and John Lynam[6]

[1]*Independent consultant, Agrotiz Limited, Uganda;* [2]*Uganda Christian University;* [3]*North America Agricultural Advisory Network and Colorado State University, United States;* [4]*Independent consultant and former head of ASTI, based in The Netherlands;* [5]*International Crops Research Institute for the Semi-Arid Tropics, Malawi;* [6]*Independent scholar, based in Kenya*

8.1 Introduction

The rapid expansion in student enrolments over the past couple of decades drew focus to teaching at the expense of research and community development. Agricultural growth and market development - with an associated need for effective public institutions - have driven the demand for graduates, particularly those with PhD degrees. In turn, these trends require significant improvement in research capacity in tertiary agricultural education (TAE). Research remains relatively ad hoc, dependent on individual researchers' motivations, networking abilities and access to grants. A more strategic approach to developing high-quality research capacity is needed in TAE institutions in order to integrate critical competencies and experience into graduate degree programmes. This chapter explores these issues, firstly, by outlining the rationale for and necessary components of strategies to transform agricultural research capacity in African countries. Thereafter, the discussion focuses on strategic approaches to the transformation through the development of graduate (i.e. master's and doctoral) programmes. Such developments form a strong argument for increasing universities' role in national agricultural innovation systems (AIS). The final section explores an effective division of labour between TAE institutions and national agricultural research institutes (NARIs) in the production of new technical advances driving agricultural development. Note that in the context of this chapter, TAE institutions refer to degree-conferring universities, schools, faculties, departments and institutes, based on their roles in both training scientists in agricultural and related disciplines - such as veterinary science, forestry, fisheries and so on - and in conducting research of relevance to agriculture.

8.2 Transforming Tertiary Agricultural Education Institutions

Building research capacity in TAE institutions is essential to the development of high-quality graduate programmes and resulting contributions

*Corresponding author: ctizikara@agrotizuganda.com; tizikara@gmail.com

DOI: 10.1079/9781789246544.0008

to technological change in the agricultural sector. Demand for agricultural graduates is increasing as universities expand, existing professional staff reach retirement age and an expanding market economy requires new skills to compete regionally and globally (see Chs. 3, 5 and 7, this volume). At the same time, university research capacity directly contributes to rural development across Africa. Commercializing African smallholder agriculture necessitates the development of a unique research agenda - quite distinct from other parts of the world - yet it also depends on academic expertise and rigor found in the world's leading universities. With their wide-ranging subject matter, recognized expertise and global and domestic experience, African universities are uniquely positioned to contribute to the multidimensional challenges of Africa's food and nutrition needs. These include: (i) simultaneously increasing yields, profitability and environmental sustainability, (ii) developing new varieties and breeds needed for sustainable food systems, (iii) decreasing food losses and wastage through more efficient distribution systems, (iv) creating and sharing knowledge, services and other resources which serve all populations, (v) ensuring inclusive and equitable food systems, (vi) addressing the dual burdens of undernutrition and obesity and (vii) ensuring a safe and secure food supply which protects and improves public health.

Nevertheless, Africa's TAE sector continues to face critical challenges which limit its effectiveness in meeting these goals. The sector has rapidly expanded, yet there are limited infrastructure, doctoral staff and research facilities to deliver quality teaching and learning, as well as research and outreach for community transformation. Secondly, teaching capacity is limited in terms of the ratios of staff to students, qualification levels (in terms of PhD training), shortages of technicians, ageing senior academics and the capacity to conduct research for development which directly affects communities and economies (as is discussed in this chapter). Training the next generation of graduate students is based on research performance, which requires: (i) faculty members with PhD degrees and a publishing record and (ii) the availability of financial support, usually linked to the ability of staff to compete for grants. Faculties of agriculture, thus, need to build research capacity to contribute both to the challenges of agricultural development and to the investment in human capital to support such transformation.

Rapidly-rising demand for agricultural science graduates has been driven by expanding demand for teaching staff and PhD graduates generally, including in civil society and private-sector organizations, and the retirement of senior staff from faculties and NARIs in many countries (see Ch. 5, this volume). Given the cost of such degree training in the North and the relative lack of support for graduate training by aid agencies, this demand will have to be met by increased capacity in African faculties of agricultural sciences, and will in turn require higher-quality research in faculties which target expanded graduate programmes. Research is central to graduate training and is a key determinant of the quality of the degree programme. It relies on research infrastructure, experienced staff, appropriate research methods and funding. Building strong graduate programmes will require a significant reallocation of both staff and financial resources, where the past focus has been on meeting the rapid expansion in student enrolments.

At the same time the supply of undergraduate students enrolling in agricultural science is not keeping pace with the sector's evolving requirements (see Ch. 5, this volume). In most countries, agriculture's share of total enrolments ranged from 0.02% to 15.4%, and averaged 4.9%. In most countries, agriculture's share of total student enrolments in higher-education institutions (HEIs) has declined. Between the early 1980s and early 1990s, for example, agriculture's share of total enrolments in African universities declined from 5.7% to 3.3% at the pre-degree (technician) level, from 5.8% to 4.6% at undergraduate degree level and from 7.6% to 3.8% at the graduate degree level (World Bank, 2007; RUFORUM, 2018a). Between 2012 and 2020, enrolments averaged less than 2% in most countries of SSA (AfDB/OECD/UNDP/UNECA, 2012; AfDB, 2020; UIS, 2020). Such enrolments do not match the countries' strategic needs, much less the requirement that all teaching staff have PhD degrees. Kurwijila (2009) noted that the numbers of candidates undertaking graduate training in agriculture in general, and livestock and veterinary science in particular, were either declining or had remained stagnant due to lack of funding and scholarships required to maintain teaching and research infrastructure. Within tertiary institutions, the World Bank recommended a strategic focus on strengthening those disciplines deemed most relevant to a country's economy and future growth prospects - in particular science, technology, engineering, and mathematics (STEM), which includes agricultural sciences (World Bank 2008a,b; Blom *et al.*, 2016).

Centres of Excellence

Approximately three-quarters of the degree-conferring TAE institutions in a sample of 37 African countries offered PhD-degree programmes, which was at least double the number observed in 1996 (see Ch. 5, this volume). Nevertheless, many of these programmes only had a small number of PhD candidates enrolled, reflecting adherence to the British system, whereby PhD training is only based on research, not course-work (see Ch. 5, this volume). Thus, only a small number of PhD-qualified faculty would supervise PhD candidates' research. The shortcomings are obvious in terms of lack of both a dedicated PhD programme and a critical mass of students to provide intellectual interaction and the foundation for a dedicated departmental research agenda. Such limitations have led to the development of centres of excellence, where particular faculties develop high-quality programmes in specific disciplines, encompassing course work, strong research competency and an evolving research agenda. As with NARIs, a critical mass is central to developing effective research capacity, which in turn provides the basis for high-quality graduate programmes. Centres of excellence are central to the World Bank's programme of strengthening higher education in STEM disciplines across Africa, e.g. Partnerships for skills in Applied Sciences, Engineering and Technology.

The above developments, however, fall short of establishing a clearly-differentiated value chain for agricultural education institutions clustered into two distinct groups: (i) technical and vocational education and training (TVETs) and undergraduate programmes designed to increase the number of field operatives and practitioners in the agricultural sectors and (ii) research-intensive, high-productivity institutions which work with research agencies and have a higher concentration of graduate enrolments for knowledge production and advancement. The centres of excellence model is key to delivering the aspects of Africa's human-resource development agenda concerned with producing adequately-trained staff, fostering capacity development, sharing expertise, promoting continuing education and training and creating cutting-edge scientific knowledge and outputs. In particular, centres of excellence play major roles in education and training for those who do not have the required facilities or resources to be self-sufficient. Graduate training should aim to solve some of the pertinent challenges in the current agricultural education system, which include (i) lack of alignment with the skills needed by the private sector, industry, nongovernmental organisations (NGOs), civil society organizations and government sectors, as well as current university programmes; (ii) inadequate focus on approaches geared to sustainable, inclusive growth (iii) inadequate opportunities for exposure to field research relevant to smallholder farmers; (iv) limited mobility - and hence intercultural exchange and integration - of staff and students across Africa; (v) agricultural education tending to be too theoretical, without a practical or specialized focus (e.g. on interdisciplinary programmes incorporating business and economics training) and (vi) curricula needing to be updated and adapted to the digital age.

The quality of programmes must not be jeopardized in the context of rapid expansion, both in enrolment numbers and diffusion across more institutions. Attention must be paid to human resources in terms of the availability of supervisors, institutions' physical infrastructure and procedures for quality assurance. The concentration of programmes in flagship universities and small numbers of centres of excellence serves to ensure high quality and creates thriving centres of specialization. Developing the necessary competencies required for Africa's development also requires attention to the entire education value chain - i.e. secondary and undergraduate study in STEM disciplines, and the linkage between TVET and university education. Disparities at lower education levels make it difficult to increase the number of PhD graduates in the short term. Moreover, enrolments are determined as much by demand as any supply-side interventions.

Differentiating Teaching and Research Universities

As noted by Lynam *et al.* (2013), the rate of growth in the number of both public and private universities, and increased competition among universities for high-quality, fee-paying students resulted in significant differentiation across universities in terms of their degree programmes and branding. Universities have undergone internal restructuring, in part to compete more effectively for private students whose tuition payments make up an increasing percentage of university budgets. This was being driven by the increasing size of the universities and the need to decentralize decision-making. Thus, many universities formed constituent colleges, which have much more autonomy, and where the principal assumed many of the responsibilities formerly held by the vice chancellor. At the same time, many larger universities began opening branch campuses in response to increasing competition from new universities in more rural areas, where a few offer agricultural courses at the BSc or equivalent level (Oanda Ogachi and Jowi, 2012). The increasing reach to more rural communities has improved rural–urban equity in access to a university education, and should be particularly important in increasing access to students interested in studying agriculture. Moreover, the decentralization to more rural areas should expand the potential for closer community outreach and research targeted to the needs of smallholders.

All universities should deliver on their respective roles so that the diverse needs of the economy and society can be addressed collectively. In differentiated systems, some institutions produce graduates with vocational and technical skills, others develop BSc-level graduates and professionals, while graduate and research-intensive institutions foster the development of master's and doctoral students, research outputs and technological innovations. Each part of the system has a specific and important role to play in a country's development. Across the world, the most successful economies have differentiated their HEIs. Efforts to differentiate Africa's HEIs - with some universities becoming research-intensive, and their resources being focused on graduate training and research - is crucial among several interlinked actions towards revitalizing African HEIs (Fonn *et al.*, 2018; Cloete *et al.*, 2018). Nevertheless, this differentiation process is only at an early stage.

One particularly compelling reason to differentiate research-intensive universities from those focused on undergraduate teaching is that Africa's population is set to double by 2050, creating continued demand for higher education. That, in turn, necessitates appropriate training of faculty with advanced degrees. It is this cohort who will be needed to staff the new universities and maintain high standards across the higher-education system. Research-intensive universities are the best place to train such people. A research university builds a culture of research which is the envy of other HEIs. Although a research university may offer both undergraduate and higher degrees, its main focus is research-based degrees and other forms of research to generate new knowledge. To this end, some universities have intentionally increased their intake of graduate students in order to increase research-based graduate training, while in other cases, some universities have been established to admit graduate students only.

Research chair programmes, whereby individual scientists lead research in relevant areas of interest, have been proposed or established in some universities (South Africa and Kenya) as strategic initiatives intended to attract talented researchers capable of (i) establishing research groups in specified thematic areas, (ii) enhancing graduate and postdoctoral research, (iii) building research capacity and (iv) fostering university collaborations with national, regional and global institutions, including industry (IAUP, 2010; NRF, 2018a,b). The research chairs concept (i) recognizes individual excellence, leadership and talent, (ii) fosters the reputation and visibility of talented individuals, (iii) catalyzes organizational expertise in a given area (in particular through the training of high-quality MSc and PhD graduates) and (iv) contributes, more broadly, to strengthening a national research system. Research chairs are a means of developing a research group and training a cadre of emerging researchers, while focusing this contribution around an individual researcher with a strong track record. The research chair and centre of excellence concepts go hand in hand in ensuring important leadership, while surrounding it with a critical mass of strong researchers and research programmes (see Ch. 6, this volume).

Current Funding and Research Patterns

Funding has been a primary constraint to the development of effective research capacity in TAE institutions. This derives from the colonial period, and the division of research between services in ministries of agriculture (MOAs) and the educational focus of the faculties of agriculture (see Ch. 3, this volume). This division persists today and is reflected in government budgets. In most African countries (with some exceptions), research is undertaken by NARIs supported through MOA budgets, and faculties must rely on quite limited research funding provided through ministries of education, where they exist. In most universities, the agricultural education and training model is based on the linkages of teaching, research and extension within a single faculty. This is patterned on the US land-grant model, promoted by the United States Agency for International Development, which supported projects across Africa in the 1960s and 1970s (Alemneh, 2014). It was not possible, however, to restructure ministerial budgets in line with this model. Consequently, even though these efforts did help to strengthen undergraduate education, the contributions to research and graduate training (Bunting *et al.*, 2014a,b; Hayward and Ncayiyana, 2014) and extension have been much weaker.

The general expectation for a university professor in most African universities is to allocate around 75% of their time to teaching and the other 25% between research and outreach/community service (Lynam *et al.*, 2013). The demand on teaching is heavy, and access to funding for research is limited. For most of the 1980s and 1990s, therefore, most university research was largely confined to on-station laboratories and field stations, allowing little direct interaction with farmers. Furthermore, most graduate training opportunities relied on overseas universities. The limited number of scholarships available to national universities rarely provided more than student tuition and living costs. In the absence of research grants, it was difficult to involve students in on-farm studies. Many universities face the problem of stagnant expertise in teaching and research. The new blood injected by graduate students provides an excellent way of stimulating new ideas, enthusiasm and different ways of doing business within a department.

In light of this, a number of private philanthropic foundations - notably the Rockefeller Foundation, Carnegie Corporation, Kellogg Foundation, Ford Foundation, MasterCard Foundation and Bill & Melinda Gates Foundation - made substantial contributions to the development of institutional research capacity based on intensified graduate training in African institutions. Between 2004 and 2018 (RUFORUM, 2018b), for example, these foundations collectively provided grants amounting to more than US$70 million to support graduate research and training through the Regional Universities Forum for Capacity Building in Agriculture (RUFORUM) network. This support provided fully-funded scholarships (i.e. tuition, research and living costs) to MSc and PhD students of agriculture in target countries. The longer-term objective was to develop a significant number of new, well-trained professionals to help intensify, scale-up and improve the quality of basic and applied research to improve the productivity of staple food crops. The emphasis on quality, the needs of the rural poor, building ownership and developing uptake pathways ensured that training and research grants left adequate 'footprints' at both university and community levels. Such external support made an important contribution to promoting and facilitating networks of specialization, funding postdoctorate fellowships, building strong and lasting connections to outreach programmes and encouraging the development of research 'stables' within the leading departments. These centres of leadership and the initiators of 'research stables' were fundamental to enhancing the quality and capacity of research and graduates from supported institutions.

Over the past four decades, Africa's TAE sector has gained substantial importance in terms of agricultural research capacity. The absolute number of agricultural researchers employed in TAE institutions (i.e. excluding those providing diploma and certificate, or equivalent, programmes) grew more than sevenfold, from approximately 500 full-time equivalents (FTEs) in 1981 to 4,210 in 2016. This represents an increase from 12% of all agricultural researchers (excluding the private-for-profit sector) to 36% (Fig. 8.1). This growth was driven by increased capacity within existing TAE institutions, as well as the establishment of many new universities, faculties, schools and institutes involved in agricultural research (see Ch. 5, this volume). Despite the high and increasing numbers of TAE institutions conducting agricultural research in sub-Saharan Africa (SSA), unlike in the NARIs, these researchers are distributed across a large number of entities, whose individual capacity, in terms of researcher numbers, largely remains small. Moreover, these regional trends mask large differences across countries (Fig. 8.2).

The prestige and better remuneration associated with the higher-education sector, coupled with the focus on teaching - which demands higher qualifications - generally translate to comparatively higher numbers of PhD-qualified researchers than are employed in the NARIs. In fact, close to three-quarters of all the FTE researchers with PhD degrees are employed in the higher-education sector (Fig. 8.3). Growth in the number of PhD-qualified researchers was higher in TAE institutions than in NARIs, but it was less effectively used to conduct research. These trends support the argument that TAE institutions should be further integrated into national AIS (which is discussed later in this chapter).

8.3 Developing High-Quality Graduate Programmes with a Research Focus

A defining characteristic of superior TAE institutions is their ability to create synergy between the production (research) and dissemination (teaching, learning, and outreach) of knowledge. The strategic plans of most African universities since the 2010s reflect a shift towards innovative research and knowledge-creation relevant for solving societal problems (Dampson and Edwards, 2019). Deepening the capacity of agricultural faculties to conduct research is essential to developing effective graduate programmes. In the context of developing smallholder production, such programmes should be structured to achieve greater balance and synergy between the research they conduct and the curricula they offer. Many

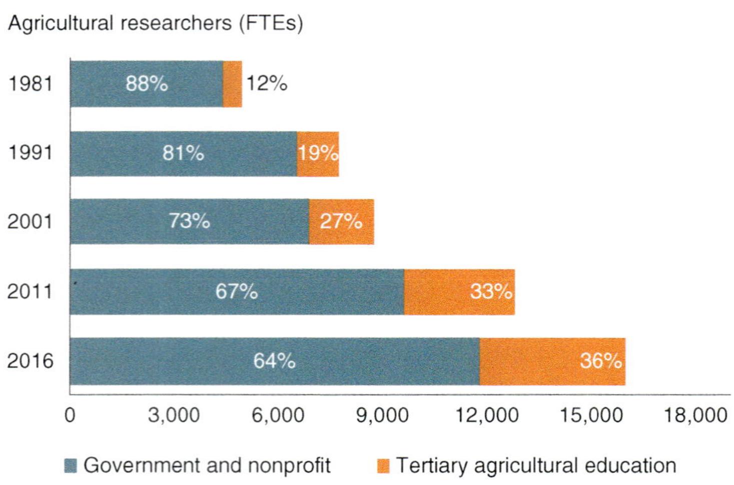

Fig. 8.1. Agricultural researchers by sector, 1981–2016
Sources: Constructed by authors based on ASTI and RUFORUM (2018) and ASTI 2019.
Notes: Data are for a sample of 39 SSA countries and indicate numbers and shares of agricultural researchers measured in full-time equivalents, which only take into account the time spent on research (as opposed to teaching and other tasks). Data for researchers employed in the government agencies only include degree-qualified scientists officially classified as researchers (as opposed to research technicians).

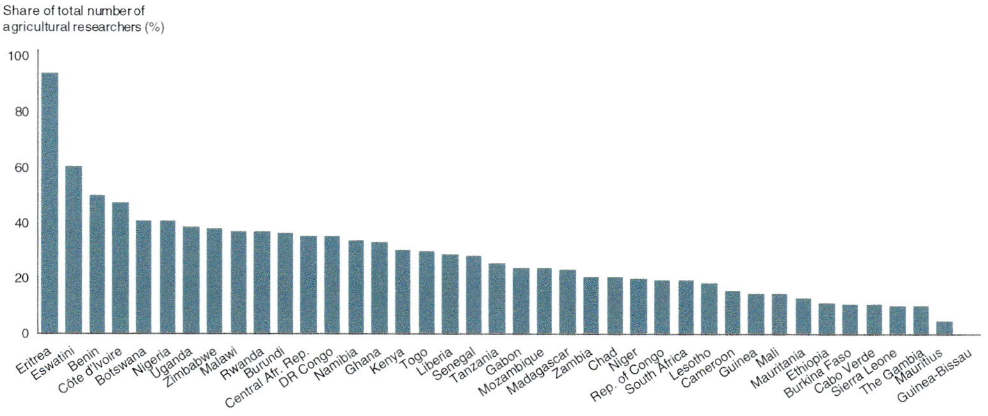

Fig. 8.2. TAE share of agricultural researchers by country, 2016
Sources: Constructed by authors based on ASTI and RUFORUM (2018) and ASTI (2019).
Notes: TAE = tertiary agricultural education. Data indicate the TAE share of each country's agricultural researchers measured in full-time equivalents, which only take into account the time spent on research (as opposed to teaching and other tasks). Data for researchers employed in the government agencies only include degree-qualified scientists officially classified as researchers (as opposed to research technicians).

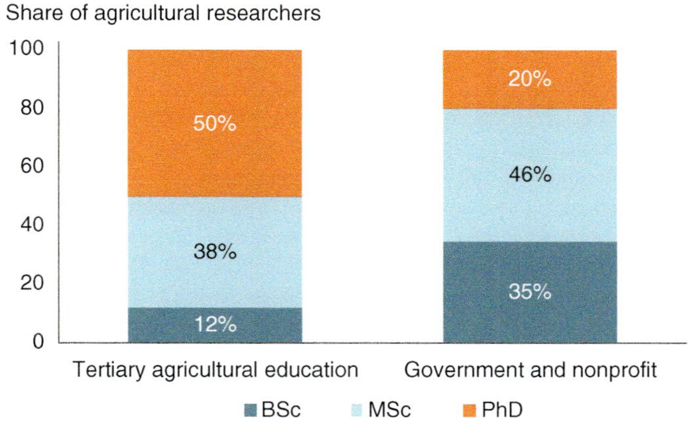

Fig. 8.3. Share of agricultural researchers by sector and qualification level, 2016
Sources: Constructed by authors based on ASTI and RUFORUM (2018) and ASTI 2019.
Notes: Data are for a sample of 39 SSA countries and indicate average shares of agricultural researchers measured in full-time equivalents, which only take into account the time spent on research (as opposed to teaching and other tasks). Data for researchers employed in the government agencies only include degree-qualified scientists officially classified as researchers (as opposed to research technicians).

African universities are responding to this demand by emphasizing experiential learning, community engagement and action research.

Strengthening Research Capacity

In general, the research undertaken by faculties of agriculture is primarily determined by a minority of faculty members who have the motivation and drive to compete for project funding to support both their research interests and the work of a few graduate students. A more strategic approach is needed in order to develop disciplinary depth beyond these ad hoc approaches. The principal (though not mutually-exclusive) pathways for defining such strategic approaches include: (i) centres of excellence, (ii) regional research networks and (iii) partnerships with universities in the Global North. The first of these pathways, as discussed above, involves building research capacity around a small number of disciplines. The crop-breeding programmes at Makerere University and the University of Ghana, for example, support the development of both high-quality graduate programmes and an evolving research agenda. Centres of excellence, using targeted eligibility criteria, can attract candidates from the local region, as the World Bank's programmes have done.

This regional focus is fundamental to the second approach and was used by international centres in the 1980s and 1990s. Their programmes were primarily organized around commodities, with plant breeding as a central organizing principle. In faculties of agriculture, such networks would be organized by discipline - e.g. with the African Economic Research Consortium, which focuses on economics and agricultural economics. In Africa, many disciplines are represented by professional associations that run conferences, e.g. in soil science and pathology, but research networks would differ in that they would be organized around specific research themes and incorporate the participation of graduate students. This kind of regional network approach could build on existing networks, such as the sub-regional organizations (SROs)[1], which - together with the Forum for Agricultural Research in Africa (FARA) - largely began as networks of NARIs. They have supported discussion, shared learning, joint research and regional priority-setting among researchers and research organizations working on commodity-specific issues. This approach could serve to strengthen and integrate collaborative research between the universities and NARIs, a theme which is further explored later in this chapter.

Examples of the third pathway include collaborations focusing on plant breeding between the University of Ghana and Cornell University, and collaborations in the area of rural innovation among the University of Wageningen (The Netherlands), Makerere University (Uganda), Sokoine University of Agriculture (Tanzania) and Egerton University (Kenya). These linkages help strengthen graduate-level curricula, along with the research methods and supervision conducted by graduate students, particularly those undertaking PhD degrees. Nevertheless, all of these pathways depend on adequate funding, which is discussed below.

Ensuring a Balanced, Multidisciplinary Research Focus

High-quality graduate programmes are built on effective research capacity, but this is currently not possible across all agricultural departments. Faculties must balance deepening the capacity of existing departments - where they may have some competitive advantage - and diversifying into new areas where there are markets for graduates. Research skills in crop and animal science are still in demand to support agricultural innovation, but as African agriculture becomes increasingly more commercial and responsive to the impacts of climate change, more specialized areas - such as natural resources, forestry and agroforestry, agricultural economics, rural innovation and aquaculture - will become more prevalent. Given the broad scope covered by the agricultural sciences, some specialization across faculties is appropriate, but not at the expense of potential multidisciplinary approaches to the challenges of smallholder development.

Crops largely remain the dominant focus of scientific research, as is evidenced by grants competitively awarded by RUFORUM to member universities during 2004–2018 (RUFORUM, 2018b) and bibliometric indicators (AOSTI, 2013; ECOWAS, 2017). Even where departments are more diversified in terms of staffing, crop science still dominates research activities. As of 2016, on average, 46% of the research conducted in faculties of agriculture focused on crop sciences, and 18% focused on animal sciences (Fig. 8.4). These data are for the more established agricultural departments, which would have greater success in competing for research funding; graduate programmes in more emerging disciplines will therefore require more strategic investments.

Using the frequency of citations in scientific literature as an indicator of the standard of research capacity, Maassen (2015) reports that the departments of African flagship universities whose work is most cited have the following characteristics: (i) a large number of research staff, (ii) a diversified portfolio of research and funding, (iii) multiple research bases across the

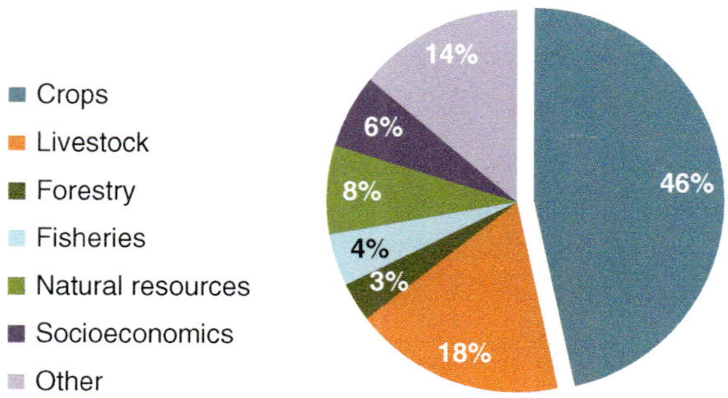

Fig. 8.4. Focus of research conducted in TAE institutions by commodity area, 2016
Sources: Constructed by authors from ASTI and RUFORUM (2018) and ASTI 2019.
Notes: TAE = tertiary agricultural education. Data are for a sample of 39 SSA countries and indicate the average share of TAE researchers measured in full-time equivalents, which only take into account the time spent on research (as opposed to teaching and other tasks).

country and (iv) well-established and productive partnerships with foreign institutions. These factors all go beyond the characteristics of individual researchers, which - once again - supports the argument for a more strategic approach to developing research capacity. Nevertheless, activities supporting the recruitment, development and motivation of researchers are critical to improving the quality of research programmes. Recruiting and retaining outstanding researchers is one of several strategies which increase research productivity - alongside setting specific goals and monitoring progress using specific external measures (Musiige, 2014; Sreeramana, 2016; Muhamad *et al.*, 2019). Finding, motivating and retaining talent were also among a number of drivers of research excellence and commercial performance, including strong leadership and strong links between theory and practice (Olsson and Meek, 2013). The most effective strategies for attracting staff included: (i) offering highly-competitive salaries, (ii) instituting proactive selection and recruitment initiatives and (iii) allocating funding for internal staff development and research (Horwitz *et al.*, 2003; Musiige, 2014).

Recognizing the Central Role of Funding

Funding for research in faculties of agriculture essentially depends on grants to individual researchers. For example, in 2013 almost 80% of research funding at Makerere University was derived from donor agencies (Musiige and Maassen, 2015). This situation creates something of a paradox in that universities and faculties have little influence on the direction of the research conducted, yet attaining research funding depends on the individual researcher's success in competing for grants. Musiige and Maasen (2015, p. 125) frame the issue as 'a loose coupling between the institutional research ambitions and strategies, and individual academic staff members' engagement in research activities'. To improve that coupling requires funding to shift the focus from individual researchers to disciplinary departments. This is the strategy of the World Bank's support for centres of excellence and also of some bilateral donors, often supported by a linkage with an agricultural faculty in the donor country.

Moving to departmental funding, however, does not negate the importance of competitive grant funding to individual faculty members. Competitive grants generally aim to enhance the quality of research through a competitive, peer-review process, which also supports graduate students in obtaining their degrees and gaining field experience. Proposal guidelines for most grants targeting agriculture and rural development call for research to improve the well-being of smallholder farmers and rural communities, with as much engagement with them as possible. This is intended to enhance the relevance of the universities' research. To foster more collaborative and synergistic relationships with other key actors, projects are encouraged to engage with other partners, such as government extension services, the private sector, NGOs and researchers from other disciplines. It is important to nurture researchers whose ideas have scientific merit but who may need help with conceptualizing and writing proposals, e.g. through targeted workshops building the skillset required. Feedback on unsuccessful proposals is also highly important. In this way, the pool of talented scientists can be retained, developed and - ultimately - expanded.

In this way, prestigious external research funding supports and stimulates the development of research capacity within universities. Success in competing for external research funds also directly correlates with academic prestige, intra-university autonomy and high research productivity (Maassen, 2015). Improvements in research productivity are particularly important because they underscore the value of collaborative arrangements. This is demonstrated, for example, through joint publications with researchers in the Global North, which indicate greater African integration within international research networks and enhanced opportunities to leverage research funding. The increasing prevalence of these collaborations contributes to universities research capacity and highlights their greater contribution to national development.

Regional Approaches to Developing Graduate Students' Skills

Rapidly-rising costs make the pursuit of graduate degrees from Northern universities increasingly infeasible for Africans, which has led SSA to assess alternative strategies, including: (i) increasing training opportunities in the South, (ii) undertaking sandwich programmes with Northern universities and (iii) improving the quality of PhD programmes within African universities (Lynam *et al.*, 2013). Graduate study is typically offered through course- and thesis-based programmes, both of which include specified disciplines, coursework and subject matter. Coursework is necessary to provide students with current knowledge in their chosen discipline, an understanding of the relevant tools to apply to their research and information on relevant thesis topics. Students must graduate with core competencies appropriate for a rapidly-changing world. This includes the capacity and confidence to be creative, cooperative and entrepreneurial, and to adapt and commit to developing their communities.

Additionally, an interactive environment supports students in formulating ideas, posing scientific questions, discussing methodological issues, exchanging ideas with colleagues and teachers and participating in a broader development community, such as through conferences and seminars. To ensure the quality of course-based programmes, funding is essential to support wider participation in graduate programmes and strong quantitative research methods. PhD programmes require a higher order of design and oversight than is required for MSc programmes. Faculties themselves need to be retooled to support the development of their programmes and their graduates. A strong focus on leadership is needed across all levels, as is building an understanding of the value of organizational change to facilitate learning, freeing staff and students to attain knowledge through a self-driven, lifelong approach to learning (see Ch. 6, this volume).

Universities need to pursue an integrated agenda, significantly expanding and strengthening regional programmes in research and learning, and allowing much more regional sharing of resources. This represents an expansion of what RUFORUM and others have already been doing. RUFORUM's collaborative PhD programmes are typically structured on the model of coursework, which requires completion of a certain number of academic courses in addition to independent scientific research culminating in an oral defence of a thesis. Implementation is linked to a series of competitively-awarded grants, either from RUFORUM or other development partners. Universities and faculties across Africa vary considerably in their respective capacities and needs. This creates different incentives and, thus, benefit streams from participation in competitive grant processes. The stronger, elite faculties are more successful in competing for grants. They also have more capacity to mount regional degree programmes, which require higher resource commitments, but offer the benefit of increased enrolment numbers. The smaller and more recently-established faculties also have the opportunity to expand their range of courses, offer higher degree training to their staff and develop innovative new courses to meet workplace demand. While the more elite faculties would be required to invest staff and resources in this kind of regional effort in order to achieve scale economies, potential exists for all universities to mutually benefit.

Student guidance is either provided by individual supervisors or, in some cases, joint supervisory committees. Different models of graduate supervision - especially of theses or dissertations - have been adopted, including: (i) the traditional 'one student-one supervisor', face-to-face interaction, (ii) reflective practice, whereby supervisors act as mentors, enabling students be more critical, creative and innovative in their research (Zhao, 2003), (iii) the collaborative cohort model, which requires students to develop greater skills through interaction and peer review by fellow students (Burnett, 1999) and (iv) the digital information-management system, which aims to help research managers, support staff, supervisors and students perform research-related activities online (Duffy, 2000; Uzoamaka and Olayiwola, 2012).

The inter-disciplinarity or thematic-framing of most programmes, in part driven by a need to pool resources for supervision, research training, general research infrastructure and research equipment is crucial to significantly expand programmes and generate socially-relevant, innovative research output. Graduates are able to utilize their skills in the various fields to address existing and emerging agricultural development challenges. Linking training and learning to working with communities is instrumental in: (i) swiftly and efficiently delivering university-generated knowledge and outputs to farmers' fields, (ii) facilitating the mutually-beneficial exchange of knowledge and resources through partnerships and reciprocity, (iii) enriching scholarship and research, (iv) enhancing curricula, teaching and learning and (v) producing well-educated, engaged graduates prepared to address critical societal issues and contribute to the public good.

E-learning enables students, staff and researchers to access high-quality, up-to-date learning materials at no cost. Old paradigms requiring students to be physically present in the classroom are being enriched with innovative pedagogic methods and remote learning, including: (i) blended learning (combining in-person and virtual methodologies), (ii) mobile learning (allowing students to work from devices such as tablets, notebooks and smart phones and (iii) 'flipped classrooms' (whereby students develop videos for fellow students as a means of improving their comprehension of a certain topic).

Ensuring the Relevance of Research Methods and Quantitative Skills

Effective research practice requires that knowledge and skills are substantively and methodologically sound. A researcher must have a deep understanding of the substantive topic being researched, be technically skilled and employ methodologically-appropriate techniques. The importance of statistics and research methods in empowering the agricultural research process and sharpening interventions cannot be over-emphasized. In a desk study involving research articles published in the *African Crop Science Journal* between 2000 and 2017, Odong *et al.* (2019) revealed that statistical ineptitude was partly the reason for the high rejection rate of up to 90% of manuscripts submitted. Although the majority of authors in the 165 published articles succeeded in describing how the studied variables were measured, key weaknesses related to poor choice and description of experimental design, data processing, data analysis and presentation of results. As pointed out by Kramer *et al.* (2016), the possible reasons for these weaknesses include: (i) researchers being unable to cope with changes in both agricultural and statistical sciences, (ii) inadequate interaction between researchers and statisticians/biometricians, (iii) inadequate teaching/learning of statistics/biometry at the universities and (iv) weakness in the journal-review process.

RUFORUM launched an MSc programme in research methods, hosted by Jomo Kenyatta University of Agriculture, which focuses on training in research methods not generally taught in the universities offering agricultural programmes. Research methods, including statistics and biometry, underpin effective applied science, yet the number of professionals in these areas has declined over time. This is a major gap in most African national agricultural research systems (NARS), including universities. The RUFORUM programme is enhanced by having linkages with research systems and experts across SSA. The programme focuses on developing the skills and aptitudes needed to support research, orient students to think creatively and value teamwork/ partnerships and prepare them to step into leadership roles upon completion of their studies.

African and International Linkages

African countries depend on international collaboration and research grants for their research output (World Bank and Elsevier, 2014; Blom *et al.*, 2016; ABC Agri-Tech Research Programme, 2016; Mouton *et al.*, 2019). In 2012, for example, of all the research output of Southern Africa, East Africa, and West and Central Africa, 79%, 70% and 45%, respectively, were produced through international collaborations. In contrast, collaboration among researchers in SSA ranged from 0.9% in West and Central Africa to 2.9% in Southern Africa. AOSTI (2013) reported an average of 4% collaboration among researchers in Africa. Between 2005 and 2013, 36.5% of West Africa's publications were produced in collaboration with Economic Community of West African States (ECOWAS) members and other countries, only 9.5% were produced through collaborations with ECOWAS exclusively, and 27.7% involved collaborations outside of Africa (ECOWAS, 2017).

Mouton, *et al.* (2019) reported that the three countries with the largest increases in international collaboration during 2000–2015 were Egypt (27–57%), South Africa (34–52%) and Botswana (32–77%). Tanzania, Kenya, Uganda, Senegal, Ethiopia and Tunisia also increased their levels of international collaboration substantially. The impact of papers authored by African researchers - as measured by the mean normalized citation score - also rose steadily for a number of countries, with Botswana, Uganda, Kenya, Ethiopia and Senegal standing out in 2015. The well-established correlation between citations and the intensity of international collaboration is the most plausible explanation for these improvements.[2] Collaboration enhances the quality of scientific research, improves the efficiency and effectiveness of that research and is increasingly necessary as the scale of research challenges and budgets grows.

The primary driver of most collaboration is the scientists themselves. In developing their research, scientists seek to work with the best people, institutions and equipment (Royal Society, 2011). Their formal and informal connections to people, diasporas, virtual networks and communities of professionals with shared interests are invaluable drivers of international collaboration. Motivated through bottom-up exchange of scientific insight, knowledge and skills, scientists are changing the focus of the research they undertake from the national to the global level. Collaboration brings significant benefits - both measurable, such as the increased impact of citations and access to new markets, and less easily quantifiable, such as broadening research horizons. The facilitation of collaboration, therefore, has a positive impact not only on the science conducted, but also on the broader objectives of any science system, be they enhancing domestic prosperity or addressing specific challenges.

The emergence of institutionalized African networks and consortia for research and higher education in the late 1980s and early 1990s is among the positive developments in enhancing Africa's research capacity and sustaining graduate research and training in African universities. These research and higher-education networks have served as focal points for partnerships and resources to support knowledge generation and dissemination. This is particularly the case because Africa's agricultural research and TAE capacity is fragmented across some 54 countries, which differ greatly in size, focus, competency and national coherence. These networks and consortia have become popular with agricultural research scientists, academia, administrators and donors as tools to strengthen academic and research capacity, and the relevance of African universities (Faris, 1991; Rockefeller Foundation, 1995; IAC, 2004a,b). The networks have played a significant role in (i) raising higher-education and research standards, (ii) generating economies of scale in the use of training facilities and resources, (iii) pioneering collaboration with new partners, (iv) sharing scarce resources through regional graduate degree programmes, (v) supporting collaborative multidisciplinary research programmes and (vi) promoting innovative community and industry engagement and the development of entrepreneurial approaches.

Partnerships between African institutions and universities and advanced research institutions have also been a strong form of capacity building for Africa. Nevertheless, several accounts point to a number of collaborations in which international partners had disproportionate influence over the activities of the African NARIs and universities (Eicher and Rukuni, 2003; IAC, 2004a; Oanda Ogachi, 2013). This undermined the ability of the African institutions to determine and act on their own priorities, and often weakened their ability to influence partners to focus on their countries' priorities. The Comprehensive African Agriculture Development Programme (CAADP) and the emergence of FARA and the SROs provided avenues for African research institutions to take dominion over their research agendas. CAADP has strengthened the technical soundness and African ownership of planning processes (for agriculture, generally, including research). This has somewhat reduced the extent to which research agendas are driven by external partners' project objectives. FARA and the SROs have provided strategic leadership on research strategies and priorities at the pan-African level, and have also designed and financed sub-regional research agendas formulated by the African NARIs themselves.

8.4 Agricultural Research within an Agricultural Innovation System

The Path to a Functional Agricultural Innovation System

African NARI and TAE configurations have been transforming over time since the 1970s to become more inclusive and pluralistic for the purpose of improving agricultural research and the delivery of training. Diversified NARS exist in each country, dominated by a NARI, which has a quasi-monopoly on research activities and mobilizes a regional average of about 50% of all the researcher capacity in the NARS (Anderson *et al.*, 2013; Beintema and Stads, 2014). Decentralization promoted vertical and horizontal linkages with other institutions, and CAADP has promoted more impact-driven approaches in the conduct of research and training. This AIS approach to research and training involves a network of diverse agents and their interactions, along with the institutions, organizations and policies which condition their behaviour and performance with respect to generating, exchanging and using knowledge. AIS require the development of new capacities and skillsets, reviewing internal organization and expanding field-level operating capacity (Spielman, 2006; Swanepoel *et al.*, 2014). Tertiary education is increasingly being set within an innovation systems framework, both generally (World Bank, 2008a) and for HEIs in particular (Maguire, 2012). Placing higher agricultural education in an AIS attempts to redress:

> . . . the declining quality of many [TAE] systems, the general failure to articulate a strong policy framework for agricultural education, and the corresponding failure of educational institutions to build a strong and vocal constituency within the agricultural sector. . . (Maguire, 2012, p. 113).

Reform of TAE institutions attempts to ensure that curricula meet the demand for graduates, that linkages to other actors in the agricultural sector exist and that competencies meet the needs of the workplace (Lynam *et al.*, 2013). However, AIS have yet to reflect significant changes in financing research, which still constrains the development of research capacity. Although some countries, such as Kenya, have instituted competitive grant programmes for agricultural research, this does not support institutional development of research capacity.

The Impact of the Comprehensive African Agriculture Programme

CAADP provides (political, as well as technical) leadership to Africa's agricultural programmes at every level. CAADP'S Agricultural Education and Skills Improvement Framework (AESIF), 2015–2025, emphasizes the importance of developing human-resource capacity in agriculture, as well as suggesting approaches to achieving this goal (NEPAD, 2015). The Framework for African Agricultural Productivity (FAAP) also emphasizes the development of human-resource capacity in African agriculture and provides general principles for designing agricultural education programmes and institutions (FARA, 2006). Nevertheless, neither AESIF nor FAAP offer the level of guidance CAADP has been able to provide other agricultural policy, programmatic and institutional topics. This may in part be because CAADP has primarily focussed on topics managed by MOAs, whereas most universities draw their funding and leadership from ministries of education (see Ch. 3, this volume). Expanding CAADP's attention and guidance to agricultural education - especially at the tertiary level - is an important priority. Otherwise, agricultural education may fall through the cracks, as it has done in the past, because neither the agriculture nor the education ministries see it as a core area of responsibility.

Differentiating Tertiary Agricultural Education Institutions and National Agricultural Research Institutes

As previously discussed, the TAE sector employs about three-quarters of Africa's PhD-qualified agricultural scientists. This share has been trending upwards for some time because more universities have made it a prerequisite, positions at universities have greater prestige and remuneration is generally better at universities than at NARIs. As a result, many of the most talented and accomplished researchers are found in TAE institutions. Despite this, however, the agricultural research and output conducted within Africa's universities - and, importantly, students' exposure to research processes - remains modest. There are many reasons for this, but primarily: (i) teaching loads are demanding, leaving most faculty members with little time to devote to research, (ii) research facilities and equipment in most TAE institutions are limited and (iii) the financial resources available for research within universities are meagre. Consequently, research and outputs at the universities continue to fall well short of their potential. Meanwhile, NARIs tend to have more resources for research, in terms of facilities and equipment, based on their research mandates, external partnerships (e.g. with international research institutions and universities) and dedicated - if still limited - external financing for research (e.g. from development banks and donors).

As discussed above, the lack of integration between the research undertaken by TAE institutions and NARIs has its historical roots in the divide in oversight by ministries of education on the one hand, and MOAs on the other. Examples of cross-linkages do exist, but this longstanding dual-track system in most, if not all, African countries represents inefficient use of both available national agricultural research capacity and available research funding and infrastructure. Greater interaction and coordination between TAE institutions and NARIs therefore deserves strong consideration. As an interim measure, most countries would benefit from developing and implementing a strategic template for jointly-assessing and planning research priorities, including promoting complementarity in the research undertaken at NARIs and within TAE institutions.

Faculties of agriculture naturally tend to emphasize more basic research, exploring the frontiers of their disciplines with a focus on the African context. NARIs, on the other hand, tend to address applied research themes and agroecologies focusing on commodities or natural resource management. NARIs also tend to have experiment stations located across a country's principal agroecologies, making it possible to conduct research relevant to those zones. A significant advantage for TAE institutions is the contribution of their graduate students, who are available and committed to completing research in order to obtain their degrees. This complementarity provides a clear rationale for building effective linkages between the two sectors, as opposed to the more common competition for funding and human resources. Such linkages could - and should - be established and nurtured through a comprehensive joint agricultural research agenda focusing on sustainably increasing the productivity and commercialization of smallholder agriculture. Such an agenda would also stipulate the related vision, priorities, workplans and budgets of both the TAE institutions and the NARIs.

Smallholder agriculture is threatened by (i) declining soil fertility, (ii) climate change and the increasing intensity of weather events, (iii) associated pest and disease attacks, including desert locust swarms and (iv) increasing human health issues, such as from aflatoxins. At the same time, with an increasingly integrated market economy, smallholder agriculture has the potential to increase marketable surpluses. Given the diversity of smallholder agriculture, technical solutions not only require disciplinary depth, but also multidisciplinary approaches, as exemplified in the joint research currently carried out with international universities or research centres. The viability of Africa's agricultural research will require broadening the scientific base (i) to include fundamental sciences within and outside of the colleges of agriculture, (ii) to enhance NARIs' ability to conduct significant research on biotechnology and emerging disciplines and (iii) to focus on the commercialization of research outputs and resulting knowledge. Advances in conventional technology will remain the primary source of growth in crop and animal production but will be inadequate to sustain future demands on agriculture. The physiological constraints to continued improvement along conventional paths are severe. The transfer of research and technology through information management technology is increasingly becoming an important source of crop and animal productivity growth.

Collaboration in these areas of research between TAE institutions and NARIs is both possible and highly desirable. Such collaboration would also generate improvements in graduate research and degree programmes by: (i) expanding the research possibilities open to graduate students, (ii) taking advantage of NARI scientists' experience and applied perspective through joint student supervision and (iii) developing students' skills and experience of working within multidisciplinary teams. Such a learning environment would allow students to develop what are termed 'T' skills - the combination of experiencing disciplinary depth through field research and participation in research teams focusing on smallholder systems. The students bring their disciplinary specialization to research problems, thereby developing an understanding of how a disciplinary issue is affected by a systems' constraints to improving productivity. Bringing new skills, methods and insights to research programmes in NARIs improves the resulting research, the graduate programmes, the graduates of those programmes and, ultimately, the overall AIS.

Rationalizing Funding of Agricultural Research

African countries have under-invested in their agricultural sectors both generally and in terms of agricultural research (Lynam, *et al.*, 2016). Although average agricultural research expenditure increased for Africa as a whole during 2000–2016, total spending averaged only 0.39% of agricultural GDP in 2016 (ASTI, 2019). This under-investment can partly be explained by (i) the lack of visibility of the benefits to agricultural research, (ii) the time lag required from the point of the investment to the point of reaping its benefits and (iii) the lower political weight given to agriculture in government budgets (Benin *et al.*, 2016). Yet, the development of a functional AIS requires minimum levels of funding. Many countries have introduced competitive grant programmes to support research in public and private universities, research agencies and other private entities. Nonetheless, if competitive grants are to make more than a marginal impact on agricultural research capacity, the core support for research in government budgets needs to be increased. The tendency also exists for the more elite universities to dominate competitive grant programmess, edging out the potential for more balanced funding allocation to build research capacity across the TAE sector. The fact that the majority of PhD-qualified agricultural scientists are employed in the university sector supports the argument for expanding MOA budgetary processes to include core support for TAE research capacity. Since the TAE institutes train the next generation of agricultural scientists, this would also contribute to agricultural research as a whole, including financially. Flexibility needs to be built into the budgeting process to address the dynamic nature of the agricultural research landscape beyond the predominate short-term, incremental approaches.

Building Linkages with Extension Services and Community Development

Another central component of organizational linkages within an AIS is downstream actors, including extension, NGOs, farmer organizations, private businesses and communities. The TAE sector does not have the capacity for widespread involvement in disseminating technologies and information at the field level, yet TAE institutions could play an important role in supporting agricultural extension and other community-level programmes. Universities could expand their research programmes to include extension and community service, particularly with regard to developing and assessing methodologies (see Ch. 15, this volume). Perhaps even more importantly, universities could take a more integral role in building capacity within agricultural extension programmes and in updating the information being disseminated by extension and community-development agents.[3] This would include playing a valuable bridging role between the research function, undertaken by NARIs, universities, and others, and the agricultural extension role. These issues are particularly relevant after the collapse of training and visit extension in the 1990s (see Chs. 3 and 9, this volume), leaving a vacuum in the dissemination of technologies which has largely been filled by NGOs. This was also associated with the development of new extension methodologies, increasingly within the framework of AIS. Innovation platforms were a particular model for fostering interaction among actors, especially within the context of improving smallholders' access to and participation in markets. TAE institutions have the capacity to continue developing such methodologies, as well as bringing rigor to the evaluation of their effectiveness.

Extension education takes different forms: community-development strategies, agricultural extension and advisory services, communication strategies and rural innovation methods. Strengthening community engagement by TAE institutions is a common feature in the capacity-development and skills-assessment plans of many African countries, e.g. South Africa, Kenya, Tanzania and Uganda. In South Africa, for example the 1997 white paper on higher education mandated that universities be more responsive to society's needs, calling for a new relationship between TAE institutions and communities (Hall, 2010). Traditional and new funders of African agricultural research and education have openly stated clear expectation that NARIs and TAE institutions make a strategic commitment to public engagement, and that researchers and academics be recognized and valued for their involvement with communities. Importantly, several networks and alliances have emerged which are significantly or exclusively focused on partnerships between universities and the broader community.

The purpose of community engagement is to cross-pollinate the knowledge, information and resources of the TAE sector with those of the public and private sectors to (i) enrich scholarship, research and creative activity, (ii) enhance curricula, teaching and learning, (iii) prepare educated, engaged citizens, (iv) strengthen democratic values and civic responsibility, (v) address critical societal issues and (vi) contribute to the public good. The practice of community engagement has taken many forms and is incorporated into university curricula in multiple ways, as is outlined below.

1. ***Learning with the community.*** Under this approach, students and teachers apply their knowledge and skills to address the challenges of a chosen community to improve the lives of its people. Linking learning with service is commonly described as 'service-learning'. The basic objective is to put students' theoretical knowledge to practical use in the field. The regional training programmes implemented by RUFORUM member universities were designed with a strong community-based learning approach facilitated through graduate research grants, field-attachment fellowship awards and community-action research grants.
2. ***Researching with the community.*** Under this approach, various university and college faculties devise joint research projects in partnership with the community, such that the community's own knowledge is integrated into the design and conduct of the research. New research undertaken by students and teachers forms the basis of theses, dissertations and research papers, which complete academic requirements while formalizing the community's knowledge.
3. ***Knowledge sharing with the community.*** Under this mode of engagement, students' and teachers' knowledge in various disciplines is made available to the local community in support of its development goals, while at the same time securing its rights and entitlements from various public and private agencies. These can take the form of enumerations, surveys, camps, trainings, learning manuals/films, maps, study reports, public hearings, policy briefs, engagement with urban homeless shelters, teaching and health services in poor communities, legal aid clinics, etc. The intention of such initiatives is to disseminate

valuable and practical knowledge to communities for use in improving their well-being.

4. *Devising new curricula and courses.* This form of engagement provides for consultations with local students, communities and community-based organizations. The resulting information is used by universities and colleges to design curricula and courses in response to specific community needs. These can take the form of short-term workshops, certificates and even degrees, which enhance existing curricula with locally-relevant content and facilitate the creation of new educational programmes to meet the needs and interests of the new generation of students. Such courses augment learners' theoretical knowledge with practical community experience.

5. *Including practitioners as teachers.* Local community elders, female leaders, indigenous peoples and civil society practitioners have enormous practical knowledge to share across a wide variety of issues. Inviting such practitioners to contribute to courses, both in the classroom and in the field, allows this valuable expertise to be captured. Such instructors should be duly recognized, compensated and respected for their knowledge.

6. *Social innovation by students.* Student-initiated learning projects - in consultation with student unions, associations and clubs - have a valuable social impact and can facilitate meaningful linkages with the community.

8.5 Conclusion

A strong argument can be made for rationalizing and integrating agricultural research capacity between TAE institutions and NARIs across African countries. TAE institutions employ the majority of PhD-qualified agricultural scientists - a shift which has occurred over the past couple of decades without a commensurate increase in research funding and capacity. This imbalance is now more pressing because TAE institutions have the responsibility to train the next generation of agricultural scientists to the doctoral level, and at the same time ensure the quality and diversity of that training. Research capacity is absolutely necessary to guarantee that quality and to meet the increasing demand for graduates. The status quo will cause the effectiveness of African agricultural research capacity to decline overall, at a time when the smallholder market economy is poised to increase its productivity through the adoption of new technologies. Correcting this imbalance will require fundamental changes to the funding of agricultural research and the division of labour between TAE institutes and NARIs, along with a restructuring of agricultural PhD programmes.

A more strategic approach is needed across TAE institutes to develop the dual mandates of effective research capacity and graduate programmes. This will entail the diversification of TAE institutions, involving the emergence of centres of excellence in graduate programmes, reinforced by national and regional networks across faculties, supported by linkages to universities in the Global North. Whether such centres emerge in a competitive TAE 'market' or are coordinated by a body - akin to the national science and technology agencies - will depend on the direction taken by such market forces. Such diversification will be reinforced by changes in research funding and will require a move towards (i) course-based PhD programmes, (ii) increased class sizes to reduce per student costs and (iii) stronger research methods and quantitative skills, often within a regional framework. Strengthening agricultural research capacity within African countries will facilitate the region's evolving research agenda, led by African researchers with the training, skills and experience to participate as equals in global research networks and international forums.

Endnotes

[1] The SROs are the Association for Strengthening Agricultural Research in Eastern and Central Africa, the Centre for Coordination of Agricultural Research and Development for Southern Africa, the West and Central African Council for Agricultural Research and Development and the Northern Africa sub-Regional Research Organisation.

[2] Various bibliometric studies have shown that citation impact increases as international collaboration increases (Mouton *et al.*, 2019).
[3] One of the systemic challenges which has hampered the effectiveness of agricultural extension programmes across Africa (and many other parts of the world) is the difficulty of providing a constant stream of sound and practical information to agricultural extension agents in order to help them, in turn, to pass on relevant and valuable information to farmers.

References

ABC Agri-Tech Research Programme (2016) *Scoping Study: Africa-Britain-China (ABC) Agricultural Technology Research Collaborations for International Development*. Available at: https://assets.publishing.service.gov.uk/media/57a08957ed915d622c000187/61579_Africa-Britain-China-Scoping-Study-Report_Final-February2016.pdf (accessed 14 April 2021).

AfDB (African Development Bank) (2020) *African Economic Outlook 2020: Developing Africa's Workforce for the Future*. Available at: https://au.int/sites/default/files/documents/38116-doc-african_economic_outlook_2020_.pdf (accessed 25 May 2021).

AfDB/OECD/UNDP/UNECA (African Development Bank, Organisation for Economic Co-operation and Development, United Nations Development Programme and United Nations Economic Commission for Africa) (2012) *African Economic Outlook 2012: Special Theme: Promoting Youth Employment*. Available at: https://www.afdb.org/en/documents/document/african-economic-outlook-aeo-2012-99865 (accessed 25 May 2021).

Alemneh, T. (2014) Agricultural higher education in sub-Saharan Africa: Partnerships and the land-grant model. Ch. 5 in Swanepoel, F. Stroebel, A. and Ofir, Z. (eds.) *Towards Impact and Resilience: Transformative Change In and through Agricultural Education and Training in Sub-Saharan AJC&PABfrica*. Cambridge Scholars Publishing, Newcastle upon Tyne, UK.

Anderson, J. Roseboom, J and Weidemann Associates, Inc. (2013) *Towards USAID re-engaging in supporting national agricultural research systems in the developing world*. Available at: https://www.agrilinks.org/sites/default/files/resource/files/NARS%20Issues%20Paper-Lit%20Rev%20revised-final.pdf (accessed 25 May 2021).

AOSTI (African Observatory of Science, Technology and Innovation) (2013) *Assessment of Scientific Production (2005–2010) in the African Union*. African Innovation Outlook, Bibliometric Series 1. AOSTI, Equatorial Guinea.

ASTI and RUFORUM (2018) Unpublished survey data. International Food Policy Research Institute, Washington, DC, and Regional Universities Forum for Capacity Building in Agriculture, Kampala, Uganda.

ASTI (2019) ASTI database. International Food Policy Research Institute, Washington, DC. Available at: https://www.asti.cgiar.org/data/overview (accessed 24 March 2021).

Beintema, N. and Stads, G.-J. (2014) *Taking Stock of National Agricultural R&D Capacity in Africa South of the Sahara*. ASTI Synthesis Report. ASTI, International Food Policy Research Institute, Washington, DC. Available at: https://www.asti.cgiar.org/pdf/AfricaRegionalReport2014.pdf (accessed 15 June 2021).

Benin, S. McBride, L. and Mogues, T. (2016) Why do African countries underinvest in agricultural R&D? Ch. 5 in Lynam, J. Beintema, N. Roseboom, J. and Badiane, O. (eds.) *Agricultural Research in Africa: Investing in Future Harvests*. International Food Policy Research Institute, Washington, DC. Available at: https://www.ifpri.org/publication/agricultural-research-africa-investing-future-harvests (accessed 14 April 2021).

Blom, A., Lan, G. and Adil, M. (eds) (2016) *Sub-Saharan African Science, Technology, Engineering, and Mathematics Research: A Decade of Development*. World Bank, Washington, DC.

Bunting, I., Cloete, N. and van Schalkwyk, F. (2014a) *An Empirical Overview of Eight Flagship Universities in Africa: 2001–2011*. Centre for Higher Education Trust, Cape Town. Available at: https://doi.org/10.6084/M9.FIGSHARE.1150442 (accessed 14 April 2021).

Bunting, I. Cloete, N. and Van Schalkwyk, F. (2014b) *Data on eight flagship universities in Africa 2009–2011*. Available at https://dataverse.harvard.edu/dataset.xhtml?persistentId=doi:10.7910/DVN/25427 (accessed 18 March 2021).

Burnett, C.P. (1999) The supervision of doctoral dissertations using a collaborative cohort model. *Counsellor Education and Supervision* 39(1), 46–51.

Cloete, N., Bunting, I. and van Schalkwyk, F. (2018) *Research Universities in Africa*. African Minds, Cape Town.

Dampson, D.G. and Edwards, A.K. (2019) Strategic thinking, leadership and governance for African universities: Lessons from successful universities. In Emeagwali, O.L. (ed.) *Strategic Management: A Dynamic View*. Available at: https://doi.org/10.5772/intechopen.81196 (accessed 25 May 2021).

Duffy, J. (2000) Knowledge management: To be or not to be? *Information Management Journal* 34(1), 64–67.

ECOWAS (Economic Community of West African States) (2017) *Scientific Research in West Africa: Bibliometric Outlook 2016*. ECOWAS, Abuja, Nigeria.

Eicher, C.K. and Rukuni, M. (2003) *The CGIAR at 31: An Independent Meta Evaluation of the Consultative Group on International Agricultural Research*. Thematic Working Paper: The CGIAR in Africa: Past, Present, and Future. World Bank, Washington, DC. Available at: https://ieg.worldbankgroup.org/sites/default/files/Data/reports/cgiar_wp_eich_ruki.pdf (accessed 14 April 2021).

FARA (Forum for Agricultural Research in Africa) (2006) *Framework for African Agricultural Productivity*. Available at: https://library.faraafrica.org/wp-content/uploads/2019/10/FAAP_English.pdf (accessed 20 April 2021).

Faris, D.G. (1991) *Agricultural Research Networks as Development Tools: Views of a Network Coordinator*. International Crops Research Institute for the Semi-Arid Tropics, Hyderabad, India. Available at: http://oar.icrisat.org/967/ (accessed 20 April 2021).

Fonn, S. Ayiro, L. Cotton, P. Habib, A. Mbithi, P. *et al.* (2018) Repositioning Africa in global knowledge production. *The Lancet* 392(10153), 1163–1166. Available at: https://doi.org/10.1016/S0140-6736(18)31068-7 (accessed 13 June 2021).

Hall, M. (2010) Community engagement in South African higher education. In CHE (ed.) *Kagisano No. 6: Community Engagement in South African Higher Education*. Council for Higher Education, Pretoria.

Hayward, F.M. and Ncayiyana, D.J. (2014) Confronting the challenges of graduate education in sub-Saharan Africa and prospects for the future. *International Journal of Higher Education* 1(1), 173–216. Available at: https://ejournals.bc.edu/index.php/ijahe/article/view/5647 (accessed 18 March 2021).

Horwitz, F. Chan, H. and, Hesan, Q. (2003) Finders, keepers? Attracting, motivating and retaining knowledge workers. *Human Resource Management Journal* 13(4), 23–44.

IAC (InterAcademy Council) (2004a) *Realizing the Promise and Potential of African Agriculture: Science and Technology Strategies for Improving Agricultural Productivity and Food Security in Africa*. IAC, Amsterdam.

IAC (2004b) *Inventing a Better Future: A Strategy for Building Worldwide Capacities in Science and Technology*. IAC, Amsterdam.

IAUP (International Association of University Presidents) (2010) Academic Chairs for Africa: A New Approach to Capacity Building: Creating a Platform for Excellence in the Knowledge Society. A 21st Century Model. Funding Proposal prepared by the International Association of University Presidents for the G8.

Kramer, M.H. Paparozzi, E.T. and Stroup, W.W. (2016) Statistics in a horticultural journal: Problem and solutions. *Journal of American Society of Horticultural Sciences* 141(5), 400–406. Available at: https://www.ars.usda.gov/ARSUserFiles/3122/KramerEtAl2016c.pdf (accessed 14 April 2021).

Kurwijila, L.R. (2009) *Status of Postgraduate Training in the Livestock Sector in East and Central Africa and Priorities for ILRI's Support.* International Livestock Research Institute, Nairobi.

Lynam, J.K., Medvecky, B. and Lyakurwa, W. (2013) *An Evolving Regional Platform for Higher Agricultural Education: A Review of RUFORUM*. RUFORUM Secretariat, Kampala.

Lynam, J.K., Beintema, N. Roseboom, J. and Badiane, O. (2016) *Agricultural Research in Africa: Investing in Future Harvests*. International Food Policy Research Institute, Washington, DC. Available at: https://www.ifpri.org/publication/agricultural-research-africa-investing-future-harvests (accessed 14 April 2021).

Maassen, P. (2015) Research productivity at flagship African universities. *University World News*. Available at: https://www.universityworldnews.com/post.php?story=20150226083557823 (accessed 14 April 2021).

Maguire, C.J. (2012) Agricultural education and training to support agricultural innovation systems. In World Bank (ed.) *Agricultural innovation systems: An investment sourcebook*. World Bank, Washington, DC.

Mouton, J., Basson, I. Blanckenberg, J. Boshoff, N. Prozesky, H. *et al.* (2019) *The State of the South African Research Enterprise*. DST-NRF Centre of Excellence in Scientometrics and Science, Technology and Innovation Policy, University of Stellenbosch, Matieland, South Africa.
Muhamad, S. Nurhattati, F. and Unifah. R. (2019) *Strategy in Improving Research Productivity of the Lecturers*. Proceedings of the International Conference on Educational Sciences and Teacher Profession. Available at: https://doi.org/10.2991/icetep-18.2019.30 (accessed 13 June 2021).
Musiige, G. (2014) *Research at Makerere University: Faculty perceptions towards the factors that influence research productivity*. University of Oslo, MSc Thesis in Higher Education.
Musiige, G. and Maassen, P. (2015) Faculty perceptions of the factors that influence research productivity at Makerere University. In Cloete, N. Maassen P. and Bailey T. (eds.) Knowledge Production and Contradictory Functions in African Higher Education. African Minds, Cape Town.
NEPAD (New Partnership for Africa's Development) (2015) *Agricultural Educations and Skills Improvement Framework (AESIF), 2015–2025*. Available at: https://www.nepad.org/publication/agricultural-education-and-skills-improvement-framework (accessed 20 April 2021).
NRF (National Research Foundation) (2018a) *Framework and Funding Guide for the South African Research Chairs Initiative*. Available at: https://www.nrf.ac.za/sites/default/files/documents/SARChI%20Framework%20and%20Funding%20Guide_%20SACN%20Research%20Chairs%20%28Final%29_0.pdf (accessed 14 April 2021).
NRF (2018b) *OR Tambo Africa Research Chairs Initiative Framework and Application Guideline, National Research Foundation*. Available at: https://www.nrf.ac.za/sites/default/files/documents/Framework%20and%20Application%20Guideline_%20OR%20Tambo%20Africa%20Research%20Chairs%20Initiative.pdf (accessed 14 April 2021).
Oanda Ogachi, I. and Jowi, J. (2012) University expansion and the challenges to social development in Kenya: Dilemmas and pitfalls. *Journal of Higher Education in Africa* 10 (1), 49–71.
Oanda Ogachi, I. (2013) Trends in internalisation of higher education and implications for research and innovation for development in African universities. In Wiseman, A.W. and Wolhunter, C.C. (eds.) *The Development of Higher Education in Africa: Prospects and Challenges. International Perspectives on Education and Society*. Vol. 21. Emerald Group Publishing, Bingley, UK. Available at: https://doi.org/10.1108/S1479-3679(2013)21 (accessed 14 April 2021).
Odong, T.L., Tenywa, J.S. and Nabasirye, M. (2019) Revisiting application of statistics in agricultural research in sub-Saharan Africa: Entry points for improvement. *African Crop Science Journal* 27(3), 529–544.
Olsson, A. and Meek, L. (eds.) (2013) *Effectiveness of Research and Innovation Management at Policy and Institutional Levels: Cambodia, Malaysia, Thailand and Vietnam*. OECD, Paris. Available at: https://www.oecd.org/sti/Effectiveness%20of%20research%20and%20innovation%20management%20at%20policy%20and%20institutional%20levels_Meek%20and%20Olsson.pdf (accessed 25 May 2021)
Rockefeller Foundation (1995) *Directory of Selected African Research Networks*. 2nd. Edition. Available at: http://www.cohred.org/downloads/564.pdf (accessed 14 April 2021).
Royal Society, R. (2011) *Knowledge, Networks and Nations: Global Scientific Collaboration in the 21st Century*. London.
RUFORUM (Regional Universities Forum for Capacity Building in Agriculture) (2018a) *Strengthening Higher Agricultural Education for Agri-Food Systems Transformation in Africa (SHAEA) Project: Preliminary Project Description*. RUFORUM Secretariat, Kampala. Available at: http://www.ruforum.org/SHAEA/sites/default/files/Project%20Description_SHAEA%20083018.pdf (accessed 25 May 2021).
RUFORUM (2018b) *Summative Outcome Evaluation Report Building Africa's Capacity to Train for Africa*. RUFORUM Secretariat, Kampala.
Spielman, D. (2006) A critique of innovation systems perspectives on agricultural research in developing countries. *Innovation Strategy Today* 2(1), 41–54.
Sreeramana, A. (2016) How to increase research productivity in higher educational institutions: SIMS model. Srinivas Institute of Management Studies, Pandeshwar, Mangalore, India. Available at: https://mpra.ub.uni-muenchen.de/71750/1/MPRA_paper_71750.pdf (accessed 25 May 2021).
Swanepoel, F., Ofir, Z. and Stroebel, A. (2014) *Towards Impact and Resilience: Transformative Change in and through Agricultural Education and Training in Sub-Saharan Africa*. Cambridge Scholars Publishing, Newcastle upon Tyne, UK.
UIS (UNESCO Institute of Statistics) (2020) Education stats database. United Nations Educational, Scientific and Cultural Organization. Available at: http://data.uis.unesco.org/ (accessed 24 March 2021).

Uzaomaka, C.G.T. and Olayiwola, U.T. (2012) Re-engineering postgraduate studies in Nigerian universities through e-supervision of research projects. In Oji, S. (ed.) *Revitalising Nigerian Education in Digital Age*. Trafford Publishing, Bloomington, Indiana, USA.

World Bank (2007) *Cultivating Knowledge and Skills to Grow African Agriculture: A Synthesis of an Institutional, Regional, and International Review*. World Bank, Washington, DC.

World Bank (2008a) *Closing the Skill Gap: The Role of Education Supporting Growth and Competitiveness in the ECA Region*. World Bank, Washington, DC.

World Bank (2008b) *World Development Report: Agriculture for Development*. Available at: https://openknowledge.worldbank.org/handle/10986/5990 (accessed 27 September 2021).

World Bank and Elsevier (2014) *A Decade of Development in Sub-Saharan African Science, Technology, Engineering and Mathematics Research*. Available at: http://documents1.worldbank.org/curated/en/237371468204551128/pdf/910160WP0P126900disclose09026020140.pdf (accessed 14 April 2021).

Zhao, F. (2003) *Enhancing the Effectiveness of Research and Research Supervision through Reflective Practice*. School of Management Faculty, RMIT University, Melbourne, Australia.

9 Transformative Outreach in Tertiary Agricultural Education in Africa

Milcah Mulu Mutuku[1]*, Nancy W. Mungai[1], Duncan Ongeng[2], Daniel Sherrard[3] and Festus Annor-Frempong[4]

[1]Egerton University, Kenya; [2]Gulu University, Uganda; [3]EARTH University, Costa Rica; [4]University of Cape Coast, Ghana

9.1 Introduction

Alongside teaching and research, outreach is considered a core responsibility of universities globally. These three university functions should cross-fertilize each other, allowing universities to contribute holistically to development (Fitzgerald *et al.*, 2012). However, many universities do not have the requisite governance and implementation structures in place. In such universities, outreach is largely implemented in a fragmented and disengaged manner, such that its effectiveness and impact on institutions, graduates and other communities (i.e. non-university stakeholders) are minimal. University outreach activities generally revolve around extension, consultancies, public lectures and dissemination of research outcomes. Universities were traditionally viewed as a hub for teaching and research (AUCEA, 2008) with outreach acting as an 'arm' connecting universities to communities. From this perspective, outreach is seen as the conduit through which knowledge produced within the confines of the university is delivered to end users, usually in a 'one-way', discipline-specific model which disregards the complexity of societal problems and needs, and communities as repositories of knowledge. As Fitzgerald *et al.* (2012) put it, higher education institutions reached out to communities in an expert model of knowledge delivery. In this case, little thought was given to the role outreach could play in shaping (i) faculties' and students' knowledge of communities and their needs, (ii) the skills required to meet community needs and, by extension, national and global needs and (iii) the requisite attitude for such work. Also overlooked was the role communities play in generating knowledge.

Unfortunately, the expert model of knowledge delivery, with its associated thinking patterns and mode of operation, have persisted in many African universities with negative results - the obvious being universities' disengagement from communities and poor integration among their teaching, research and outreach functions. A typical negative outcome is the criticism levelled against universities as 'churning out graduates who are not fit for purpose' and research products which are 'out of sync with society's needs'. Many newer and growing universities are keen to experiment with pedagogical and community-engagement approaches designed to educate market-oriented graduates and create focused research and outreach programmes. On the other hand, some old, esteemed and established universities may be confident in their hallowed positions and hence may not be as innovative in seeking new ways of approaching

*Corresponding author: mmutuku@egerton.ac.ke

 Transforming Tertiary Agricultural Education in Africa (eds D. Kraybill, J. Lynam and Adipala Ekwamu)

DOI: 10.1079/9781789246544.0009

teaching, research and outreach. The expert model of knowledge delivery is strongly entrenched. While it may have many gatekeepers, including some political elites, it is likely to change in the advent of reliable online resources which are easily accessible to students.

Universities have been excoriated for conducting research which is disengaged from societal needs and aspirations, and for not extending their significant resources to relevant community, social and national issues (O'Meara *et al.*, 2011). A majority of stakeholders nowadays view graduates as ill-prepared and incapable either of solving societal problems or addressing societal aspirations. In addition, even when the results of research conducted in universities are relevant to the needs of the community, they are largely shelved and seldom trickle down to end users. Universities are rarely held accountable for the fitness of their graduates as long as they continue to produce an ever-increasing supply. Moreover, to the extent that faculty promotion continues to rely heavily on the volume of peer-reviewed publications - rather than also considering other criteria, including teaching and outreach - it is unlikely that significant change will occur. Nevertheless, higher education can be a critical driver of national development and competitiveness through the systematic generation of useful, high-quality knowledge (Kivati, 2017; Muriisa, 2014). Consequently, there is a clarion call to reorient universities' outreach missions towards community engagement, whereby the knowledge extended is generated in collaboration with communities to maximize gains to all parties. A well-planned and carefully executed community-engagement programme can act as a pivot, focusing university teaching and research firmly on societies' needs and aspirations, an alignment currently lacking in many parts of the world, and in developing countries in particular.

Outreach encompasses all university activities designed to support society in obtaining, understanding and applying information, whether at the individual, group, organizational, business, governmental or civil society levels, and whether in urban or rural settings (Muse, 2000). Outreach is a university's way of using its expertise to identify, understand and address community-based issues (Muse, 2000). This chapter explores models of university–community interactions intended to build knowledge-driven partnerships and yield mutually-beneficial outcomes by aligning university strengths with relevant local contexts. The discussion focuses on the ethos of community engagement, through which teaching, research and outreach in universities have the potential to be transformed. Within the context of this chapter, 'community' is defined as non-university stakeholders who directly or indirectly benefit from the university and its activities, and vice versa. In the words of Byrne (2016, p. 55):

> Engagement involves transfers in two directions: a partnership of exchange between the university and its constituents. Engagement is mutually beneficial to the university and to society, and frequently involves shared goals, agendas, and measures of success. It includes working together to solve problems and share expertise.

This chapter advocates for partnerships between universities and communities which drive the processes of generating and transmitting knowledge, and promoting constructive change targeting societal needs and aspirations, with the capacity to reinvigorate development, particularly in the agricultural sector.

9.2 The Evolution of University Outreach in Africa

From the independence period (see Ch. 3, this volume), many universities were established to serve as critical drivers of national development and to immediately have an impact on social welfare and national identity after years of colonization. The core mandate of these universities was to produce a knowledgeable and skilled workforce to occupy positions formerly held by colonial masters and to bring about rapid development. As would be expected, colonialism played a great role in shaping the educational philosophy, organization and structure of African universities, with many being modelled after higher education systems of their colonial masters (Alemu, 2014; Mackatiani *et al.*, 2016). In fact, even universities established after independence typically reproduced colonial ideologies, contributed to the selection of dominant elites and dedicated themselves to the training of a select group of civil servants (see Zeelen, 2012; Mosweunyane, 2013; Mackatiani, *et al.*, 2016;

Nampala, *et al.*, 2017). Furthermore, local communities, especially peasant farmers, were considered 'primitive' and underdeveloped, with nothing to offer universities. Consequently, African university education systems lacked philosophical and organizational structures to support reciprocal relationships with communities and to hinge university education on local and national needs and aspirations.

As such, universities were perceived as 'ivory towers', set apart from the societies which established them. Knowledge was considered the preserve of the university, which periodically disseminated it, largely to the community of university academics in the form of peer-reviewed publications and symposia (see Nampala *et al.*, 2017). This original 'closed system' of university operation limited engagement with non-university actors in the generation and transfer of knowledge to the wider community of end users. This situation elicited pertinent questions on the relevance of universities to the broader societal development beyond academics. The development of the German university model as a driver of development led to emergence of the one-way engagement model (O'Boyle, 1983), which emphasized the delivery of knowledge and services to the society with little community influence on university teaching and research (Tornatzky and Rideout, 2014; Ntimi *et al.*, 2016). Over time, however, a shift has occurred to a two-way model which focuses on the interactive exchange of knowledge between universities and their communities in the context of partnership, reciprocity and mutual learning (Ntimi *et al.*, 2016).

A majority of African universities have largely been funded by government, but economic pressure and structural adjustment programmes in the 1980s caused university funding to drop sharply. At the same time, student enrolments were increasing, leading to cutbacks in research, outreach activities, staff development, library acquisitions and infrastructure development and maintenance. Universities were required to seek alternative sources of funding to continue their operations (New African Magazine, 2017). Funding alternatives were limited to research grants and tuition fees for privately-sponsored studies, mostly charged below market rates. Universities were, therefore, unable to raise adequate funding for their programmes, and this problem has persisted (see Wangenge-Ouma and Nafukho, 2011; Gudo, 2014). With dwindling financial resources, the outreach mission of universities was the easiest to sacrifice as available resources were directed towards teaching and, to a lesser extent, research. This led to neglect and stagnation of outreach activities in many universities, with little innovation occurring in the organization, governance and execution of outreach. Additionally, there was a lack of research on the true benefits of outreach (McCann *et al.*, 2015) leading to a perception of its being 'an extra burden' on an already overworked faculty.

In the 1980s, the search for a development role for East African universities focused on several aspects of (i) community involvement and service, (ii) interactions with government, including contributions to policy, (iii) professional and vocational issues, (iv) cost-sharing, (v) equitable admissions policies and (vi) multidisciplinary research and teaching (Court, 1980). However, the response and outcome for each university was influenced by the political environment and prevailing ideologies of the time. Universities were challenged to convince their governments and populations at large that their contributions to national development were the formation of trained minds and a thinking society (Court, 1980). In the 1990s, debates emerged as questions arose regarding the role of universities in national and global development. This debate was fuelled by the contention of the World Bank that investments in university education generated a limited return on investment (Muriisa, 2014). These debates focused on changing processes of teaching and learning within universities to promote civic awareness among students and to contribute towards communities' improved quality of life. As a result, new and relevant forms of teaching, learning and transmitting knowledge in partnership with communities emerged, emphasizing experiential learning and a heightened degree of community engagement.

A survey of experiential learning activities in institutions of higher learning offering agriculture and related subjects was conducted in East Africa in November 2015. Ten institutions were covered: Kenya's Egerton University, Jomo Kenyatta University of Science and Technology, the University of Nairobi and Baraka Agricultural College; Tanzania's Sokoine University of Agriculture, Livestock Training Institute, Morogoro

and Ministry of Agriculture Training Institute, Ilonga; and Uganda's Makerere University, Gulu University and Bukalasa Agricultural College. The survey, which covered 58 undergraduate and 5 certificate programmes over the 2011–2015 period, documented three main areas: (i) field attachment, (ii) community engagement and (iii) internship programmes. Results confirmed that university students' participation in community engagement was less than their counterparts in middle-level colleges in all three countries. However, of the three Kenyan universities, Egerton university and the University of Nairobi ran community-engagement programmes involving smallholder farmers, and both Makerere University and Gulu universities also undertook community-engagement activities (Mungai and Ombati, 2016; Kisiangani, 2019).

As African economies become incrementally more knowledge-based, lifelong learning is the new reality, and societies are assuming many characteristics of a learning organization, which requires universities to engage effectively with them (Byrne, 2016). Fortunately, a number of universities are enthused about community engagement. As Byrne (2016, p. 55) points out,

> Engagement involves transfers in two directions: a partnership of exchange between the university and its constituents....both outreach and "in-reach" into the university.

Community engagement holds the key for effective university outreach involving shared goals, agendas, measures of success and cultural changes inside and outside the university (Byrne, 2016). In particular, engaged university outreach is significant in transforming higher agricultural education through knowledge-driven partnerships capable of jointly generating, sharing and utilizing knowledge aimed at promoting constructive change in the agricultural sector. Inclusion and integration of experiential and service-learning components in agricultural education curricula not only provide students and faculty with opportunities for reaching out to communities, but also exposes them to community needs and aspirations. This, in turn, can lead to an increased commitment to the public good (Harden and Loving, 2015). Such universities as Gulu University (Uganda), EARTH university (Costa Rica), Egerton University (Kenya), University of Cape Coast (Ghana), Mohammed VI Polytechnic University (Morocco), University of Development Studies (Ghana), Kabasa University (Uganda) and University of Venda (South Africa) have implemented engaged outreach. Additionally, the research model of the Regional Universities Forum for Capacity Building in Agriculture (RUFORUM) emphasizes community action research as a good example of engaged outreach programmes.

Feedback from students, faculty and community members attests to the power of such engagement in transforming mindsets and developing a culture of social responsiveness and civility among students. In addition to the direct contribution towards solving community problems, lessons learned through community engagement have the potential to transform university programmes and activities to contribute meaningfully to development. While recognition of the benefits of robust university-community engagement is growing, there are also many constraints to its widespread adoption. Costs, lack of support from some senior administrators and the inherent complexities of development work in general are all factors. The need for more data on the number of institutions developing innovative models of community engagement across Africa is urgent, as is evaluations of the efficacy of such models. This chapter is founded on the premise that engaged university teaching, research and outreach, and learning pathways which support partnerships between universities and communities operate as an ecosystem which is beneficial to all stakeholders (Fig. 9.1). Community engagement is seen as the nerve centre of teaching, research and outreach which influences and is influenced by the social, economic and political developments within communities.

9.3 The Emergence of the Innovation-Systems Approach

African societies are increasingly demanding a higher level of engagement by universities. While the call is for universities to engage across multiple sectors, given that two-thirds of Africa's population depend on farming for their livelihoods, agriculture and rural development are clearly among the highest priorities. The need

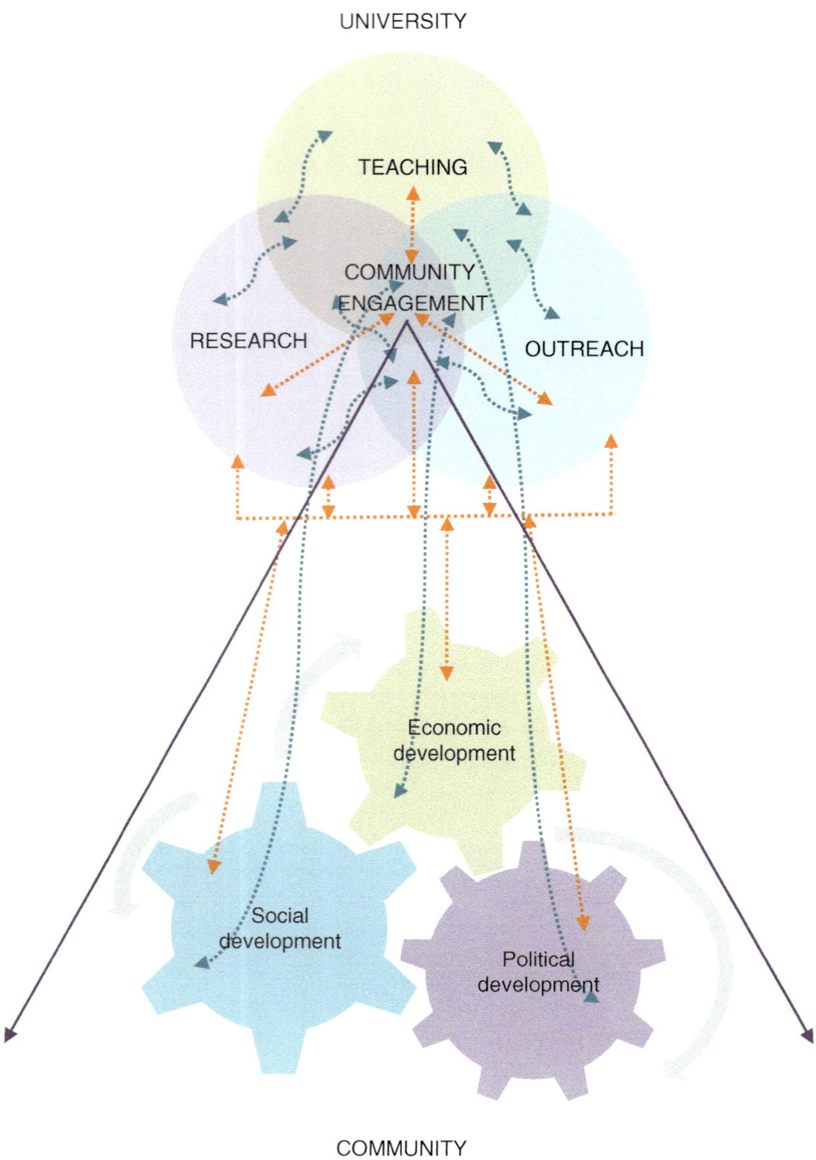

Fig. 9.1. A model for engaged university teaching, research and outreach
Source: Devised by authors.

for profitable and sustainable technologies capable of driving agricultural development and ensuring effective global competition compels African nations to transform their systems of agricultural education and training to strengthen the capacity of actors in the agricultural sector to innovate (Spielman *et al.*, 2008). Engaged university outreach plays a crucial role in this regard. In addition, the need to transform the agricultural sector obliges nations to train a workforce capable of offering demand-driven and targeted agricultural advisory services to all actors in the agricultural innovation system (AFAAS, 2011). Policy reviews and development of new agricultural advisory services models are therefore inevitable. Consequently, many African nations are transforming agricultural advisory policies to focus on offering services suitable for linking farmers with markets within the framework of value chains (AFAAS, 2011).

The goal is to develop agricultural advisory services capable of supporting the commercialization of agricultural outputs in a sustainable, pro-poor manner (Neuchâtel Group, 2008). This requires major institutional reform to accommodate and facilitate innovation (Rivera and Sulaiman, 2009) and to ensure diversity in the delivery of services. This reality provides rich opportunities for Africa's agricultural universities to play the guiding role by providing evidence-based practicable knowledge and expertise. The emergence of the innovation-systems approach provides the impetus for the continuous innovation of novel development interventions for African agriculture. In embracing the innovation-systems approach, it is important to acknowledge the ineffective and under-staffed state of the largely government-driven extension systems in many parts of Africa (RUFORUM, 2016). Consequently, as key component of the agricultural development 'ecosystem', agricultural universities need to review their role in the agrifood system, along with their contribution to transforming agricultural advisory services across Africa. This is a defining moment for those universities able to embrace transformative change and implement engaged outreach programmes - especially in the agriculture sector - which complement the efforts of agricultural advisory service providers. It is an opportunity for universities to differentiate themselves by aligning their teaching, research and outreach activities with community needs and aspirations. Stronger connections and collaborations are needed between universities and their communities, with faculty members focusing on studying and responding to social problems and public issues (Woods *et al.*, 2013).

As RUFORUM (2016) states, universities are well placed to support rural transformation as they employ a large number of highly-trained scientists (i) engaged in the continuous generation and conservation of knowledge, (ii) sharing knowledge with subsequent generations of scholars and (iii) training a workforce which can support agricultural innovation, even at the farm level. Nevertheless, agricultural universities need to re-examine their strategies and orientation towards (i) jointly generating knowledge with communities, (ii) providing information which leads to effective and sustainable agricultural advisory services models and approaches, (iii) preparing personnel with the right attitude, knowledge, and skills for such reforms and (iv) building and strengthening collaborations among diverse actors in agricultural value chains and actively participating in innovation platforms. All of this requires innovative financing strategies and high levels of investment in terms of intellectual resources and time. Neuchâtel Group (2008) noted a major gap in the systematic documentation of lessons which have the power to guide the policy and institutional development necessary for establishing and strengthening market-oriented agricultural advisory services. Experiences and lessons learned from projects remain largely undocumented or scattered, providing only limited access to those who need them. Through engaged, well-structured and coordinated outreach programmes, universities can efficiently and continuously document and disseminate these lessons in a useful form to support sustainable development. Further, through engaged outreach, universities can play the role of facilitating multistakeholder innovation platforms in agricultural value chains.

9.4 Challenges to Effective and Engaged University Outreach

Effective and engaged university outreach requires decisive planning in terms of human and financial capital in order to foster its contribution to national development. As it is, agricultural education and training systems in sub-Saharan Africa have not kept pace with demand for technically- and professionally-qualified human resource capacity capable of producing applied knowledge and technologies, and conducting outreach and dissemination of research results (Chimpololo, 2017). These shortcomings stem from numerous factors, key among them being education and training programmes which lack a grounding in society's actual needs. Engaged university teaching, research and outreach has the potential to solve this problem. While partnerships with communities would reinvigorate university outreach activities and redirect efforts to serve real community needs, overall lack of funding or financial incentives for community partnerships prevail for most African universities (Zeelen, 2012). Reliance on government funding and student fees may not adequately serve the financial requirements of engaged outreach. Innovative and entrepreneurial funding strategies, coupled with the establishment of systems of institutional incentives and support, are required if such programmes are to grow and be sustainable. EARTH University, for example, exports over one million 18 kg boxes of bananas every year from its commercial farm, not only providing a living laboratory for students to analyze agribusiness operations, but also generating important resources to support student scholarships.

In addition, the great socioeconomic divide, characteristic of many African societies, and the range of partners with which universities should engage to play a role in development poses a challenge for engaged outreach. These partners include private firms in various industrial sectors, actors in the informal sector, cooperatives, rural communities, small-scale farmers, social movements and, even, individuals and households all requiring different modes of engagement. University policy and guidelines for engaging with the community should accommodate such diversity. Lack of community interest in university partnerships, ambivalent or 'fluid' academic culture, a mismatch between university research strengths and community needs are additional barriers to engagement requiring a complete about-face if transformative university outreach is to be realized. Stakeholders' involvement in the design and implementation of university curricula must not only contribute to cultivating community interest in universities, but also to increasing the responsiveness of university programmes to meeting societal needs and aspirations. Nevertheless, efforts to solve society's problems require new approaches to generating knowledge, which is generally understood to mean partnerships, collaborations, knowledge exchange and co-created solutions (Fitzgerald *et al.*, 2012). Important regional problems also pose challenges for engaged outreach programmes. These include: (i) weak integration structures, (ii) economic, social and political instability and (iii) diverse contexts and policies, e.g. the massification of university education.

9.5 Successful Models of Engaged University Outreach

Numerous examples of successful models of engaged university outreach exist, fully embedded within communities, dictated by community needs and fully integrated with the teaching mission of the university. Four such models are discussed in detail in the sections that follow: (i) Gulu University's student-attachment and enterprise model, (ii) EARTH university's community-engagement model, (iii) Egerton University's student farm-attachment programme and (iv) the University of Cape Coast's supervised enterprise projects in collaboration with the Sasakawa Fund for Extension Education.[1]

Gulu University's Community-Engagement Model

The Government of Uganda established Gulu University in 2003 as a strategic response to the need for social transformation and economic progress in Northern Uganda, which had undergone protracted armed conflict from the mid-1980s to mid-2000. This had led to massive displacement of the population representing significant humanitarian and economic crises, in turn shaping the university's strategic focus on community development - as evidenced by its motto 'for community transformation'. Community engagement therefore underpinned all of the university's academic programmes, teaching, research and training as a means of contributing to the country's development challenges. The 2005 establishment of the Faculty of Agriculture and Environment was instrumental in championing Gulu's community transformation agenda. The flagship bachelors degree in agriculture strongly emphasized student–farmer outreach, experiential learning and social entrepreneurship. This worked well operationally because the university had neither an experiment farm nor the land to develop one. The university's rural setting - surrounded by a community of smallholder farmers who desperately needed the university's services - provided an ideal opportunity for students to work closely with local farm families. This engagement provided mutual benefits, including assisting farmers in identifying and resolving farm-related issues, and providing opportunities for students to gain hands-on agricultural and outreach experience (Kalule and Ongeng, 2016). The arrangement involved students regularly visiting farmers using bicycles (Kalule *et al.*, 2016; Kalule and Ongeng, 2016; Odongo *et al.*, 2017). As of 2017, all undergraduate and graduate academic programmes embraced community engagement, drawing on useful lessons from this initial implementation of the undergraduate programme.

Many lessons emerged from the connections forged between the university's curricula, students and staff and the local farming community:

1. ***Growing realization of the viability of an undergraduate programme founded on community outreach.*** Across all undergraduate curricula, content was devised, structured and assessed to reflect cumulative learning and academic rigor. In the earlier years students focused on experiential learning, needs assessment and problem identification, then in later years, they became more involved in developing training programmes, facilitating learning and sharing experiences with the farmers and peers. All curricula articulated mechanisms for student assessments, along with a separate manual to guiding students in undertaking their attachments.

2. ***Increased interest on the part of farm families to host university students.*** Farmers interest in hosting students on their farms for longer periods was taken as an indicator of their appreciation for the knowledge they gained from interacting with students, and of the students' attitudes while on the farms.

3. ***Changes in the attitudes of students, faculty members and administrators.*** Students, staff and faculty administrators came to perceive their experiences in the community as highly positive and useful, especially in terms of the 'feel good' factor in having an impact. Behaviourally, learning practices evolved to include journal publications, brochures and banners.

4. ***Differentiation in the approaches to community engagement.*** The different approaches to community engagement included: (i) student-to-farmer attachments, largely for smallholder farmers within a short (10 km) distance of the university, (ii) model-farmer attachments, mainly for more progressive and commercially-oriented farmers, (iii) farmers' associations, mainly employed at the graduate-student levels and (iv) industry attachments, involving larger scale farming operations. These approaches complement each other, placing the student at the centre of activities supported by faculty members who provided supervision and technical support.

5. ***Misalignment of expectations between the students and host farmers.*** Despite the positive lessons learned, the attachments - also known as student-centred outreach - were not free from challenges. Many farmers tended to perceive students as an additional source of labour, which led to some complaints from students and, at times, loss of morale. Faculty members were sensitive to this issue, clarifying arrangements with farmers and reinforcing the value of active participation in farm work to students as an important aspect of their learning.

EARTH University's Community-Engagement Model

EARTH University - a private, international institution, located in Costa Rica - was created with the goal of preparing leaders to promote sustainable development and build fair and prosperous societies. The university's creation grew out of the conviction that the best response to the region's endless cycle of poverty, conflict and environmental destruction was education. The university's more than 400 students come from over 45 countries in the Global South, including 16 African nations. While this mission is primarily achieved through the university's four-year undergraduate degree programme, EARTH is actively involved in efforts to empower rural communities by building their capacity to promote rural development. The university seeks to build a two-way bridge to the larger community, while providing students direct experience of and exposure to the complexities of rural development.

EARTH University has developed a distinctive and novel curriculum with a strong focus on experiential learning. Coursework includes work experience, a community-development programme, a third-year internship and entrepreneurial projects. While the work experience course begins in the first year of study with routine and often labour-intensive activities related to crop and animal production on the university's farms, from the second year, students spend a full-day each week living and working directly with rural families and communities. Through this experience, they not only gain a better understanding of the realities of smallholder agricultural production and community development, but also have the opportunity to work with farm families promoting sustainable production practices, as well as with local organizations to build stronger communities. Students in their fourth and final year function as leaders in these activities and are responsible for developing strategies and activities to accomplish established goals. Community-development staff provide organizational and logistical support, while faculty members maintain responsibility for providing the programme's strategic vision and encouraging students to critically reflect on their experiences.

In their third year, students live with rural families for a seven-week period, allowing them to engage in community-development activities and undertake short-term internships with local companies. In the final trimester of their third year, students undertake a 15-week internship with a host organization, such as a business, non-governmental organization or farm, often outside Costa Rica. Using the knowledge and skills they have acquired in their first three years at EARTH, students obtain real-world practical experience upon which they can reflect during their fourth and final academic year. Students are required to fulfil the requirements of their full-time internship and must also analyse the community's situation to identify a local need and develop a programme or project to meet that need. There is tremendous diversity in the projects they design, from mobilizing the community to build a playground for children, to organizing a waste-management programme.

In addition to these courses students and faculty have numerous other opportunities to engage with the community. Many courses include community-based projects, including opportunities for students to provide advisory services to farmers to improve their production. Each year, students at each level of the programme dedicate a week in which classes are cancelled, and students and faculty decide to focus on an issue of particular interest. In first and second years, in particular, teams of students might spend the week in a rural community collecting information on and assessing social, economic and agricultural conditions. Activities include interviews and conversations with local residents, farmers, authorities, etc. At the end of the week, students present their findings and reflect on what they learned. Providing meaningful volunteer opportunities is another important tool that EARTH University uses to promote community engagement. Such activities benefit the community, provide essential experience for students and - perhaps most importantly - contribute to developing lifelong habits of positive citizenship.

EARTH University's community-engagement efforts demonstrate how meeting community needs can complement academic programmes and provide valuable experience for students, faculty members and the community. Over the past 15 years, EARTH University has collaborated with African partners to share experiences and lessons learned. EARTH's faculty and staff have conducted workshops and participated in numerous regional events in Africa. In addition, many university leaders and faculty members from African universities have visited EARTH's campus in Costa Rica to learn from their experiences and approaches.

Egerton University's Farm-Attachment Model

Egerton University, which was founded as a farm school in 1939, has evolved over the years to offer diverse diploma, undergraduate and graduate programmes and to become a centre of excellence in agricultural sciences. Known for its practical training orientation, the university has various programmes which expose students to practical learning, including laboratory sessions, field trips and demonstrations, clinical and community-based practicums, community-outreach activities and attachments. Over 60% of Egerton's undergraduate diploma and bachelor's degree programmes require students to undertake field attachments as course units for a minimum of eight weeks in the latter years of their studies in their areas of specialization. Over the years, student attachments have focused on government ministries and parastatals, and on agribusinesses with little interaction with farmers (Edziwa *et al.*, 2012). Consequently, eight-week farm attachments were introduced in 2014 as a variant of field attachments. Farm attachments provide students with an opportunity to interact directly with farmers, learn and utilize theoretical knowledge acquired in class, understand the opportunities and challenges farmers deal with and propose mitigation strategies to some of these challenges. The goal of the farm attachment programme is to facilitate students' experiential learning by creating linkages between the university and the farming community, and in doing so enhance Kenya's agricultural productivity and competitiveness. Specific objectives are: (i) to produce undergraduate students who will effectively contribute to the agricultural development agenda through experiential learning with farming communities, (ii) create and strengthen linkages between Egerton University, farming communities and service providers to improve service delivery to rural communities, (iii) provide empirical and socially acceptable solutions to farmers, (iv) develop an information and communication technology mechanism to enable farmers to access agricultural information, (v) facilitate the transition of small-scale agriculture from subsistence to commercial farming and (vi) obtain feed-back from farmers and other stakeholders to support the redesign and revision of university curricula, research activities and subsequent extension messages.

Farm attachments are designed such that different cohorts of students remain with the same farm(s) for at least three consecutive years. Each cohort of students builds on and follows-up on the recommendations of its predecessors. The first cohort focusses on making a general appraisal of the farm - i.e. identifying strengths, weaknesses, opportunities and threats, and making proposals for improvement. The students and farmers are supported by project coordinators, lecturers and existing extension officers in synthesizing reports and proposing recommendations to the farmers. Meanwhile, students gain experience and are supported in becoming analytical observers, allowing them to contribute to solutions which then accumulate into knowledge. The second and third cohorts of students are sent to the same farm(s) to oversee the implementation of agreed interventions (Fig. 9.2). Preliminary evaluations of the programme have assessed students', lecturers' and farmers' perceptions of the farm-attachment model. Each eight-week learning cycle brought new challenges and solutions for incorporation into future research and inclusion in future dissemination messages (Mungai *et al.*, 2016).

Rural communities are keen to engage with the university, particularly in sharing unbiased agricultural knowledge. While farmers have numerous sources of information, some are inclined towards influencing them to purchase products or services. A public university actively engaging with the rural community can contribute to closing this information gap. The students appreciated the fact that farmers had experience which enriched their understanding and application of the theoretical principles they learned through their university studies. Notably, community engagement through student attachments requires more time and resources than the traditional attachment model, but the potential benefits are immense. Further examples include:

1. ***Productivity impacts***. In addition to increased awareness and use of agricultural technologies

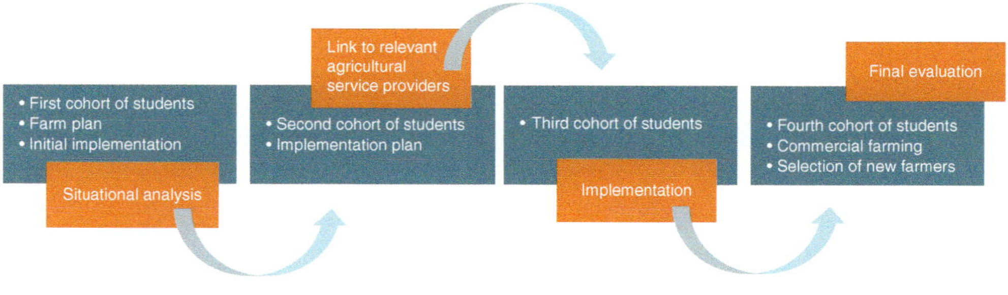

Fig. 9.2. Egerton University's farm-attachment model
Source: Mungai *et al.* CC BY 4.0 (2016, Fig. 1).

by farmers and students, one Kenyan farm increased milk production (from 100 to 170 kg day). This was achieved through students' contributions to (i) formulating a new feed ration, which included protein supplementation, (ii) modifying the timing of feeding to allow for milk synthesis, (iii) modifying the timing of milking to establish a fixed routine and (iv) improving the hygiene of the milking shed and feeding area.

2. *Record keeping.* Over 95% of the students attached to farms helped to organize farm operations and initiated record keeping. Students also helped farmers undertake farm planning and budgeting in addition to their routine activities. As a result, farmers began viewing their farms as businesses, rather than simply as a subsistence activity.

3. *Improved family nutrition and incomes.* Students assisted farmers in establishing vegetable gardens and fruit orchards, implementing soil-conservation measures and using fuel-saving stoves. They assisted small-scale farmers in adding value to their products, e.g. making ghee, refining honey and preserving fruits and vegetables through sun-drying. Students encouraged female farmers to have their children registered and immunized, and provided information on balanced diets and family budgeting. Some farmers recorded increased farm income as a result of the advice and contribution of the students.

4. *The need for curriculum review.* Students undertaking Egerton's BSc Degree in Community Development do not take any agricultural courses, although they do participate in agricultural practicums, such as composting and multistorey gardens. One student was quoted saying, 'limited knowledge on farm activities, including farm inputs and outputs, dwarfed my expertise in my relevant field'. Since over 60% of the Kenyan population is based in rural areas, a great need exists for the inclusion of basic agricultural skills and knowledge in the BSc Community Development programme. In undertaking their farm attachments, students were confronted with questions which underscore importance of capacity building in new approaches to extension service delivery providing prompt, reliable and unbiased agricultural information. A call centre would be a useful resource for students on attachment. More general community-based capacity building is also needed given the large number of new farmers who have developed interest in the farm-attachment programme.

The University of Cape Coast's Supervised Enterprise Projects

The University of Cape Coast, Ghana, was established as University College in 1962. It became a fully-fledged university in the early 1970s, and evolved to become the University of Cape Coast in 1992. Its mandate is to educate a highly-qualified and skilled workforce, across numerous professions, including the agricultural sciences for the purpose of social and economic transformation. Over the years, the University has built relationships with the community to fulfil its vision of benefiting the broader society. Community engagement in training students has been part of the university's practice since its inception. It is therefore not surprising that in 1993 the University of Cape Coast - in collaboration with the Sasakawa Fund for Extension Education, an international NGO, and Ghana's Ministry of Food and Agriculture - instituted 'supervised enterprise projects' as an approach to training mid-career extension personnel and ultimately improving the livelihoods of rural communities.

Supervised enterprise projects require a student to work with stakeholders to identify the needs of a group of people in a community, design a project to effectively address those needs, and implement the project in a way that generates the dual outcomes of development and learning. The development outcomes address the needs of target communities by improving the productivity and quality of farm produce, reducing postharvest losses and generating income for beneficiaries. The learning outcomes take the form of mini research projects, which enable students to attain competencies and training methods relevant to their future work, and develop or adapt technologies to suit the local environment (Annor-Frempong, 2013). The success of this approach has resulted in its replication in 19 programmes in the universities and agricultural colleges of 9 African countries. The approach has also been adopted in two other programmes in Ghana.

The philosophical basis of supervised enterprise projects is experiential learning, which holds that the combination of theory, experience and critical reflection provides learners with the opportunity to develop lifelong learning skills, while building their confidence in and commitment to the application of learning to bring about planned change (Kolb, 1984; Carr and Kemmis, 1986). The strategies and extent of engagement involve students undertaking relevant courses, conducting participatory needs assessment with project beneficiaries and developing a successful proposal under the guidance of assigned lecturers and other personnel. Fieldwork involves direct interactions with beneficiaries to apply academic subject matter, appropriate technologies and indigenous knowledge to solve problems. The projects are evaluated by university lecturers, community development experts, students' employers and beneficiaries, who compare students' proposed ideas and field observations and offer suggestions for improvements. The university supports community engagement by providing logistics and lecturers who monitor the projects. Knowledge, skills and technologies are provided to students for transfer to beneficiaries. Approximately $1,000 is invested in each project, from students' own sources, donations (from non-governmental organizations, banks, municipalities or districts), beneficiary communities or the Ministry of Food and Agriculture.

This approach has also contributed to human resource development for the agricultural sector and the mobilization of resources for community development. Beneficiary communities have been empowered and provided with appropriate technologies which have raised the productivity and quality of agricultural products. Furthermore, the approach has led to rural infrastructure development and has galvanized commitment to agricultural development on the part of support institutions. The key constraints are the costs associated with supervised enterprise projects, both to students and the university, and weaknesses in the sustainability of projects after students graduate. The supervised enterprise project approach to community engagement has demonstrated that university programmes can be designed to promote long-term sustainable community development.

9.6 Conclusion

Developing transformative and sustainable engaged university outreach programmes requires significant transformation of university priorities and paradigms. The results would elevate outreach to a significantly higher status than it is currently afforded in many university agendas. Society is increasingly demanding this transformation, particularly in developing nations where competing demands for scarce resources require clear articulation of the benefits derived from investment in higher education. It is therefore essential that the impact and contributions of higher education institutions are not only substantial, but also benefit the greatest number of citizens. Simply put, as the third branch of a university's mandate, outreach deserves to be prioritized alongside teaching and research. University engagement with the community involves many distinct activities, most often falling into four categories:

1. ***Research.*** Investigation should go beyond simply sharing results to involve direct engagement with external stakeholders such as through action research and community-based participatory research.
2. ***Knowledge transfer.*** The mutual sharing of knowledge with stakeholders typically goes beyond the more traditional forms of outreach and extension to involve two-way and tacit knowledge sharing.
3. ***Teaching and learning engagement.*** Students should be actively engaged in the community in service learning activities linked to their academic endeavours.
4. ***Engagement through service.*** Faculty members, staff and students should apply their expertise to address community needs.

A number of preconditions need to be met in developing effective community engagement programmes which cross-fertilize teaching, research and outreach and ensure universities' holistic contribution to development (Fitzgerald *et al.*, 2012). Perhaps most important is the existence of an institutional culture which values engagement with external stakeholders. Since attitudes shape behaviour (Velasco and Harder, 2014), university leadership, policy and decision-makers must adopt a positive attitude towards engaged outreach. Byrne (2016) suggests that attitudinal shifts within universities are necessary for effective community engagement, including the formal recognition of the value of community-engaged outreach activities and the institutionalization of rewards for those involved (Woods *et al.*, 2013). Faculty evaluation, promotion and reward schemes should value community engagement and provide incentives for innovative and intensified community-based activities. Similarly, operating policies and guidelines are necessary to provide a framework for universities to engage with communities. Many universities engage without such frameworks, resulting in duplication of effort, poor resource management and difficulty in monitoring or documenting the successes or challenges which accrue.

Commitment on the part of university leadership is another essential element of ensuring strong community linkages, as is the commitment of other internal stakeholders. Abrahamsson (2001) argues that commitment represents a conscious and overt act which binds or obliges a person to some future action. Viewed from this perspective, commitment is a function of the right attitude, evidenced by favourable behaviour. While the commitment of all stakeholders is paramount to the success of community engagement, university leadership must set the tone (Muriisa, 2014). Another key component of success is an entrepreneurial orientation and mindset, characterized by proactive innovation, risk-taking, autonomy, and competitiveness, which support creativity and the active pursuit of opportunities for community engagement. Faculty and students require a reasonable level of autonomy to seek out opportunities proactively (Clark, 1998). Significant positive shifts in universities' organizational structure and policy are vital for committed implementation of community engagement. Finally, funding for effective community engagement requires innovative, entrepreneurial and sustainable strategies potentially involving private-sector linkages. This calls for widened and diverse resource mobilization strategies that involve cost sharing among stakeholders in mutually beneficial knowledge-driven partnerships.

Endnote

[1] Links to other community-engagement programmes, along with RUFORUM's research model, can be found on RUFORUM's website: www.ruforum.org.

References

Abrahamsson, P. (2001) Rethinking the concept of commitment in software process improvement. *Scandinavian Journal of Information Systems* 13, 35–59.

AFAAS (2011) *Market-Oriented Agricultural Advisory Services (MOAAS): Guidelines for Setting Up MOAAS Pilots*. African Forum for Agricultural Advisory Services, Kampala, Uganda, and Forum for Agricultural Research in Africa, Accra, Ghana.

Alemu, S.K. (2014) An appraisal of the internationalisation of higher education in Sub-Saharan Africa. *Center for Educational Policy Studies Journal* 4(2), 71–90.

Annor-Frempong, F. (2013) *Supervised Enterprise Projects as a Community Development Tool. Case Study Series 4 on Human Resource Development in Agricultural Extension*. Feed the Future. Available at: https://meas.illinois.edu/wp-content/uploads/2017/02/MEAS-HRD-CS-4-SEP-Community-Development-Jan-2013.pdf (accessed 23 April 2021).

AUCEA (Australian Universities Community Engagement Alliance) (2008) *Universities and Community Engagement*. Position Paper (2008–2010). Available at: https://engagementaustralia.org.au/uploads/universities_CE_2008_2010.pdf (accessed 23 April 2021).

Byrne, J.V. (2016) Outreach, engagement, and the changing culture of the university. *Journal of Higher Education Outreach and Engagement* 20(1), 53–58.

Carr, W. and Kemmis, S. (1986) *Becoming Critical: Education, Knowledge and Action Research*. Falmer Press, Basingstoke, UK.

Chimpololo, A. (2017) Glaring mismatch in the demand and supply of technical, entrepreneurial, vocational education and training (TEVET) skills in the agricultural sector: An analysis of the oil seeds and sugar industries in Malawi. *International Journal of Vocational and Technical Education* 9(7), 62–75.

Clark, B. R. (1998) Entrepreneurial pathways of university transformation. Ch. 1 in *Creating Entrepreneurial Universities: Organizational Pathways of Transformation*. Issues in Higher Education. Pergamon Press, Oxford, UK.

Court, D. (1980) The development ideal in higher education: The experience of Kenya and Tanzania. *Higher Education* 9, 657–680.

Edziwa, X. Chakamba, J. Jomo, C. and Chiweshe, M. (2012) Experiential learning in agriculture education: A Zimbabwean experience. *International Journal of Humanities and Social Science* 2(21), 280–285.

Fitzgerald, H.E. Bruns, K. Sonka, S.T. Furco, A. and Swanson, L. (2012) The centrality of engagement in higher education. *Journal of Higher Education Outreach and Engagement* 16 (3), 7–27.

Gudo, C. (2014) Financing higher education in Kenya: Public-private partnership approach. *International Journal of Educational Policy Research and Review* 1(1), 1–5.

Harden, S.B. and Loving, K. (2015) Outreach and engagement staff and communities of practice: A journey from practice to theory for an emerging professional identity and community. *Journal of Community Engagement and Scholarship* 8(2), 7–15.

Kalule, W.S. Odongo, W. Kule, E. Ndyomugyenyi, K.E. Omara, P. and Ongeng, D. (2016) Conceptualizing the student-centered outreach model for experiential learning and community transformation. *African Journal of Rural Development* 1(3), 219–227.

Kalule, W.S. and. Ongeng, D. (2016) *Embedding Community Engagement in University Training Curricula: Experiences from Gulu University in Uganda. RUFORUM Working Document Series* 14 (1), 409–413.

Kisiangani, B.W. (2019) *Effect of farm attachment program on farm technology adoption among small holder farmers in Baringo and Nakuru counties, Kenya*. MSc thesis in Agricultural Economics, Egerton University, Kenya.

Kivati, G. (2017) The role of Kenya's formal higher education in sustainable development within the context of globalization. In Leal Filho, W. Skanavis, C. do Pasco, A. Rogers, J. Kuznetsova, O. *et al.* (eds.), *Handbook of Theory and Practice of Sustainable Development in Higher Education*. World Sustainability Series, Springer International Publishing. Available at: https://doi.org/10.1007/978-3-319-47889-0_2 (accessed 23 April 2021).

Kolb, D.(1984). *Experiential Learning: Experience as a Source of Learning and Development*. Prentice Hall, Englewood Cliffs, New Jersey, USA.

Mackatiani, C. Imbovah, M. Imbova, N. Gakungai, D.K. (2016) Development of education in Kenya: Influence of the political factor beyond 2015 MDGs. *Journal of Education and Practice* 7(11), 55–60.

McCann, B.N. Cramer, C.B. and Taylor, L.G. (2015) Assessing the impact of education and outreach activities on research scientists. *Journal of Higher Education Outreach and Engagement* 19(1), 65–78.

Mosweunyane, D. (2013) The African educational evolution: From traditional training to formal education. *Higher Education Studies* 3(4), 50–59.

Mungai, N.W. Ambula, M. Chege, N.W. Mberia, P. Ombati, J.M. *et al.* (2016) Experiential learning for agricultural students in institutions of higher learning: The case of Egerton University. *African Journal of Rural Development* 1(3), 229–237.

Mungai, N.W and Ombati, J.M (2016) *Survey of Experiential Learning Activities for Agricultural-Based Programmes in Institutions of Higher Learning in East Africa*. Report submitted to RUFORUM for Project No. RF 2014 NG 12. RUFORUM, Kampala, Uganda.

Muriisa, R.K. (2014) Rethinking the role of universities in Africa: Leadership as a missing link in explaining university performance in Uganda. *Journal of Higher Education in Africa* 12(1), 69–92.

Muse, W.V. (2000) University outreach in the twenty-first century. In *University Outreach: University Connections to Society*. Auburn University, Alabama, USA. Available at: http://www.auburn.edu/outreach/publications/auoutreachbook.pdf (accessed 23 April 2021).

Nampala, P. Kityo, R. Makuma-Massa, H. and Adipala, E. (2017) Tracing the evolution of higher education institutions and linkage to rural development in Africa. *African Journal of Rural Development* 2(2), 143–151.

Neuchâtel Group (2008) *Common Framework on Market-Oriented Agricultural Advisory Services*. Technical Centre for Agricultural and Rural Cooperation (CTA), Wageningen, The Netherlands.

New African Magazine (2017) *Higher education in Africa: Start of a new golden age?* May 17. Available at: http://newafricanmagazine.com/higher-education-africa-start-new-golden-age/ (accessed 23 April 2021).

Ntimi, N.M. Samuel, N. Fongwa, S.N. Wangenge-Ouma, G. (2016) The scholarship of university-community engagement: Interrogating Boyer's model. *International Journal of Educational Development* 49, 126–133.

O'Boyle, L. (1983) Learning for its own sake: The German University as nineteenth-century model. *Comparative Studies in Society and History 25*(1), 3–25. Available at: http://www.jstor.org/stable/178570 (accessed 23 April 2021).

Odongo, W. Kalule, W.S. Kule, K.E. Ndyomugyenyi, E.K. Omara, P. and Ongeng, D. (2017) Responsiveness of agricultural training curricula in African universities to labour market needs: The case of Gulu University in Uganda. *African Journal of Rural Development* 2(1), 67–76.

O'Meara, K.A. Sandmann, L.R. Saltmarsh, J. and Giles Jr., D.E. (2011) Studying the professional lives and work of faculty involved in community engagement. *Innovative Higher Education* 36, 83–96. Available at: https://doi.org/10.1007/s10755-010-9159-3 (accessed 4 June 2021).

Rivera, W.M. and Sulaiman, V.R. (2009) Extension: Object of reform, engine for innovation. *Outlook on Agriculture* 38(3), 267–273.

RUFORUM (Regional Universities Forum for Capacity Building in Agriculture) (2016) *Strengthening University Engagement for Community Transformation and Impact*. Concept note for plenary session at the Fifth African Higher Education Week and RUFORUM Biennial Conference, Cape Town, South Africa, 2 July.

Spielman, D.J. Ekboir, J. Davis, K. and Ochieng, C.M.O. (2008) An innovation systems perspective on strengthening agricultural education and training in sub-Saharan Africa. *Agricultural Systems* 98(1), 1–9. Available at: https://doi.org/10.1016/j.agsy.2008.03.004 (accessed 3 March 2021).

Tornatzky, L.G. and Rideout, E.C. (2014) *Innovation U 2.0: Reinventing University Roles in a Knowledge Economy*. Available at: https://www.Innovation-U.com (accessed 23 April 2021).

Velasco, I. and Harder, M.K. (2014) From attitude change to behaviour change: Institutional mediators of education for sustainable development effectiveness. *Sustainability* 6, 6553–6575. Available at: https://doi.org/10.3390/su6106553 (accessed 23 April 2021).

Wangenge-Ouma, G. and Nafukho, F.M. (2011) Responses to conditions of decline: The case of Kenya's public universities. *Africa Education Review* 8(1), 169–188.

Woods, L. Willis, J. Wright, D.C. and Knapp, T. (2013) Building community engagement in higher education: Public sociology at Missouri State University. *Journal of Public Scholarship in Higher Education* 3, 67–90.

Zeelen, J. (2012) Universities in Africa: Working on excellence for whom? Reflections on teaching, research, and outreach activities at African Universities. *International Journal of Higher Education* 1(2), 157–165.

10 Quality Assurance in Tertiary Agricultural Education in Africa

Violet Makuku[1]* and Young Kafui Abel Etsey[2]

[1]*The Association of African Universities, Accra, Ghana;* [2]*University of Cape Coast, Ghana*

10.1 Introduction

The quality of higher education has become a global concern in recent decades (Salmi, 2015). In Africa, widespread decline in the quality of institutions and programmes is an oft-mentioned theme in papers and presentations on higher education (Materu and Righetti, 2010; Shabani *et al.*, 2014). Among the key reasons are rapid increases in the number of higher-education institutions (HEIs) and student enrolments, along with the decline in funding by national governments. Other reasons include the poor quality of education at primary and secondary levels, inadequate higher-education infrastructure, insufficient training of academic staff, poor institutional governance, absent or weak national quality-assurance agencies (QAAs), weak institutional quality-assurance mechanisms and frequent university closures due to strikes (Teferra and Altbachl, 2004). In response to these concerns, many African countries established mechanisms and procedures to maintain and improve the quality of higher education, mainly by establishing 16 QAAs in some African countries, including South Africa and Zimbabwe, mostly during 1998–2007 (Materu, 2007). According to Okebukola (2012), 21 QAAs had been established by 2012; he predicted that number would reach 50 by 2020.

This chapter focuses on the quality of higher education as a means of transforming learning and knowledge outcomes, measured in terms of graduates, research products and outreach activities. Learning outcomes include students' competencies - i.e. knowledge, skills and attitudes - and the degree to which they are relevant to workplace needs. Knowledge outcomes take the form of ideas, analyses, inventions and publications; they can be measured based on the extent to which they address current and emerging problems of economic and social importance. The chapter begins by defining key terms and explaining how quality assurance fits within the management and operation of universities. The chapter then explores the role of organizations, networks, associations, activities and initiatives in developing and implementing sound quality-assurance practices across institutional, national, regional and pan-African levels.

*Corresponding author: vmakuku@gmail.com

 Transforming Tertiary Agricultural Education in Africa (eds D. Kraybill, J. Lynam and Adipala Ekwamu)
DOI: 10.1079/9781789246544.0010

10.2 Defining Quality Assurance and Related Terms

Quality

In a comprehensive report on quality assurance in Africa, Materu (2007) noted the difficulty of defining quality in the context of HEIs given a lack of common concepts and standards (Table 10.1). Definitions range from quality as a measure of excellence, to quality as perfection, as value for money, as customer satisfaction, as fitness for purpose and as transformation of learners (Smout, 2002). Employers expect HEIs to produce graduates with knowledge and skills fit for the purpose of problem-solving and innovation. Governments, donors and parents expect HEIs to deliver value for money by providing graduates with knowledge, skills and experiences which result in employment or self-employment. The academic community, alumni, students and parents expect HEIs to deliver exceptional programmes, produce exceptional graduates and create and maintain a reputation of distinction and excellence. Education ministries and agencies expect HEIs and related systems to adhere to government-mandated academic and administrative standards. Citizens, governments and donors expect HEIs to be accountable in the use of funds, the management of assets and the maintenance of academic standards. All stakeholders expect HEIs to have a transformative impact on their constituents, including the students, graduates, businesses and communities which rely on their education programmes, research outputs and outreach activities.

Quality Assurance

Quality assurance has been defined in numerous ways by numerous authors (Vlăsceanu, Grünberg and Pârlea, 2007; Materu and Righetti, 2010; and Ryan, 2015). In the context of this chapter, quality assurance refers to a planned, systematic and continuous process of monitoring, evaluating and improving or enhancing the quality of higher-education systems, services, programmes and graduates in order to meet or surpass stakeholders' expectations. This is done by maintaining relevant education systems which minimize or solve contemporary socioeconomic challenges. In the context of tertiary agricultural education (TAE), these challenges are linked to the Sustainable Development Goals (SDGs). While university administration is an important dimension of quality assurance, to be effective, quality assurance must be factored into all aspects of a university's functioning. In fact, it should be a widely-held cultural value by the whole university community, including academic support services and staff, such as laboratory technicians, examination officers, secretaries, maintenance workers, etc. This is why more recent definitions of quality assurance include elusive aspects, such as an institutional quality

Table 10.1. Conceptions of quality from the various perspectives of higher-education stakeholders

	Conceptions of quality	Stakeholders
1.	Fitness for purpose	Employers
2.	Value for money	Governments, donors, parents/guardians
3.	Exceptional programmes and graduates	Academic community, alumni, students, parents/guardians
4.	Adherence to set standards	Education ministries/agencies, accreditation organizations
5.	Accountability	Citizens, governments, donors
6.	Transformation	All stakeholders
7.	Excellence	Employers, students, institutions, quality assurance agencies, professional bodies
8.	Brand	Institutions, students, employers
9.	Value addition	Students, employers, donors, governments
10.	Meeting or surpassing set standards	Employers, students, quality assurance agencies, professional bodies
11.	'Doing right when no one is looking' (Henry Ford)	All stakeholders

Source: Constructed by authors.

culture (Hayward, 2017), defined by Bendermacher *et al.* (2016, p. 39) as 'structural/managerial and cultural/psychological elements [acting] in synergy to continuously improve education'.

It has been noted with concern that most institutions concentrate on, invest more in and pay special attention to quality assurance in academic aspects at the expense of all other university activities. Consequently, the need to pay attention to all facets of university services must be emphasized and acknowledged. This encompasses both academic and supporting departments, including such functions as staff recruitment, student and staff support services, student enrolments and examinations, the library, accommodation, catering, building works and maintenance, and so on. University personnel should also consider their colleagues to be their customers in order that they, in turn, can improve their services to students - who in most cases are considered the university's most important customers.

Internal and External Quality Assurance

Accreditation is an aspect of external quality assurance and is mostly undertaken by national QAAs through institutional audits, making institutions accountable to their stakeholders and helping to improve and enhance the quality in these institutions. According to Materu (2007), accreditation is a process of self-study, and external quality assurance in higher education is used to scrutinize an institution and its programmes to uphold quality standards and identify the need for quality improvements. External reviews typically involve site visits. The purpose of accreditation is to determine whether an educational institution meets the minimum standards set by an external body, such as a government, national QAA or professional association. Internal quality assurance is carried out by HEIs through systems, tools, processes and procedures under the guidance of the government, relevant ministry and professional bodies, which set regulations and standards beyond the purview of the higher-education sector itself. Data for internal quality assurance are primarily generated from sources within the institution, although they may be supplemented by audits or reviews conducted by external quality-assurance experts. While formal quality-assurance processes are administrative, the term 'quality culture' encompasses both the day-to-day and long-term priorities associated with quality assurance across all levels of an institution.

A quality culture can be promoted top-down by university administration, but it cannot be enforced because it is founded in the attitudes and behaviours of the people who make up the institution. A quality culture can only be encouraged through education, awareness and lobbying to promote buy-in by constituents. A quality culture is cultivated through clear and open communication, shared ownership and the empowerment of staff at all levels to enable them to carry out their duties to a high standard. When a quality culture exists, it is evident in a widespread commitment to quality by staff and through pervasive adoption of quality-enhancing practices in all aspects of university business, including teaching and learning, research, administration, student services and stakeholder engagement. No common blueprint exists for a quality culture; it should be based on the institutional values and vision of being and doing the best within given circumstances, national needs and global best practices.

Building a quality culture involves institutionalizing and adopting a culture which makes quality assurance a powerful tool for raising the level of performance of HEIs to meet the expectations of stakeholders. A quality culture in higher education is a way of life in a university community, whereby staff continuously and systematically strive to improve governance and management, teaching, research, the campus environment, punctuality, efficiency, effectiveness and service delivery, among other things. A quality-assurance unit (or department or directorate) can serve as a change agent to aid the transformation of university personnel's attitudes and foster a mindset which embraces change and new ways of doing things. Institutions are advised to establish appropriately-sized units based on their student populations and available resources, and especially their financial resources. Quality-assurance education and awareness campaigns can be mounted through special events, competitions, posters and billboards and branded clothing, among other initiatives.

10.3 The Role of Quality Assurance

Education systems and institutions engage in cycles of planning and doing (implementation), checking and adjusting (monitoring and evaluation for improvement), which are repeated in response to a calendar, a perceived need or the availability of resources. The quality of services, infrastructure, staff and student support, programme implementation, curricula, teaching methods, research or outreach may improve or worsen over time. Quality assurance builds feedback and self-correction into the cycle of planning and doing (Fig. 10.1). As a plan is put into action, outcomes are checked and implementation is adjusted to increase the likelihood of achieving goals. Even the best plans require adjustments during implementation; foresight is never perfect, and circumstances or conditions are likely to change. The 'plan-do-check-adjust model' - developed by the management scholar, W. Edward Deming - is commonly applied to the management of industry and education (Koiesar, 1994). This simple model integrates quality assurance as a core component of managing and operating an institution. A first step is to incorporate a quality-assurance strategy in an institution's strategic plan, as was done at the University of Mauritius (Materu, 2007). Identifying quality assurance as central to the institution's administrative strategy facilitated the inclusion of appropriate structures, processes and activities across all levels of university operations, which, in turn, have positive impacts on the quality of governance, teaching, research, curricula, outreach, infrastructure and so on. Such positive changes are possible within a framework of transparency, accountability, efficiency and communication and documentation.

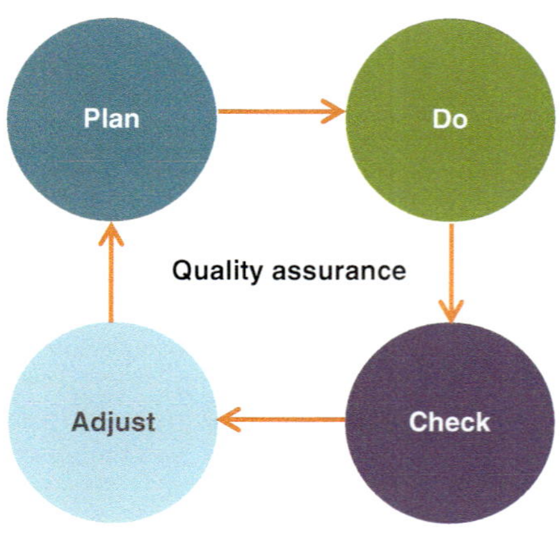

Fig. 10.1. Integrating quality assurance into higher education planning and implementation
Source: Adapted by editors based on Koiesar (1994).

10.4 Conceptual Framework

Quality-management mechanisms can and should help educational institutions become more purposeful, accountable, transformative and successful in achieving overall excellence (Schindler *et al.*, 2015). This begins with a clear and transparent vision and mission, which the institution strives to fulfil. Thereafter, standards - typically set by national or international accrediting and regulatory authorities - must be adopted and pursued via specific sets of goals with associated targets, time periods and resources. An accountable institution focuses on the requirements and expectations of its stakeholders in the design and delivery of educational products and services. The institution procures high-quality human and physical resources, including facilities, and uses them responsibly to meet those requirements and expectations. Teaching is a core service of HEIs, and accountability requires that teaching prepare students for employment. The accountable institution should strive continuously to improve the quality, relevance and usefulness of all its activities. An exceptional institution gains distinction by maintaining high-quality standards relevant to its purpose. Each of the various kinds of institutions within the higher-education sector has its unique purpose and stakeholders. Excellence is achieved, not by aspiring to reach the top of the value chain, but by establishing credibility, legitimacy, reputation, ranking and prestige in relation to peer institutions of the same type (Fig. 10.2).

A transformative institution creates a learning environment in which students acquire both technical and 'soft' skills and knowledge, and are able to actualize their professional and personal potential. This can be enhanced by incorporating time for community engagement, allowing students to learn by carrying out research activities in actual communities (see Ch. 9,

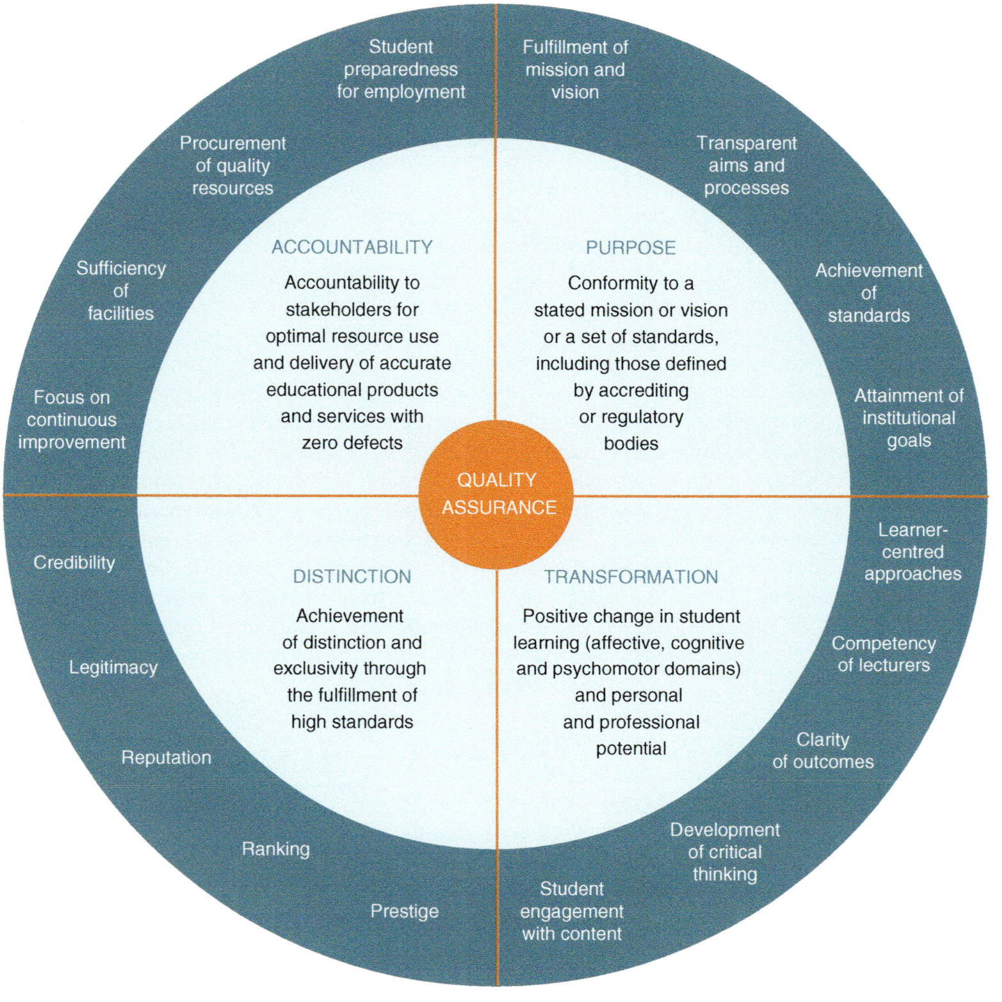

Fig. 10.2: Indicators of quality in the higher education sector
Source: Adapted by authors from Schindler *et al.* (2015, Fig. 1).

this volume). In this way, students are transformed while directly contributing to the transformation of the communities they work with, as well as achieving SDG 1 (no poverty) and SDG 2 (zero hunger). A transformative institution (i) sets clear outcomes in terms of learning goals, (ii) ensures that lecturers are competent through rigorous hiring processes and in-service training and (iii) uses student-centred curricula and teaching methods. Student-centred approaches supplement the traditional lecture format with experiential and hands-on activities in which students engage with the content and learn how to think critically and independently.

10.5 Pan-African Quality Assurance

The Association of African Universities (AAU) is an apex organization and forum for consultation, information exchange and cooperation among African HEIs. AAU interacts with regional and international bodies and supports networking among institutions in the areas of teaching, research, quality assurance and information exchange and dissemination. AAU also serves as the coordinating body for the African Union's Continental Education Strategy for Africa, 2016–2020 (African Union, 2015). In this regard, AAU sponsors activities to enhance institutional quality, including live discussion forums, webinars and workshops on such subjects as pedagogy, designing and implementing quality-assurance mechanisms, relevant research, information and communications technology, university management and so on. The African Quality Rating Mechanism (AQRM) is an institutional method for self-assessment, evaluation and improvement. AQRM is administered by AAU, and institutions participate on a voluntary basis. AQRM aims to bring institutional transformation in such areas as university governance, finance, teaching, research, industry linkages, community engagement and global networking. AQRM evaluations focus on an institution's strengths, areas needing improvement and recommendations at departmental, faculty and institutional levels. This mechanism is a valuable aid for enhancing institutions' quality-assurance mechanisms, thereby strengthening their quality cultures. Activities include self-assessments, site visits and verification by international quality-assurance experts, who also interview different university groups, including students, and document their analyses over a longer time frame compared with national QAAs (i.e. three days or more).

To this end, AAU is part of a consortium of at least four members in the Harmonisation of African Higher Education Quality Assurance and Accreditation (HAQAA) Initiative, which was conceived by the African Union Commission and is funded by the European Union Commission.[1] A mapping study undertaken by AAU under the HAQAA Initiative in 2016 revealed that both internal and external quality assurance were unevenly developed throughout Africa at institutional, national and regional levels. At that time, some countries - e.g. Zimbabwe, Ghana and Uganda, among others - had established national quality-assurance agencies (or units, directorates or departments) within their HEIs. The HAQAA Initiative focuses on harmonizing different academic programmes and durations, credit systems, procedures and evaluation mechanisms across countries to establish a minimum body of knowledge in all areas of specialization under the Pan-African Quality Assurance and Accreditation Framework. The tools adopted for this process include: (i) the African Standards and Guidelines for Quality Assurance, known as ASG-QA (African Union, 2018b), (ii) AQRM, as discussed above (African Union, 2018a) and (iii) the Continental Qualifications Framework, which aims to facilitate student and staff mobility across institutions and countries by standardizing national entry requirements and establishing comparable qualification levels.

The Continental Strategy for Technical and Vocational Education and Training (African Union, 2018c) was developed by the African Union Commission in conjunction with the ministries of higher education and ministries responsible for agricultural and technical vocational education and training. This strategy aims to reshape agricultural education curricula to ensure alignment with the needs of the workforce (see Ch. 13, this volume). The African Qualifications Verification Network coordinates qualifications authorities across Africa by promoting institutional linkages in order to streamline the process of accessing and hence verifying students' records, which in turn harmonizes the process of verifying students' and graduates' qualifications when they pursue further study or seek employment.

10.6 Regional-Level Quality Assurance

Several educational organizations, associations and networks in Africa provide leadership on quality assurance at the regional level:

1. **The Inter-University Council of East Africa**, which serves five East African countries, developed a quality-assurance manual and tools and procedures for use throughout the East African Region and beyond.
2. **Conseil Africain et Malgache de l'Enseignement Superieur** promotes quality assurance in 19 francophone HEIs in West Africa and Malagasy.
3. As part of its mission, the **Regional Universities Forum for Capacity Building in Agriculture** addresses quality assurance issues in HEIs focusing on agriculture and the STEM disciplines (science, technology, engineering, and mathematics).

A number of other regional organizations exclusively focus on quality assurance, aiming to assist institutions in building a quality culture and adopting quality-assurance mechanisms, along with harmonizing standards to facilitate staff and student mobility across instititions. Among the regional bodies are the Southern Africa Quality Assurance Network, which has developed a common credit system and is developing minimum knowledge standards for all academic programmes offered by multiple institutions within a country. The ultimate intention is to harmonize educational programmes across member countries of the Southern Africa Development Community. The Network for Higher Education Excellence in West Africa recently finalized its efforts (i) to harmonize programme content in agricultural sciences and related disciplines and (ii) to eliminate disparities within specialities, in graduation requirements and in the duration of different programmes in order to further enhance and promote staff and student mobility by reinforcing the comparability of qualifications and skills.

10.7 National-Level Quality Assurance

National QAAs promote awareness of quality issues and train, regulate, monitor and enforce quality standards in higher education. National QAAs typically develop and enforce standards for staff qualifications, staff-to-student ratios, curricula, Internet access, governance, infrastructure and facilities and other issues identified in ASG-QA. Typically, national QAAs also set the procedures for external quality assurance and external monitoring and enforcement carried out through audits and accreditations. Recommendations on the structure and functions of QAAs were developed under the HAQAA Initiative and are presented in the ASG-QA (African Union, 2018b), noting that a QAA should:

1. be an autonomous legal entity constituted at the national level with a clearly-defined mandate, scope and powers;
2. be independent in its operations, outcomes, judgements and decisions;
3. have adequate and appropriate human, financial and material resources to carry out its quality-assurance mandate effectively and efficiently;
4. have written and published vision and mission statements or objectives which take the national or regional higher-education context into account;
5. have clearly-defined structures which ensure sound and ethical governance and management, including good quality-assurance practices which support its mission and legal mandate;
6. have policies and processes for its own internal quality-assurance related to defining, assuring and enhancing the quality and integrity of its activities;
7. undergo periodic internal and external reviews for continuous improvement; and
8. be subjected to peer review, adopting the same ASG-QA guidelines.

The structure and function of national QAAs across Africa demonstrate great diversity, especially in terms of levels of autonomy, relationships with the government and the type of HEIs they regulate. In some countries, the national quality-assurance and accreditation bodies only regulate private institutions. However, if public institutions are not subjected to the same stringent measures as those of private universities to maintain quality, educational quality may become compromised. In Namibia, for example, the focus of the country's QAA was initially on private institutions only. As the quality of education offered by private institutions improved, however, public institutions realized they were being left behind and asked the national QAA agency to treat them as they treated private universities. Zimbabwe and Ghana provide examples of how national QAAs operate and how diverse their structures and functions can be. Zimbabwe has a single quality-assurance and accreditation agency, the Zimbabwe Council for Higher Education, which accredits both private and public institutions and their programmes, as well as carrying out institutional audits and training of higher-education personnel. In Ghana, the National Accreditation Board is charged with quality assurance and accreditation for both private and public institutions, while the National Council for Tertiary Education sets norms and standards and approves institutional budgets. Once the Board receives Council approval, it proceeds with accrediting institutions and their programmes. This links accreditation with funding, which complicates the issues involved, lengthens the accreditation process and encourages conflicts and contradictions among the QAAs.

10.8 Institutional-Level Quality Assurance

Materu (2007), noted that success at the national level cannot be achieved if HEIs do not have a strong culture of quality. HEIs have a primary responsibility for ensuring the quality of their own services, programmes and activities (Box 10.1). This can be executed and implemented efficiently through a well-established institutional quality-assurance unit, which spearheads the implementation of institutional and programmatic quality measures. Such a unit should provide technical support for quality-assurance procedures across the institution's various departments. This requires capacity building, education and awareness and publicizing institutional best practices, as well as designing, systematizing and institutionalizing quality-assurance tools, mechanisms, procedures and processes. Through training and internal communication, a quality-assurance unit can help foster a quality culture within an institution. To improve quality, HEIs engage in self-evaluation, submit to academic audits by external reviewers, conduct internal programme reviews and design and implement new programmes. Institutions should also undertake education and awareness campaigns on an annual basis to build their quality culture. Quality can be improved by shifting from teacher-centred to learner-centred pedagogical approaches, developing updated curricula, reviewing old curricula, adopting new and better teaching methods, expanding and improving classroom facilities, outfitting laboratories with new equipment and software, investing in staff development - including the induction of new staff and re-tooling of old staff - and providing improved student services - e.g. academic

Box 10.1. Examples of Institutional Quality-Assurance Practices

The University of Cape Coast (UCC) in Ghana undertakes quarterly consultative meetings between staff and students for the purpose of discussing issues related to teaching and learning. The dean and heads of department maintain an open door policy, and students are encouraged to discuss matters critical to their performance. At Botswana University of Agriculture and Natural Resources (BUAN), each department has a quality-assurance committee and a performance-management system. Information is gathered on how each department is performing, and awards are given to those who perform well. At both institutions, students assess lecturers and provide comments and suggestions for continuous performance improvement. Both institutions also use the results of accreditation reviews to ensure further improvements. An important aspect which has enhanced the quality of agricultural education at both institutions is participation in agricultural extension services, providing students with hands-on experience and greatly enhancing the quality of their education. Both institutions use external examiners for academic audits, which they found to be highly effective based on feedback, although some challenges arose in relation to the costs associated with engaging external examiners (not only in terms of remuneration, but also for transportation, accommodation and food, especially for non-local examiners). In some cases the timely availability of experienced external examiners was also an issue.

Quality-assurance practices at UCC's School of Agriculture have led to improvements in the quality of graduates as indicated by employer attitudes and graduate employment rates. At BUAN, the majority of graduates obtained government employment. BUAN's quality-assurance practices also resulted in improved research, indicated by the adoption and commercialization of technologies developed by faculty members. Both institutions undertook tracer studies to elicit information on employer attitudes towards their agriculture graduates. At both institutions, quality-assurance practices resulted in improved research outputs, indicated by the rate of articles published and the adoption and commercialization of technologies developed by faculty members. Staff-development practices were in place at both institutions. BUAN encouraged its staff to undertake PhD-level studies by providing paid study leave, and additional opportunities for short-term training, both locally and outside the country, were available. The rate of staff promotion had increased over time at UCC, and PhD and MSc studies had been developed for some programmes. In addition, UCC collaborates with other universities, and the number of research grants won by the School of Agriculture had increased. The quality and quantity of UCC's research publications had been highly rated based on the number of the papers accepted in well-known, high-impact journals.

Source: Compiled by authors based on information provided by UCC and BUAN.

remediation, counselling, career development and so on (see Ch. 7, this volume).

Guidance on institutional quality assurance is provided by ASG-QA, as previously described (African Union, 2018b). The purpose of ASG-QA is to support HEIs and QAAs in member states in Africa to (i) have a common framework and understanding of quality-assurance systems for teaching and learning among all stakeholders at pan-African, regional and national levels; (ii) develop mutual trust, thus facilitating the mobility of students and university personnel within and across countries; (iii) enhance and improve the quality of higher education through self-assessment, external peer review and continuous monitoring and evaluation; (iv) promote transparency and accountability by providing appropriate quality-assurance information to the public; (v) support HEIs in developing a sustainable quality culture; (vi) promote international competitiveness within Africa's higher education system and (vii) support the production of relevant teaching and learning resources, as well as student assessment mechanisms. As previously described, ASG-QA provides guidance on both internal and external quality assurance (African Union, 2018b). For internal quality assurance, the ASG-QA identifies 13 standards with associated guidelines, which HEIs are expected to implement through internal administrative and academic processes (Table 10.2).

Table 10.2. Guidelines for internal quality assurance in higher education

No.	Standard
1.	Vision, mission and strategic objectives
2.	Governance and management
3.	Human resources
4.	Financial resource management
5.	Infrastructure and facilities
6.	Student recruitment, admission, certification and support services
7.	Design, approval, monitoring and evaluation of study programmes
8.	Teaching, learning and assessment
9.	Research and innovation
10.	Community engagement
11.	Information management system
12.	Public communication
13.	Collaboration, staff and student mobility

Source: African Union (2018b).

ASG-QA (African Union, 2018b) is one of the most important reference documents guiding institutions in devising and implementing AAU's quality-assurance strategies; this is mainly because it was derived through a mapping study of the standards used by all African countries.

10.9 Conclusion and Recommendations

Through a survey of HEIs, Loukkola and Zhang (2010) identified procedures and processes used in European universities which are relevant to African HEIs. Results indicated that student assessment procedures should be clearly outlined within the institutional framework and be designed to measure the attainment of intended learning outcomes. The procedures should (i) have clear and publicly-available criteria for grading, (ii) have pre-defined examinations or other assessment methods in place, (iii) have regulations covering student absences, illness and other mitigating circumstances and (iv) ensure that assessments are conducted securely in accordance with the institution's stated procedures. Students should be informed of assessment procedures either through student portals, course outlines, study guides or websites, and checks should also be in place to ensure that procedures are followed. Internal evaluation processes at institutional, faculty and departmental levels should provide feedback for strategic planning at both short (i.e. quarterly) and longer term (i.e. annual) intervals. Strategic gaps should be identified, and action plans formulated to address the gaps. The AQRM and ASG-QA guidelines could be used for this purpose. It is also important that institutions conduct regular surveys of staff, students and external stakeholders to analyze their perceptions of the institution's quality-assurance strategy and implementation. Institutional quality assurance should also include course reviews through collaborative and collegial faculty-centred processes, augmented by external examiners' feedback. Reviews should focus on improving course content and teaching methods, rather than on individual instructors, the goal being to ensure that courses remain up-to-date.

Institutions must also have processes in place to ensure the quality of research activities.

These processes include: (i) internal seminars where research projects and ideas can be discussed, (ii) internal and external peer-reviews of research projects, (iii) quality-checking of scientific articles sent for publication in scientific journals, (iv) statistical assessments of articles published, including impact monitoring and (v) performance indicators defined for each research group, department or faculty. Institutions must also serve the needs of their communities and society in general. This requires that - in addition to learning, undertaking research and participating in internships and farm attachments - students should also spend time working with their local communities. Processes and tools need to be established to ensure the quality of these activities, including: (i) defining performance indicators for each service, (ii) monitoring the impact of these services and activities in terms of research, teaching and learning, (iii) monitoring the number of patents, technology transfer agreements and cooperation agreements stemming from these activities, (iv) monitoring interactions with external stakeholders (including alumni) and (v) eliciting feedback from stakeholders, as well as institutional responses to this feedback, e.g. through questionnaires. The COVID-19 outbreak raised the urgent need to ensure universities are equipped to offer virtual teaching and learning environments. Institutions should be heavily guided by ASG-QA Standard 5 in this regard, which encourages institutions to have adequate and appropriate infrastructure, facilities and resources to support teaching, learning and research, as well as other services and activities.

HEIs are the custodians of institutional quality, which accumulates to national, regional and pan-African levels. Institutional quality-assurance systems should be guided by the ASG-QA standards outlined in this chapter. Organizational structures and activities must be available to support internal quality-assurance processes. These include having: (i) a central quality-assurance unit, (ii) a university-wide quality-assurance committee which represents all categories of employees and (iii) additional faculty-level, school and departmental committees. In addition to internal quality assurance, structures should facilitate compliance with external quality-assurance processes. The institution's overarching quality-assurance structures should be under the office of the vice chancellor, rector or president of the institution, or a specially-assigned vice-president or vice-chancellor. The centralized quality-assurance unit (or department or directorate) should have a full-time director and specialized staff. Faculty and departmental quality assurance committees throughout the institution should carry out assigned quality-assurance duties. Personnel working in the centralized quality-assurance unit require a strong background in qualitative, quantitative and mixed research methods such that they can design appropriate data collection tools and analyze and interpret data. These personnel should be well versed in information and communications technologies so that the quality-assurance system can be run efficiently and effectively. As previously mentioned, this is even more imperative in a post-COVID-19 era. Quality-assurance personel must be professional, fair and possess excellent communication skills so they can garner the respect of their institution's community and become effective promoters of a culture of quality.

Endnote

[1] The initiative was guided by the Addis Convention (2014), which is the former Arusha Convention (1981). The consortium of partners included the AAU, German Academic Exchange Services, European University Association, University of Barcelona and European Network on Quality Assurance.

References

African Union (2015) *Continental Education Strategy for Africa 2016–2025*. Available at: https://au.int/sites/default/files/newsevents/workingdocuments/27671-wd-continental_strategy_for_education_cesa_16-25_english_0.docx (accessed 28 April 2021).

African Union (2018a) *African Quality Rating Mechanism: Consolidated Evaluation Report. (Fifteen Higher Education Institutions)* Available at: https://haqaa.aau.org/wp-content/uploads/2019/03/AQRM-CONSOLIDATED-REPORT_AAU_FINAL-.pdf (accessed 28 April 2021).

African Union (2018b) *African Standards and Guidelines for Quality Assurance in Higher Education*. Available at: https://haqaa.aau.org/wp-content/uploads/2018/07/ASG-QA_2018_edited_EN.pdf (accessed 28 April 2021).

African Union (2018c) *Continental Strategy for Technical and Vocational Education and Training (TVET): To Foster Youth Employment*. Available at: https://au.int/sites/default/files/pressreleases/35308-prtvet-english_-_final_2.pdf (accessed 28 April 2021).

Bendermacher, G.W.G. oude Egbrink, M.G.A. Wolfhagen, I.H.A.P. and Dolmans, D.H.J.M. (2016) Unravelling Quality Culture in Higher Education: A Realist Review. *Higher Education* 73 (1), 39–60. Available at: https://doi.org/10.1007/s10734-015-9979-2 (accessed 28 April 2021).

Hayward, F.M. (2017) Reflections on two decades of quality assurance and accreditation in developing economies. *Planning for Higher Education Journal* 46(1), 21–32.

Koiesar, P.J. (1994) What Deming told the Japanese in 1950. *Quality Management Journal* 2 (1), 9–24.

Loukkola, T. and Zhang, T. (2010) *Examining Quality Culture. Part 1: Quality Assurance Processes in Higher Education Institutions*. European University Association, Brussels.

Materu, P. (2007) *Higher Education Quality Assurance in Sub-Saharan Africa: Status, Opportunities, and Promising Practices*. World Bank, Washington, DC. Available at: https://doi.org/10.1596/978-0-8213-7272-2 (accessed 28 April 2021).

Materu, P. and Righetti, P. (2010) Quality assurance in Sub-Saharan Africa. *Research in Comparative and International Education* 5 (1), 3–17.

Okebukola, P. (2012) *Quality Assurance in Higher Education: The African Story of Achievements and Outstanding Needs*. Washington, DC. Available at: https://www.academia.edu/5158108/Quality_Assurance_in_Higher_Education_The_African_Story_of_Achievements_and_Outstanding_Needs (accessed 28 April 2021).

Ryan, T. (2015) Quality assurance in higher education: A review of literature. *Higher Learning Research Communications* 5 (4). Available at: https://doi.org/ 10.18870/hlrc.v5i4.257 (accessed 28 April 2021).

Salmi, J. (2015) *Is big brother watching you?* Available at: https://www.chea.org/big-brother-watching-you-evolving-role-state-regulating-and-conducting-quality-assurance (accessed 28 April 2021).

Schindler, L. Welzant, H. Puls-Elvidge, S. and Crawford, L. (2015) Definitions of quality in higher education: A synthesis of the literature. *Higher Learning Research Communications* 5 (3): 3. Available at: https://doi.org/10.18870/hlrc.v5i3.244 (accessed 28 April 2021).

Shabani, J. Okebukola, P. and Oyewole, O. (2014) Quality assurance in Africa: Towards a continental higher education and research space. *International Journal of African Higher Education* 1 (1), 139–171. Available at: https://doi.org/10.6017/ijahe.v1i1.5646 (accessed 28 April 2021).

Smout, M. (ed) (2002) *Quality Assurance in South African Universities: Views from SAUVCA's National Quality Assurance Forum*. SAUVCA, Pretoria, South Africa.

Teferra, D. and Altbachl, P.G. (2004) African higher education: Challenges for the 21st century. *Higher Education* 47(1), 21–50. Available at: https://doi.org/10.1023/B:HIGH.0000009822.49980.30 (accessed 29 June 2021).

Vlăsceanu, L., Grünberg, L. and Pârlea, D. (2007) *Quality Assurance and Accreditation: A Glossary of Basic Terms and Definitions*. Revised and Updated Edition. United Nations Education, Scientific and Cultural Organization, Bucharest.

11 Gender Considerations and Practices for Transforming Tertiary Agricultural Education in Africa

Jemimah Njuki[1]* and Salome Bukachi[2]

[1]*International Food Policy Research Institute, Africa Region;* [2]*Institute of Anthropology, Gender and African Studies, University of Nairobi*

11.1 Introduction

Education is hailed as a key driver for economic, social and sustained development, as well as an effective way to reduce poverty and promote prosperity. It is also listed as one of the priorities in various global development agendas, such as the United Nations' Sustainable Development Goals (SDGs) of the 2030 Agenda for Sustainable Development (United Nations, 2015). Tertiary agricultural education (TAE) contributes to agricultural development by strengthening individuals' capacity to innovate - i.e. to introduce new or existing products and processes of social or economic relevance to smallholder farmers, in particular, as well as other agents along agricultural value chains (Davis *et al.*, 2008). Given that agriculture is the backbone of many African economies, it needs significant attention to ensure maximum performance and output. Nevertheless, it is not appropriate to focus on agricultural education and outputs without considering gender issues. Women play numerous key roles in agriculture, especially in Africa, yet they face multiple barriers to effective participation in the sector. They need gender-responsive agricultural products, services and systems. Gender equality in education is an integral aspect of rural development and economic growth. When included as a tenet of development, gender mainstreaming requires urgent attention in universities and colleges because it goes beyond social justice and equity (Aissetou *et al.*, 2015) to promoting economic growth.

Demystifying the concept of gender equality in education through a realistic framework can support the design, management and evaluation of education programmes and investments. UNESCO (2003) outlines three gender concepts in education: (i) gender parity, (ii) gender equity and (iii) gender equality. Each concept involves different metrics and different types of interventions (Box 11.1). This chapter explores gender considerations and practices in the context of TAE in terms of the issues involved, the strategies needed and the barriers to be overcome.

11.2 Key Outcomes of Gender Equality in Education

UNESCO outlines four key outcomes of integrating gender issues into education equality: (i) equality of access, (ii) equality in the learning process, (iii) equality of educational outcomes and (iv) equality of external results (UNESCO, 2003; EQUATE, 2008).

*Corresponding author: j.njuki@cgiar.org

 Transforming Tertiary Agricultural Education in Africa (eds D. Kraybill, J. Lynam and Adipala Ekwamu)

DOI: 10.1079/9781789246544.0011

Box 11.1. Definitions of key terms

Gender parity is attained when the same proportion of male and female students - relative to their respective age groups - enter the education system, achieve educational goals and advance through the different educational cycles . Reaching parity in enrolment is necessary, but not sufficient, for achieving equality and should be considered a 'first stage' measure of progress towards gender equality in education.

Gender equality in education means that male and female students have equal opportunities to realize their full human rights and contribute to and benefit from economic, social, cultural and political development.

Gender equity is the process of treating women and men fairly. To ensure fairness, measures must be available to compensate for the historical and social disadvantages which prevent girls and boys from engaging on a level playing field. Equity does not imply treating all learners the same because many factors could disadvantage students in having a chance to achieve equitable outcomes.

Source: Compiled by authors from Subrahmanian (2005) and UNESCO (2003).

These expected outcomes aid in reviewing gender interventions and progress within TAE institutions. The framework, described below, is a useful guide (Fig.11.1). It should be systematically employed in initiatives seeking to transform TAE in Africa, and is a useful lens for formulating or reviewing programmes intended to achieve gender parity and equity in education.

Equality of Access

This refers to equitable opportunities for women and men to gain admission to formal, non-formal or alternative approaches to basic and tertiary education, and is a principle measure for gender parity. Actual attendance, rather than enrolment, is a better indicator of whether access has been achieved (EQUATE, 2008). Programmes or policies addressing this issue in tertiary education should include the provision of scholarships, equitable entry requirements, a gender-sensitive and safe accommodation and learning environment, mentorship programmes and gender-sensitive codes of conduct to provide a safe environment. A monitoring framework which entails frequent data collection disaggregated by sex is also critical to demonstrate actual attendance.

Equality in the Learning Process

This refers to equitable treatment and attention given to both women and men such that they have equal opportunities to learn (EQUATE, 2008). This means that women and men are exposed to the same curricula, although the coursework may be taught differently to accommodate male and female students' different learning styles (EQUATE, 2008). Equality in the learning process also means that all learners should be exposed to teaching methods and materials which are free of stereotypes and gender bias. In addition, it means that male and female students should have the freedom to learn, explore and develop skills across all academic and extracurricular offerings (EQUATE, 2008). According to Mangheni *et al.* (2010), in terms of course content and the learning experience, formal and informal delivery of TAE curricula often include gender bias, which can affect gender-specific student attrition and retention rates. Classroom practices, course materials and course content reflect the underlying values of institutions and wider society, incorporating gender biases and stereotypes, which can hinder gender-sensitive learning (Mangheni *et al.*, 2010). University curricula and knowledge production are heavily biased towards male preferences (Rathgeber, 2013), hence this needs to be addressed through a gender-equity lens.

Equality of Educational Outcomes

Aside from ensuring that outcomes are based on students' individual talents and efforts, equality of educational outcomes refers to equal opportunities for fair academic achievement by both male and female students. Yet biases exist, especially when it comes to women in the agricultural sciences. Useful metrics include the rates of

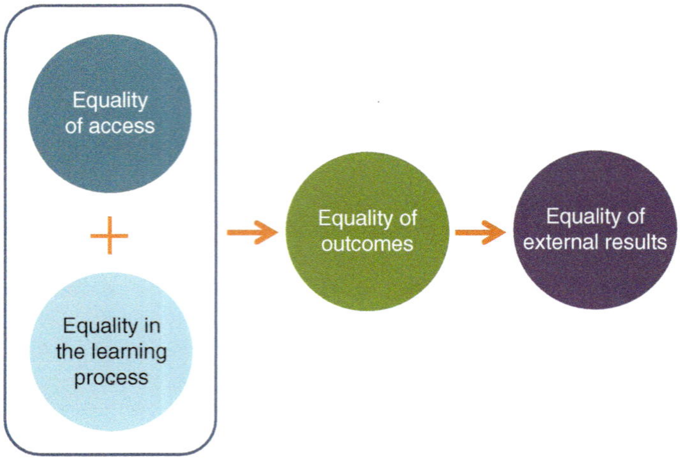

Fig 11.1. Key outcomes of gender equality in education
Source: Devised by authors based on UNESCO (2003) and EQUATE (2008).

transition of male and female students from one level to the next, along with other metrics of educational performance. To ensure fair chances for achievement, other factors - such as the length of school careers and academic qualifications and diplomas, including mechanisms for evaluating individual achievement - should also not differ based on a person's sex (EQUATE, 2008). TAE-related policies and programmes should aim to ensure equality of educational outcomes by instituting teacher trainings on (i) gender and its implication in teaching and fairness in student assessments, (ii) the development of gender-responsive student assessments and (iii) mechanisms for frequent reviews of examinations and assessments to ensure the language and examples used are free of gender stereotypes and bias.

Equality of External Results

This aspect is a product of equality in access, learning and educational outcomes. Ideally, it should reflect equality in female and male students' (i) post-college status; (ii) career opportunities, (iii) access to employment and remuneration once they complete their schooling; (iv) access to resources and (v) ability to contribute to, participate in and benefit from social, cultural and political activities. And while tracer studies have been done in several countries, including Kenya, Ghana and Uganda (Mugisha and Nkwasibwe, 2014; Kiliswa *et al.*, 2016), most of the data are not disaggregated by sex, making it difficult to determine the equality of external results. A tracer study of students of agricultural education and extension in Kenya, for example, found that 61% of 2008 graduates were employed by the Kenya Teachers' Service Commission as teachers of agriculture and biology, 14% were employed in the private sector and 10% were employed by the Ministry of Agriculture (Kiliswa *et al.*, 2016). It would be easy to disaggregate these data to determine whether women or men were more likely to follow certain career paths or were represented in some and not in others, but this continues to be a key gap in the research. These outcomes of gender integration as applied in education, though related, are not linear; they do not always work together to produce the required end in themselves, but require other interventions, external to education, to ensure that male and female students from TAE institutions get the maximum benefit from their education. Organizational policies to support both male and female students, and the pedagogy, all contribute to determining the eventual outcomes.

11.3 Barriers to Achieving Gender Equality

The gap in gender parity seems to widen as one moves along the different levels of education from the lower level-primary to the higher levels as disparities emerge in secondary school, then the gap widens at tertiary level. These issues are further exacerbated as women progress in their careers in agricultural research. Barriers women face in accessing and benefiting from institutions of higher learning can be classified as sociocultural, institutional or economic, as is discussed below.

Sociocultural Factors

In most of Africa, traditional and cultural norms and practices continue to define the roles of men and women in public life, and what men and women can and cannot do. The preference for boys' education and not girls' education when resources are scarce continues to force many young girls and women out of the school system. Cultural beliefs and society's perceptions about the roles women should predominantly play (e.g. caregiving) have led to girls and young women being taken out of school before they reach university. Santos and Rubiano-Matulevich (2019) found a correlation between female enrolment levels and the gender gap in the proportion of time spent on unpaid domestic work. In some communities, the desirability of women for marriage is pegged to their education levels, but in some countries, daughters are withdrawn from school at puberty to be married off early, often for fear of unwanted pregnancy.

Nevertheless, some of these cultural and societal perceptions are changing as more women become educated, enter the job market and become role models for young girls and women. This increases the emphasis placed on educating girls. UNICEF (2021), for example, observed widespread agreement that the education of girls is one of the most important investments a developing country can make in its own future. In the long term, almost every other aspect of progress - from nutrition to family planning and from child health to women's rights - is profoundly affected by whether a nation educates its girls. Further cultural shifts in local and national systems and structures could potentially enhance the education of girls and increase the likelihood of attaining gender equality in education. Another challenge faced by women is balancing work and family demands, which women are more likely to face than their male counterparts (Prozesky and Beaudry, 2019). Addressing sociocultural factors may seem unrealistic short-term, which is why it is important to institute strategies, such as flexible graduate programmes and gender quota systems in higher education.

Institutional Factors

Papadópulos and Radakovich (2005) note that higher education can potentially be an environment which perpetuates gender bias, based on low female representation. Internal institutional practices - such as biased admission processes, non-flexible times and schedules and lack of facilities for women with children - can drastically affect whether women enrol and, when they do enrol, whether they remain in institutions of higher learning. Mukhwana *et al.* (2020) document several institutional barriers, such as stereotypes women face in the work environment as they pursue careers in traditionally male-dominated disciplines (science, technology, engineering and mathematics). Lack of physical infrastructure to meet the needs of female students - such as childcare facilities, restrooms and breastfeeding rooms - has a significant impact on the admission and retention of female students in higher education. Unfriendly work environments coupled with unsupportive team leaders and gender-blind policies have been reported as contributors to the gender gap in higher education. Female students also report various forms of sexual harassment, as well as lack of sexual harassment policies, as critical barriers to their participation in higher education (Daddieh, 2003; Mukhwana *et al.* (2020).

Economic Factors

Economic considerations also play a key part in contributing to lower female enrolment levels. For most households, the opportunity cost of sending young women to school is the labour forgone in the form of field work, care giving, market activities, and so on. For households living on the margin, lower economic power necessitates choices about who does and does not attend school. Families often need greater inducements - e.g. in the form of scholarships - to keep girls in school. This also applies to enrolment in tertiary education because, in the face of economic challenges and competing priorities, paying for graduate studies may be considered a luxury compared with employment or marriage, especially for young women.

11.4 Strategies to Achieve Gender Equality in Education

Strategies to Achieve Equality of Access

One of the key inclusion strategies not only involves ensuring that female students are represented in approximately equal numbers (i.e. parity), but also that policies, systems and infrastructure support the inclusion and retention of students of both genders. For example, while enrolment may be at parity, unless specific measures are put in place to support female students, they may be more likely to drop out or not graduate due to some of the constraints mentioned above. Africa has the lowest university enrolment rates compared with the rest of the world, but massive growth has occurred in the past two decades driven by such factors as: (i) liberalization of the global economy, (ii) peoples' awareness of the value of university education, (iii) favourable institutional and national policies, (iv) universal secondary school education and (v) expansion of universities and increased investments and funding.

As discussed in several chapters in this volume, despite significant growth in tertiary enrolment levels and an increase in the provision of higher education to traditionally underserved groups, systems of tertiary education in sub-Saharan Africa (SSA) have yet to be transformed to provide universal education accessible by all. Moreover, gender-based disparities persist in tertiary enrolment across SSA. Across SSA's tertiary education systems in 2013, on average, approximately 72 female students were enrolled for every 100 male students. This lower enrolment rate is attributed to such factors as: (i) unfriendly environments for women, (ii) insufficient female role models and (iii) societal values, beliefs and practices which mitigate against higher education for women (Bunyi, 2003). The disparity is even greater in science-related subjects, which are skewed towards male students (UNESCO, 2010). This disparity is ultimately reflected in the professional arena, where women are under-represented in key positions in the agricultural sector (Blackie *et al.*, 2009). For African countries, other than South Africa, the proportion of female students enrolled in agriculture remained under 50% as of 2007, with only 5 countries (South Africa, Nigeria, Botswana, Malawi and Kenya) reporting shares above 30% (Fig. 11.2). On average 34% of the students enrolled in agricultural science in 2007 were female. Most of these women were enrolled in BSc-level studies (83%), whereas only 13% and 4% were enrolled in MSc and PhD studies, respectively (Beintema and Di Marcantonio, 2010). According to Beintema and Di Marcantonio (2010), few female graduates are likely to progress to graduate levels, which leads to their under-representation in leadership positions in the agricultural sector.

Strategies shown to have successfully-increased women's enrolment levels can be categorized into broad - 'lift all boats'- strategies, which also benefit women, and strategies specifically targeting women. Examples of both of these are discussed below.

Expansion of university education. Tertiary education has significantly expanded in the past few decades in many countries in Africa. This is taking various forms, including increases in the number of government-funded universities, increases in the number of private universities, the introduction of fee-paying programmes in public universities and the introduction of flexible (evening, weekend and part-time) programmes at both public and private universities. According to Gandhi (2018), SSA underwent the fastest growth in tertiary gross enrolment ratio (GER) during 1970–2013 (4.3% annually, compared with the global average of 2.8%). Much of this can be attributed to university expansion. Africa was home to only about seven universities in the 1960s, but by 2005 this number had expanded to 316 public and 85 private universities (Kihara,

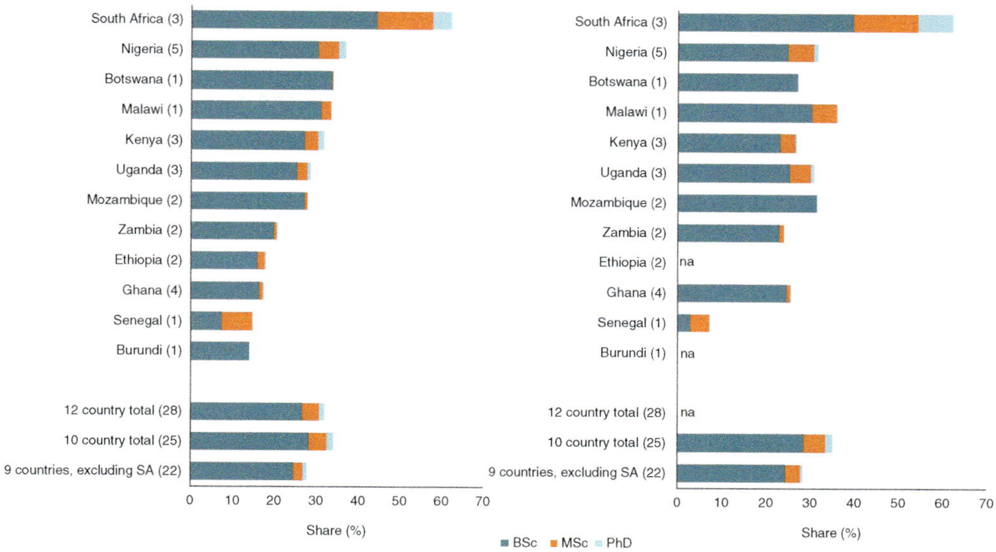

Fig 11.2. Female students enrolled versus those who graduate by qualification level, 2007
Source: Beintema and Di Marcantonio (2010, Fig. 9).
Note: Values in parentheses show the number of institutions for each country.

2005). In Kenya, for example, the higher education sector comprised 70 institutions as of 2016, making it one of Africa's largest. This expansion has been accompanied by an increase in the share of female students, but gender disparities remain (Fig. 11.2).

Government sponsored university programmes. The provision of government loans and sponsorship programmes can increase the number of students pursuing higher education, which includes women and especially those from economically marginalized groups. A study of the Government of Kenya's higher-education loan programme showed that female and male students were likely to receive equal amounts of government loans, despite the fact that female students required higher allocations (Odebero *et al.*, 2007).

Lower admission cut off points for women. To increase the number of women who enrol in tertiary institutions, some countries - e.g. Ghana, Kenya, Tanzania, Uganda, and Zimbabwe, among others - have articulated and implemented affirmative-action policies. Such policies allow female candidates who have attained the minimum required marks to enter public universities at between 1 and 1.5 points (2 points in the case of Zimbabwe) below their male counterparts (Bunyi, 2003). The rationale for this is the need to account for the social and cultural barriers female students face, which might hinder their performance. Many countries have achieved parity in undergraduate admissions through these measures. While lowering cut-off points for women has increased female enrolment levels, it has its detractors. Critics argue that tertiary institutions are meritocracies and that affirmative action dilutes educational standards. They further argue that these concessions foster the notion of women being intellectually weaker than men.

Scholarships targeting women. Many organizations have recognized the gender gap in tertiary enrolment levels and seek to close this gap by offering scholarships either specifically targeting women or with quotas for female applicants. While gender quotas do serve the purpose of increasing opportunities for women, as well as improving the gender balance, they are insufficient to address persistent inequalities. The rationale for women-only scholarships is twofold: (i) that broad scholarship programmes do not attract enough women to address the persistent institutional and structural causes of inequality and (ii) that women have specific needs which can be met by scholarships specifically targeting them.

Universities for women only. A few countries have tested the idea of universities for women only in order to increase women's enrolment levels and provide a conducive learning environment. The private university, St. Lucie Kiriri Women's University of Science and Technology, was established in Kenya in 2003. Zimbabwe's Women's University in Africa was established as a private university in 2004 with the goal of closing the gender gap in higher education; as of 2005, however, 15% of students in all of the university's programmes were male. While these universities have succeeded in increasing female enrolment levels in universities, critics argue that it is important for women to learn in a real-world environment which also includes men.

Gender mainstreaming in recruitment procedures. A broader approach to ensuring gender parity is introducing measures to avoid bias in the selection of students. Decades of research has made it clear that gender biases are nearly always present in recruitment decisions, subtly influencing our assessments about who is the 'right' or 'best' person. Although it is unlawful in many countries to advertise for or select candidates based on gender, research has shown that the language used in advertisements can subtly filter men and women out or cause certain groups not to apply or be selected. Advertisements for engineering, for example, are more likely to contain stereotypically male-oriented terms like 'competitive' or 'dominant' compared with advertisements for humanities-based courses (Gaucher *et al.*, 2011). When presented with such advertisements, women are more likely than men to rate these courses as unappealing and to perceive that women are not well represented in these courses and careers. Such language is unlikely to be a deliberate attempt to exclude women, but gender-coded language may not only deter individual women from applying for certain courses, but also reinforce wider social beliefs about who 'belongs' in those courses or institutions and who does not (Gaucher *et al.*, 2011). Using gender-neutral language eliminates these biases and potentially encourages more women to apply. Many organizations have also adopted anonymous recruitment procedures to minimize conscious and unconscious biases, but this approach has had mixed results. Although some studies have shown that anonymous recruitment procedures can increase the proportion of women receiving positive responses in selections, this has not always been the case for a number of reasons, such as unconscious biases and sociocultural issues which favour the qualities of self-confidence, self-assertion and competition (Foley and Williamson, 2018).

Strategies to Achieve Equality in the Learning Process

Once women have been accepted into institutions of higher learning, the institutional and learning environment needs to be conducive to women's participation. Critical components include, for example, gender-sensitive and safe accommodation, a learning environment free from violence and sexual harassment, mentorship programmes, gender-sensitive codes of conduct and gender-sensitive course content. As Rathgeber (2002) notes, African tertiary institutions were established and organized to meet the needs of male students. Consequently, the culture and, in some cases, the infrastructure of many of them do not meet the needs of female students. Early studies and more recent reports in several universities in East and Southern Africa showed that sexual harassment - ranging from constant derogatory verbal remarks to rape and other types of gender-based violence - is perpetrated on female students by academic and non-academic staff members and by students (Gitobu, 1994; Ndinda, 2001).

Strategies which have worked to address sexual and gender-based violence include (i) having a sexual harassment policy, (ii) infrastructure improvements which guarantee student safety, (iii) accountability mechanisms for dealing with perpetraters of gender-based violence and (iv) training on gender sensitivity for students and staff, particularly to remove the stigma associated with reporting of sexual violence. For example, while over 40,000 rapes are reported to the police in South Africa per annum, it is estimated that only 14% of perpetrators face conviction (Kapps, 2006). In fact, evidence suggests that statistics are much higher, but fear, shame, re-traumatization and lack of trust in the criminal justice system discourage women and girls from reporting sexual violence. In the context of abuses within institutions like schools and at universities, in around 50% of cases, reports of sexual violence are accompanied by reports of institutional betrayal, whereby women are not supported or where no action is taken. In their study on South Africa, Treffry-Goatley *et al.* (2018) suggest that, if sexual violence is to be addressed, it is necessary to challenge gender norms, stop trying to control the behaviour of girls and women and start investigating the sexually-violent behaviour of boys and men. The authors recommend participatory approaches to understanding women's and girls' experiences and discussion forums as a prevention measure for engaging men and boys.

Another area in which tertiary institutions fail to meet the needs of female students is in employing too few women in teaching and management positions, both to provide role models (for both women and men) and a more balanced culture. The few women who hold these positions are concentrated in the lower ranks of academia and management, generally in humanities-based disciplines which tends to promote a cycle of under-representation in science, technology, engineering and mathematics. In his review of women's leadership in higher institutions in Kenya, Odhiambo (2011) argues for the need to develop national and local-level policies and strategies geared towards increasing women's participation in decision-making and leadership in higher education. In South Africa, national data on gender representation from the Higher Education Management Information System showed that the share of women in faculty positions declined significantly with seniority: as of 2016, women represented 56.6% of junior lecturers, 53.3% of lecturers, 45.1% of senior lecturers, 39.5% of associate professors, but only 27.5% of professors (Naidu, 2018). And of the country's 26 vice chancellors, only five were female.

While policies supporting gender-equity are necessary, they are insufficient to ensure gender parity. For example, despite gender-neutral recruitment practices in tertiary education, huge gender gaps remain across student enrolment, staff recruitment and staff promotion (Aina *et al.*, 2015). Eboiyehi *et al.* (2016) demonstrated the low representation of women in senior management positions in the selected universities in Nigeria, irrespective of the gender policies in place. This was attributed to inadequate numbers of qualified women, competing domestic and family demands, patriarchal and sociocultural beliefs that leadership positions are the domain of men, lack of encouragement and support, and lack of leadership skills and poor mentorship, among other factors. Ultimately, the organizational culture of most tertiary institutions needs to change in order for policies to take hold. Women require leadership training, and programmes need to be developed to eliminate the barriers to women's participation in non-traditional sectors. This includes targeted recruitment, training and initiatives to support women. Existing management also requires ongoing leadership and management on gender issues, and together appropriate administrative structures need to be established so that gender policies can be acted upon.

Another factor influencing learning outcomes is the lack of gender integration in pedagogy. Gender-responsive pedagogy considers the needs of both men and women in the design of teaching and learning approaches to enhance women's and men's different learning styles. Recognition emerged in the late-1990s that science needs to be taught in ways which accommodate the diversity of students' needs, interests, experiences and cultures. Gender-inclusive pedagogy acknowledges that teaching does not occur in a vacuum, but rather under complex cultural, economic, political, social and historical interactions (Barton, 1998). In addition, teachers and students are themselves subject to all these differences and hence may resist or reinforce outmoded power dynamics, and thereby affect the learning process (Mossman, 1995). Gender-responsive pedagogy calls for teachers to adopt an all-encompassing gender approach to lesson planning, teaching, classroom management and performance evaluation, and to respond to the specific needs of their students, ensuring equal access and encouraging equal participation. This also includes students' specific needs in terms of facilities (e.g sanitation and seating infrastructures). A key element is ensuring that learning materials are free of gender stereotypes

(FAWE, 2020), such as scientists being portrayed as male only.

Strategies to Achieve Equality in Educational Outcomes

Low numbers of female students in TAE institutions lead to low representation of women in agricultural research and development institutions (Fig. 11.3). As a result, women's potential is rarely fully tapped in international agricultural research (Blackie *et al.*, 2009; Forysthe *et al.*, 2010). Female scientists remain grossly under-represented in agricultural R&D, even though they are in a unique position to address the pressing challenges facing African farmers, the majority of whom are female. Even with adequate male representation in research institutions, a significant gap remains in adequately representing women's specific needs and having their voices heard in policy debates and decision-making processes. The expansion of TAE in recent years, with its accompanying increase in numbers of female students, offers institutions a great opportunity to attract higher numbers of female staff if the barriers to female participation described in this chapter can simultaneously be addressed. In Ethiopia, for example, as of 2007 the share of women studying agricultural science was three times higher than the share of female (BSc, MSc and PhD) graduates employed in agriculture (18% compared with 6%).

As previously discussed, women's participation in science and technology (S&T) generally, and TAE specifically, diminishes with career advancement - a phenomenon known as 'the leaking pipeline'. Huyer and Westholm (2007) point out that women entering S&T systems encounter two levels of segregation. Firstly, horizontal segregation involves a considerably higher concentration of women in 'softer' fields of science, such as biology, other life sciences and social sciences, rather than 'harder' fields, such as engineering and physics. Notably also, women's involvement in agricultural research and training is more skewed to agricultural disciplines related to life and social sciences, such as food and nutrition sciences, molecular biology and agricultural economics (Forsythe *et al.*, 2010). Women remain under-represented in engineering courses, such as irrigation, natural resource management,

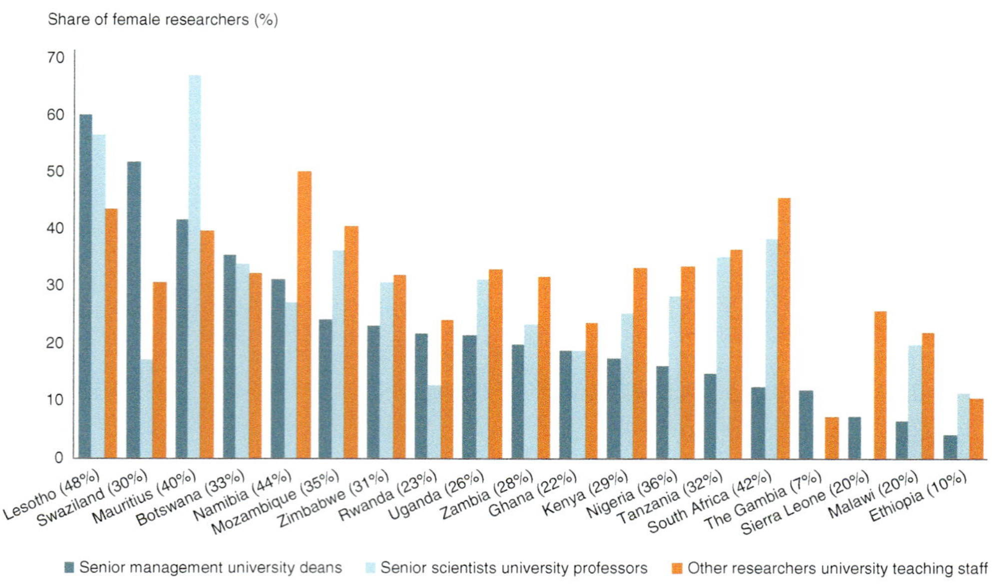

Fig 11.3. Total agricultural researchers by seniority level, 2014
Source: Beintema and Stads (2017, Fig. 20).
Notes: Data in parentheses indicate the country's average share of female agricultural researchers. Data for Nigeria, Sierra Leone and Uganda exclude the government sector. Senior management includes university deans; senior scientists include university professors; and other researchers include teaching staff.

water management and soil science (Beintema and Di Marcantonio, 2010). Secondly, vertical segregation involves the over-representation of women at lower levels of the professional hierarchy, and this is true in industry and education, as well as in S&T. Hence, women are less represented in high-level research and management positions compared with their male colleagues (Fig. 11.4). As a result, they have less influence in policy- and decision-making processes. This phenomenon begins at the lower levels of schooling, where perceptions persist that science subjects are for boys not girls (Forysthe *et al.*, 2010).

Progress has been made in recent years, however. In a sample of 15 SSA countries, the share of female agricultural graduates employed in agricultural research rose from 18% in 2008 to 24% in 2014 (with shares being lower in Ethiopia and West African countries, and higher in Southern African countries). This is in itself a positive indicator for the future, but it does not mean the proper incentives are in place to encourage BSc graduates to pursue careers and graduate studies in agriculture and related fields (Beintema and Di Marcantonio, 2010). Available data also show that, in many countries, improvements in the shares of female scientists employed in TAE and agricultural research institutions were more prevalent among those qualified to the BSc-degree level than those with MSc or PhD degrees (Beintema and Di Marcantonio, 2010; Beintema, 2014). Various RUFORUM and other initiatives - such as the leadership and mentoring programme, African Women in Agriculture and Rural Development - are promising approaches to overcoming the challenges associated not only with achieving gender equality in TAE institutions, but also overcoming the social alienation many women face in the workplace. These initiatives provide much-needed mentorship, along with the development of key competencies, such as leadership and networking ability.

Women's participation in research systems is important for two major reasons. Firstly, more equal representation of women is needed in the conduct of research and formulation of policy. Secondly, women represent a valuable resource in terms of the opportunity to increase human resource capacity to fill serious shortages in many countries (IAC, 2004; Beintema and Di Marcantonio, 2010). Various studies have argued that higher rates of female participation in S&T could improve the quality and competitiveness of research and innovation (Huyer and Westholm, 2007; European Commission, 2008;

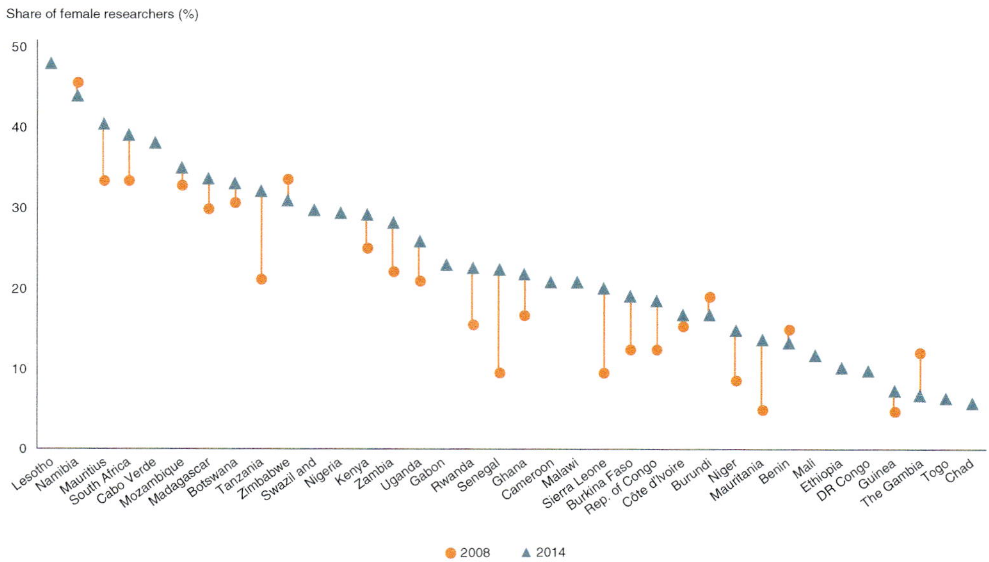

Fig. 11.4. Growth in the share of female agricultural researchers, 2008–2014
Source: Beintema and Stads (2017, Fig. 19).
Note: Data for 2008 were not available for Chad, DR Congo, Ethiopia, Gabon, Lesotho, Malawi, Mali, Eswatini and Togo.

Beintema and Di Marcantonio, 2010). Greater participation by female agricultural scientists would expand the diversity of skills, experiences, subject matter and perspectives in the workforce. Improving the gender balance in TAE would support a more inclusive culture and encourage more women to pursue careers in agricultural research, thereby sharing their unique perspectives and focusing more attention on the needs of female farmers. Importantly, greater gender balance would ultimately contribute to achieving broader national and international food security and development goals.

It is important to note, however, that a key aspect in achieving greater equality in educational outcomes is the way organizations are structured. Historically, institutions of higher learning have been conceptualized and controlled by men to reflect their interests and needs (Britton, 2000; Rathgeber, 2013). These one-sided perspectives have narrowed the scope of leadership, power, beliefs, symbols, structures, ceremonies and information flows (Rathgeber, 2013), which has fostered gender inequity and contributed to women being denied the full range of opportunities in the higher education sector. In order for corrective gender policies to be effective, however, they must be accompanied by investment in strong structural frameworks to support change, rather than simply entrenching the beliefs and behaviours that need to be transformed. This is a challenge that Africa's tertiary institutions appear to be facing, based on case studies from some of the public universities in Kenya and Nigeria.

11.4 Incorporating Gender into Agricultural Training Programmes

Gender equality is a basic human right - one that has value in and of itself. Nevertheless, ministries of agriculture and development agencies continue to devise and implement programmes which fail to tackle the structural barriers to women's full participation in agricultural development. This perpetuates the marginalization of female farmers, even though they constitute the majority of farmers in many African countries. Structural and institutional limitations - together with harmful gender and social norms - also persist in creating barriers to women's participation and empowerment. Most graduates of agriculture end up either working with smallholder farmers and agribusiness entrepreneurs or training the next generation of agricultural specialists. Successfully integrating gender issues into graduates' perspectives is critical. Evidence shows that gender has an impact on agriculture in terms of both efficiency and equality. Gender inequality in the access to and control over productive resources results in poorer agricultural and human development outcomes. A World Bank study conducted in Burkina Faso, Kenya, Tanzania and Zambia showed that providing female farmers with the same quantity and quality of inputs that men typically receive (such as fertiliser, land and labour), as well as improving their access to agricultural education, could increase national agricultural output and incomes by an estimated 10–20% in each country (World Bank, 2005). And FAO (2011) states that giving women equal access to resources can increase agricultural productivity by 20–30%.

For specialists working in agriculture, their ability to understand these gender barriers, ground them in societal and structural inequalities and facilitate the social change required to achieve gender equality in the agriculture sector will be determined by the skills they acquire through their education and training, including how well they are socialized to gender issues. Because of the lack of gender integration in agricultural training, most agriculture scientists, academics and extension staff resort to short-term training to acquire necessary skills on gender equality in agriculture. Various (often donor-funded or project-related) trainings have been developed[1], but - while useful - these short trainings have limited reach and often don't cover both the theoretical and practical aspects of gender inequality in agriculture. The need, therefore, exists to build gender content into agricultural curricula in universities, which can be done in two ways, as described below.

1. ***Gender as a module in agricultural training.*** For students of agriculture and related disciplines, integrating courses on gender into agricultural science programmes has the twofold impact of introducing students to the basic concepts of gender and gender analysis, and - importantly - raising their awareness of context-specific gender issues relevant to African agriculture. This level of gender integration does not, however, build a cadre of gender experts with the analytical grounding necessary for an in-depth understanding of gender issues as they relate to agriculture. It merely helps raise a cadre of gender-sensitive agricultural academicians, researchers and practitioners.

2. *Specialized degrees in gender and agriculture.* Building a cadre of agricultural researchers and development experts specifically targeting the study of gender issues requires the development of fully-fledged undergraduate and graduate programmes focusing on gender and agriculture. These programmes would provide the theoretical foundation for addressing gender issues within Africa's agricultural sector. This type of expertise is greatly lacking, and a majority of those working as gender and agriculture experts are either gender and development students completing mandatory agricultural units, or agricultural graduates who have undertaken short courses on gender. The danger of not having appropriately-trained experts with knowledge of both gender *and* agriculture is that the field will remain under-developed and critical opportunities to empower women in the agricultural sector and food system - thereby contributing, not only to equity, wellbeing, educational and health goals, but to overarching economic and development goals - will be foregone. Schools and programmes of agriculture could forge linkages with those of gender and women's studies to offer MSc or PhD programmes bridging the two disciplines. Students could spend one year on gender studies and the rest of their time applying their theoretical training to the agriculture sector. An example of an existing programme is the dual-title graduate degree programme in international agriculture and development at Penn State University in the United States, which is jointly offered by the university's College of Agricultural Sciences and the International Development Program. In Africa, the Royal Tropical Institute of the Netherlands and the University of Eldoret (Kenya) developed the first MSc-level programme on gender and agriculture.

11.5 Education as a Force for Changing Social and Gender Norms

Gender education has the potential to generate important knowledge and facilitate the transformation and modernization of society to achieve social justice, equity and a better quality of life for all (Aina *et al.*, 2015). Creating a gender-sensitive/gender-responsive academic environment enhances (i) the organizational culture of institutions, (ii) the social and professional interactions and productivity of men and women and (iii) the process of much-needed social change. Education can be leveraged in three key ways to support transformative social change in the realm of gender equity, as described below.

1. *Transforming women's social status.* The education of women has been shown to enhance their social status in society and to have a multiplier effect in terms of resulting higher educational outcomes for children, shifts in how a society perceives women and changes in the structure of the economy and the labour force. Husseini (2017), for example, documents how growth in the number of women in higher education in Iran to 65% of the student population during 1979–2015 led to women's greater economic empowerment and social development, as well as changes in how women are viewed, despite repressive laws in the country. The content and environment in which education occurs can, however, reinforce norms and stereotypes. For example, the norm that young women should defer to men feeds a stereotypical view that women are less capable academically, and expectations that their home-making role is more important than their participation in the labour market can lead teachers to prioritize the young men in their classes. Achieving gender equality in education would give men and women equal opportunities to realize their full human rights and contribute to and benefit from economic, social, cultural and political development (EQUATE, 2008).

2. *Transforming gender norms.* Educational programmes incorporating gender dynamics represent a critical step in confronting outmoded cultural norms and behaviours which perpetuate inequality. Education can be a force for positive change in terms of challenging socialized norms and supporting students in 'unlearning' related behaviour. Higher education has the potential to raise the aspirations of young women, build their knowledge and skills and foster an equitable environment for learning, ultimately changing how young men and women view their roles in society. Markus (2017) describes what she calls an accelerated route through which education can lead to positive change in social norms, thereby empowering women. Critical components of this accelerated approach include gender-responsive teaching, equitable opportunities for leadership by young men and women, explicit gender-equality in the content of curricula, activities which compensate for gender inequalities and an egalitarian school environment, including management.

3. ***Understanding the cause of gender inequality.*** Gender is a necessary component of transformative change in rural development. Agricultural experts who are appropriately educated on the gender issues associated with TAE, agricultural research and the agricultural sector overall have the power to raise social consciousness towards eliminating inequality and its underlying causes, not only in the context of rural and economic development, but also in terms of the broader social fabric. Given TAE institutions' critical role as pioneers of social transformation, it behoves them to embrace this role by mainstreaming gender in their educational programmes, management structures and organizational cultures to achieve equality of access, equality in the learning process, equality of educational outcomes and equality of external results.

In conclusion, and following Subrahmanian (2005), in the context of TAE, gender equality needs to be understood as *the right to education* - i.e. access and participation; *rights within education* - i.e. gender-aware educational environments, processes and outcomes and *rights through education* - i.e. meaningful educational outcomes which link education equality with wider processes of gender justice.

Endnotes

* The views, opinions, interpretations and conclusions expressed by the authors in this chapter do not necessarily reflect the policies and opinions of their institutions or their partners or collaborators.

[1] Examples include: (i) 'Gender-Responsive Researchers Equipped for Agricultural Transformation' by Makerere and Cornell Universities, (ii) gender and leadership training offered by the African Women in Agricultural Research and Development and (iii) the AgriProfocus training on Gender and Value Chains.

References

Aina, O.I. Ogunlade, I. Ilesanni, O.O. and Afolabi, C. (2015) Institutionalization of gender mainstreaming in Nigeria's tertiary. *European Scientific Journal* Special Edition (November), 314–339.

Aissetou, D.Y. Akinnagbe, O.M. Ochola, A. Chakeredza, S. and Hien, M. (2015) Assessment of gender policy in selected tertiary agricultural education institutions in Africa. *Agricultural Sciences* 6, 1039–1047. Available at: https://doi.org/10.4236/as.2015.69099 (accessed 11 May 2021).

Barton, C.A. (1998) *Feminist Science Education*. Teachers College Press, New York.

Beintema, N. (2014) Enhancing female participation in agricultural research and development: Rationale and evidence. In: Quisumbing, A.R. Meinzen-Dick, R. Raney, T.L. Croppenstedt, A. Behrman, J.A. *et al.* (eds.) *Gender in Agriculture: Closing the Knowledge Gap*. Springer, Dordrecht, The Netherlands.

Beintema, N. and Di Marcantonio, F. (2010) *Female Participation in African Agricultural Research and Higher Education: New Insights*. Synthesis of the ASTI–Award Benchmarking Survey on Gender-Disaggregated Capacity Indicators. IFPRI Discussion Paper 957. International Food Policy Research Institute, Washington, DC. Available at: https://www.asti.cgiar.org/pdf/ifpridp00957.pdf (accessed 11 May 2021).

Beintema, N. and Stads, G-J. (2017) *A Comprehensive Overview of Investments and Human Resource Capacity in African Agricultural Research*. ASTI Synthesis Report. International Food Policy Research Institute, Washington, DC. Available at: http://ebrary.ifpri.org/cdm/ref/collection/p15738coll2/id/131191 (accessed 11 May 2021).

Blackie, M. Mutema, M. and Ward, A. (2009) *ASARECA/ RUFORUM Thrust 4. A Study of Agricultural Graduates in Eastern, Central, and Southern Africa: Demand, Quality and Job Performance Issues*. A Report to ASARECA and RUFORUM. Available at: https://repository.ruforum.org/documents/

study-agricultural-graduates-eastern-central-southern-africa-demand-quality-job (accessed 11 May 2021).

Britton, D.M. (2000) The epistemology of the gendered organization. *Gender & Society* 14(3), 418–434. Available at: https://doi.org/10.1177/089124300014003004 (accessed 3 June 2021).

Bunyi, G.W. (2003) Interventions that increase enrollment of women in African tertiary institutions. A case study prepared for a Regional Training Conference on Improving Tertiary Education in Sub-Saharan Africa: Things That Work! Accra, 23–25 September. Available at: https://www.academia.edu/9001916/Interventions_That_Increase_Enrolment_Of_Women_In_African_Tertiary_Institutions (accessed 11 May 2021).

Daddieh, C.K. (2003) *Gender Issues in Ghanaian Higher Education*. Occasional Papers 36. Institute of Economic Affairs, Accra, Ghana.

Davis, K.E. Ekboir, J. and Spielman, D.J. (2008) Strengthening agricultural education and training in sub-Saharan Africa from an innovation systems perspective: A case study of Mozambique. *Journal of Agricultural Education and Extension* 14(1), 35–51. Available at: https://doi.org/10.1080/13892240701820371

Eboiyehi, C.O. Fayomi, I. and Eboiyehi, F.A. (2016) From exclusion to discrimination: Gender inequality in the senior management of Nigerian universities. *Issues in Educational Research* 26(2), 182–205. Available at: https://search.informit.com.au/documentSummary;dn=132106210446235;res=IELAPA (accessed 11 September 2019).

EQUATE (2008) *Education from a Gender Equality Perspective*. A report for USAID's Office of Women in Development. Available at: https://files.eric.ed.gov/fulltext/ED511674.pdf (accessed 11 May 2021).

European Commission (2008) *Mapping the Maze: Getting More Women to the Top in Research*. European Commission, Brussels. Available at: https://op.europa.eu/en/publication-detail/-/publication/1f690dc0-7606-48d2-a74e-6b07d15dbb26 (accessed 11 May 2021).

FAO (Food and Agriculture Organization of the United Nations) (2011) *The State of Food and Agriculture 2010–2011: Women in Agriculture: Closing the Gender Gap for Development. Rome*. Available at http://www.fao.org/3/i2050e/i2050e00.htm (accessed 19 September 2019).

FAWE (Forum for Women Educationalists) (2020) *Gender Responsive Pedagogy: A Toolkit for Teachers and Schools*. Second edition. FAWE, Nairobi.

Foley, M. and Williamson, S. (2018) Does anonymising job applications reduce gender bias? Understanding managers' perspectives. *Gender in Management: An International Journal* Available at: https://doi.org/10.1108/GM-03-2018-0037 (accessed 11 May 2021).

Forsythe, L. Mangheni, M. and Martin, A. (2010). Attracting women into agricultural education: Constraints and best practices. Paper presented at the Second RUFORUM Biennial Meeting, 20–24 September, Entebbe, Uganda.

Gandhi, D. (2018). Figures for the week: Higher education enrollment grows in sub-Saharan Africa along with disparities in enrollment by income. *Africa in focus* (January 10). Brookings Institution, Washington, DC. Available at: https://www.brookings.edu/blog/africa-in-focus/2018/01/10/figures-of-the-week-higher-education-enrollment-grows-in-sub-saharan-africa-along-with-disparities-in-enrollment-by-income/ (accessed 2 June 2021).

Gaucher, D. Friesen, J. Kay, A.C. (2011) Evidence that gendered wording in job advertisements exists and sustains gender inequality. *Journal of Personality Social Psychology* 101, 109–128. Available at: https://doi.org/10.1037/a0022530 (accessed 11 May 2021).

Gitobu, J. (1994) *Report on Sexual Harassment of Female Students at Kenyatta University*. Kenyatta University, Nairobi.

Husseini, N. (2017) *Higher education, gender and social change in Iran (1979–2015)*. (PhD thesis), Kingston University. Available at: https://ethos.bl.uk/OrderDetails.do?uin=uk.bl.ethos.733457 (accessed 11 May 2021).

Huyer, S. and Westholm, G. (2007) *Gender Indicators in Science, Engineering, and Technology: An Information Toolkit.* United Nations Educational, Scientific and Cultural Organization, Paris.

IAC (InterAcademy Council) (2004) *Inventing a Better Future: A Strategy for Building Worldwide Capacities in Science and Technology*. IAC, Amsterdam. Available at: https://www.interacademies.org/publication/inventing-better-future (accessed 11 May 2021).

Kapps C. (2006) Rape on trial in South Africa. *Lancet* 367(9512), 718–719. Available at: https://doi.org/10.1016/S0140-6736(06)68285-8 (accessed 11 May 2021).

Kihara J (2005) New departure in provision of higher education. In *University Journal*, a publication of the *Daily Nation Newspaper*, 28 July, Nation Media Group, Nairobi.

Kiliswa, I. Udoto, M. and Konyango, J. (2016) Tracer study of Bachelor of Science in Agricultural Education and Extension - Graduate class of 2008 - Egerton University. *Imperial Journal of Interdisciplinary Research* 2(8), 1080–1085. Available at: http://www.onlinejournal.in/IJIRV2I8/179.pdf (accessed 3 June 2021).

Mangheni, M.N. Ekirikubinza-Tibatemwa, L. and Forsythe, L. (2010) *Gender Issues in Agricultural Education within African Universities*. Gender Background Paper for the Ministerial Conference on Higher Education in Agriculture in Africa, Kampala, Uganda, 15–19 November 2010. Available at: https://citeseerx.ist.psu.edu/viewdoc/download?doi=10.1.1.718.3160&rep=rep1&type=pdf (accessed 11 May 2021).

Markus, R. (2017) *Education and Gender Norm Change*. Advanced Learning and Innovation on Gender Norms (ALIGN), London. Available at: https://www.alignplatform.org/sites/default/files/2019-12/education_guide_edited_-_final.pdf (accessed 11 May 2021).

Mossman, M.J. (1995) Gender issues in teaching methods: Reflections on shifting the paradigm. *Legal Education Review* 6(2), 129–152. Available at: https://core.ac.uk/display/232639214 (access 3 June 2021).

Mugisha, J. and Nkwasibwe, A. (2014) *Tracer Study of Agricultural Graduates in Uganda*. Food Security Collaborative Working Paper 183862. Department of Agricultural, Food, and Resource Economics, Michigan State University. Available at: https://doi.org/10.22004/ag.econ.183862 (accessed 11 May 2021).

Mukhwana, A.M. Abuya, T. Matanda, D. Omumbo, J. Mabuka, J. (2020) *Factors which Contribute To or Inhibit Women in Science, Technology, Engineering, and Mathematics in Africa*. African Academy of Sciences, Nairobi. Available at: http://www.aasciences.africa/sites/default/files/Publications/Women%20in%20STEM%20Report_Final.pdf (accessed 3 June 2021).

Naidu, E. (2018) Universities body to probe gender imbalance at the top. *University World News* 27 July, South Africa. Available at: https://www.universityworldnews.com/post.php?story=20180725103923330 (accessed 11 May 2021).

Ndinda, C. (2001) *Black female student's response to their problems at the University of Natal*. University of Natal, Durban, South Africa.

Odebero, S., Bosire, J. Sang, A. Ngala, F. and Ngware M. (2007) Equity in the distribution of HELB loans in Kenya in relation to students' characteristics: An empirical analysis. *Educational Research and Review* 2(8), 209–219.

Odhiambo, G. (2011) Women and higher education leadership in Kenya: A critical analysis. *Journal of Higher Education Policy and Management* 33(6), 667–678. Available at: https://doi.org/10.1080/1360080X.2011.621192 (accessed 11 May 2021).

Papadópulos, J. and Radakovich, R. (2005) *Estudio comparado de Educación Superior y género en América Latina y el Caribe*. [Comparative study of Higher Education and gender in Latin America and the Caribbean]. Available at: https://www.cned.cl/otros-estudios/estudio-comparado-de-educacion-superior-y-genero-en-america-latina-y-el-caribe (accessed 3 June 2021).

Prozesky, H. and Beaudry, C. (2019) Mobility, gender and career development in higher education: Results of a multi-country survey of African academic scientists. *Social Sciences* 8(188). Available at: https://doi.org/10.3390/socsci8060188 (accessed 11 May 2021).

Rathgeber, E.M. (2002) Women in universities and university-educated women: the current situation in Africa. In: Teferra, D. (ed.) *African Higher Education: The international reference book*. University of Indiana Press, Bloomington, Indiana, USA.

Rathgeber, E.M. (2013) *Gender Barriers Faced by African Women in Graduate Programmes and Research in the Social Sciences*. A Partnership for African Social and Governance Research Scoping Study. Available at: https://www.pasgr.org/wp-content/uploads/2017/01/Gender-Barriers-Scoping-Paper-Final.pdf (accessed 11 May 2021).

Santos, I. and Rubiano-Matulevich, E. (2019) Minding the gender gap in training in sub-Saharan Africa: Five things to know. August 12. Available at: https://blogs.worldbank.org/africacan/minding-gender-gap-training-sub-saharan-africa-five-things-know (accessed 24 February, 2021).

Subrahmanian, R. (2005). Gender equality in education: Definitions and measurements. *International Journal of Educational Development* 25, 395–407. Available at: https://doi.org/10.1016/j.ijedudev.2005.04.003 (accessed May 11, 2021).

Treffry-Goatley, A. de Lange, N. Moletsane, R. Mkhize, N. and Masinga, L. (2018) What does it mean to be a young African woman on a university campus in times of sexual violence? A new moment, a new conversation. *Behavioural Sciences* 8(8), 67. Available at: https://doi.org/10.3390/bs8080067 (accessed May 11, 2021).

UNESCO (United Nations Education, Scientific and Cultural Organization) (2003) *Gender and Education for All: The Leap to Equality*. UNESCO, Paris. Available at: https://www.sel-gipes.com/uploads/1/2/3/3/12332890/2003-4_-_unesco_-_the_leap_to_equality.pdf (accessed 11 May 2021).

UNESCO (2010) *The Central Role of Education in the Millennium Development Goals*. MDG Summit High-Level Round Table. UNESCO. Available at: http://www.unesco.org/new/fileadmin/MULTIMEDIA/HQ/ED/ED_new/images/education_for_all_international_coordination_new/PDF/analyticalnote.pdf (accessed 2 June 2021).

UNICEF (United Nations Children's Fund) (2021) Girls' Education: *Gender Equality in Education Benefits Every Child*. UNICEF/UN0284179/LeMoyne. Available at https://www.unicef.org/education/girls-education (accessed 2 June 2021).

United Nations (2015) Sustainable development goals. In: *Transforming Our World: The 2030 Agenda for Sustainable Development*. United Nations, New York.

World Bank (2005) *Gender and 'shared growth' in sub-Saharan Africa*. Briefing notes on critical gender issues in sub-Saharan Africa 2005-1. World Bank, Washington, DC.

12 Entrepreneurship and the Role of Universities in Generating Youth Employment in Africa

Anthony Egeru[1,2*], Sylvanus Mensah[1], Serge Abihona[3] and Adnane Alaoui Soulimani[4]

[1]*Regional Universities Forum for Capacity Building in Agriculture, Uganda;* [2]*College of Agricultural and Environmental Sciences, Makerere University, Uganda;* [3]*University of Abomey Calavi Foundation, Benin;* [4]*Mohammed VI Polytechnic University, Morocco*

12.1 Introduction

Entrepreneurship offers great potential as a strategic response to Africa's glaring youth unemployment and is strongly advocated by governments and development partners. In fact, Africa has the world's highest rate of entrepreneurship, with an estimated 22% of its working-age population starting their own businesses as of 2017 (AfDB, 2017). Yet the journey is short-lived for many, given that approximately 8% of business owners are estimated to discontinue their endeavours within 12 months of operation. As has been discussed in numerous chapters in this volume, the higher-educaton sector is increasingly being challenged to provide leadership in promoting entrepreneurship. This has put universities, in particular, at the forefront of wider societal demand for entrepreneurial productivity, beyond their traditional mandates of teaching, research and outreach. As a result, Africa's higher-education institutions are under increasing pressure to do much more to foster students' potential for innovation and prepare them with the skills needed in the marketplace. Universities have responded by: (i) redesigning and developing their curricula, (ii) establishing incubation hubs and accelerator centres promoting innovation and startup businesses, (iii) forging linkages with the private sector to provide students with real-world lessons and experiences (iv) reforming policy and related frameworks to strengthen the enabling environment for entrepreneurship and (v) putting these changes into effect using mixed models.

This chapter explores the role of Africa's universities in facilitating the proliferation and success of entrepreneurship as a development strategy. In particular, the discussion is intended to articulate how Africa's universities can support the generation of much-needed employment opportunities by providing education and training in entrepreneurship. The next section examines the prevailing entrepreneurial ecosystem, (i) by providing context from the available literature on the demographic challenge Africa faces in light of its rapidly-growing youth population and associated high unemployment rates and (ii) by exploring the means by which entrepreneurship can be fostered. Thereafter, the chapter focuses

*Corresponding author: a.egeru@ruforum.org

 Transforming Tertiary Agricultural Education in Africa (eds D. Kraybill, J. Lynam and Adipala Ekwamu)
DOI: 10.1079/9781789246544.0012

on working models of education and training in entrepreneurship, including intentions and adaptations, real-world successes, lessons learned and potential future directions.

12.2 The Entrepreneurial Ecosystem

Context

Africa will account for more than half the growth in the world's population by 2050 (Fig. 12.1), with the addition of 1.3 billion people (Kazeem, 2017), and by 2100, almost half the world's youth will be African (Rabana, 2019). This growth trajectory presents governments and development experts with an overwhelming conundrum as they seek to lay the foundational conditions to allow all Africans to thrive. Yet youth employment opportunities remain severely limited. Of the 74 million jobs created in Africa between 2000 and 2008, only one-third were applicable to the youth demographic - i.e. those aged 15 to 24 years (Devlin, 2013). As a result, the absolute number of working poor is rising, based on the projected estimate that by 2030 the employment market will include 198 million youth (ILO, 2018). As of 2016, Glassco and Holguin (2016) estimated that 78% of Africa's youth were living on less than US$2 per day. Employment projections indicate limited growth and expansion in youth employment despite an acceleration in economic activity, which to date has been too small to generate the number of jobs need to absorb the fast-growing labour force (ILO, 2019).

Given Africa's unemployment levels, thousands of self-starters are taking their futures into their own hands and embarking on entrepreneurial endeavours, with all the associated risks. In large part, their success depends on the environment - or ecosystem - within which they can develop their innovative capacity and exploit available market advantages and opportunities (AfDB, 2011). Institutions, such as the African Development Bank (AfDB), recognize the need to provide intentional investments and support for entrepreneurs (AfDB, 2011). Moreover, African governments have begun to support universities in establishing the necessary infrastructure and facilities to promote and commercialize innovative entrepreneurial endeavours, e.g. through technology development centres and science parks. Through these actions, universities are regaining public trust as active stakeholders in

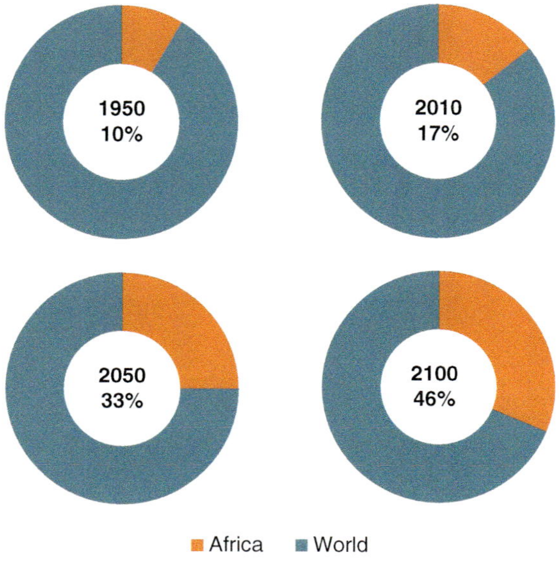

Fig. 12.1. Projected growth in Africa's share of the world's youth, 1950–2100
Source: Adapted by authors from Mo Ibrahim Foundation (2013, p.8).
Note: Youth refers to population aged 15–24 years.

addressing the youth unemployment challenge. This trust stems from early evidence of successful job creation by graduates who have received practical and experiential training and mentoring in entrepreneurship. If universities are to suceed in fulfilling this new role, the ecosystem for entrepreneurship in Africa needs to be strengthened and mainstreamed into broader systems of governance and education.

Fostering Entrepreneurship

It is important for novice entrepreneurs to approach their journey as a marathon, not a sprint (Ocici, 2017). Numerous hurdles must be overcome, requiring endurance. Extrapolating from this metaphor, entrepreneurship can generally be classified as being based either on necessity, wherein entrepreneurs seek to overcome unfavourable circumstances (Haas, 2013), or opportunity, which derives from a more deliberate intention to build on prevailing opportunities (Calderon *et al.*, 2016). Regardless of an entrepreneur's intentions at the outset, all of them begin by identifying and harnessing a given opportunity which propels them forward. A survey of youth entrepreneurship revealed varying rates of activity across selected countries of sub-Saharan Africa (SSA), ranging from high (56%) in Uganda to low (13%) in South Africa (Fig. 12.2). Ultimately, the growth, success and sustainability of entrepreneurs depend on what has come to be known as the underlying 'entrepreneurship ecosystem' (Table 12.1).

The entrepreneurship ecosystem can generally be conceptualized as the system through which successful entrepreneurship is facilitated (Stam and van de Ven, 2021). The 'systems' focus highlights the interdependence of a range of actors and factors at the community level, coordinated in such a way as to foster productive entrepreneurial enterprise (Stam, 2015). Put simply, successful entrepreneurship typically goes hand-in-hand with a supportive entrepreneurial ecosystem. Isenberg (2010, 2011a) identified six interactive domains of a self-sustaining entrepreneurship ecosystem: (i) policy, (ii) markets, (iii) capital, (iv) human capacity, (v) culture and (vi) support (Fig. 12. 2). These domains equate with natural ecosystems. For example, educators and people are within the biotic environment,

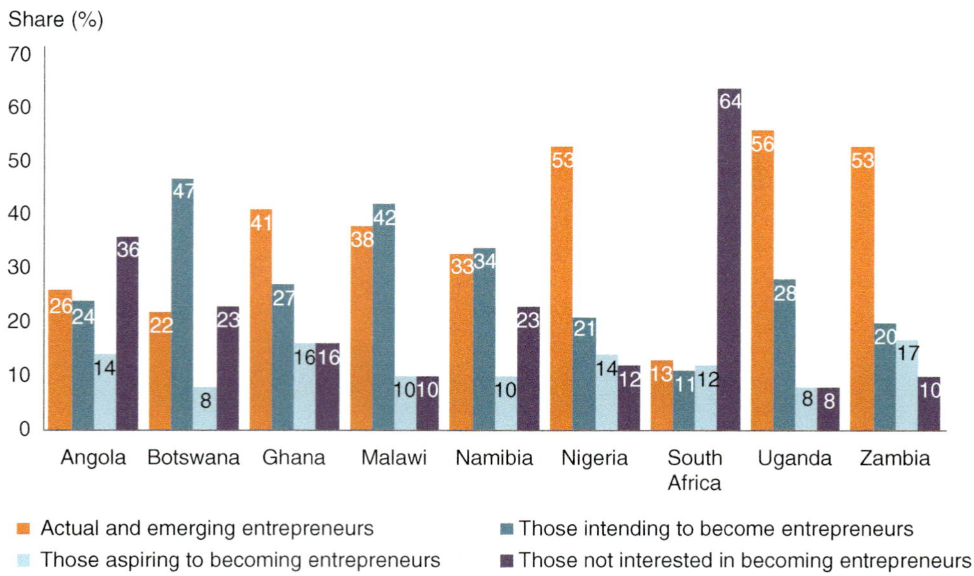

Fig. 12.2. Youth by level of entrepreneurial activity for select sub-Saharan African countries, 2013
Source: Adapted from Kew (2015, Fig. 2.2).
Note: Youth refers to the population aged 15–24 years. Actual and emerging entrepreneurs are those with either established or new businesses; those intending to become entrepreneurs expressed fixed intentions towards entrepreneurship; those aspiring to become entrepreneurs believed they had the potential to become entrepreneurs in the future.

Table 12.1. Pillars of the entrepreneurial ecosystem

1. Markets	2. Policy	3. Finance	4. Human capital	5. Culture	6. Support
EARLY CUSTOMERS • Early adopters for proof of concept • Expertise in creating products • Reference customer • First reviews • Distribution channels NETWORKS • Entrepreneurs' networks • Diaspora networks • Multi-national corporations	LEADERSHIP • Unequivocal support • Social legitimacy • Open-door advocate • Entrepreneurship strategy • Urgency, crisis and challenge GOVERNMENT • Institutions (e.g. investment, support) • Financial support (e.g. for R&D, jumpstart funds) • Regulatory incentives (e.g. tax benefits) • Research institutions • Venture-friendly legislation (e.g. bankruptcy, contract enforcement, property rights and labour	FINANCIAL CAPITAL • Micro-loans • Angel investors, friends and family • Zero-stage venture capital • Venture capital • Private equity • Public capital markets • Debt	LABOUR • Skilled and unskilled • Serial entrepreneurs • Later generation family EDUCATIONAL INSTITUTIONS • General academic and professional degrees • Specific training in entrepreneurship	SUCCESS STORIES • Visible successes • Wealth generation for founders • International reputation SOCIETAL NORMS • Tolerance of risk, mistakes and failure • Innovation, creativity and experimentation • Social status of entrepreneur • Wealth creation • Ambition, drive and hunger	INFRASTRUCTURE • Telecommunications • Transportation and logistics • Energy • Zones, incubation, centres and clusters SUPPORT PROFESSIONS • Legal • Accounting • Investment bankers • Technical experts and advisors NON-GOVERNMENTAL INSTITUTIONS • Promotion of entrepreneurship in non-profits • Business plan contests • Conferences • Entrepreneur-friendly associations

Source: Adapted by authors from Isenberg (2011b, p. 7).

while infrastructure, culture, policy and markets are part of the abiotic environment (McKague *et al.*, 2017). Given the interactions at play, an entrepreneurial ecosystem supports knowledge transfer across various actors, e.g. at market, business, community, government, banking and professional levels (Purbasari *et al.*, 2019). The economic exchange of knowledge, materials products, etc., are influenced by: (i) organizational experiences and capacities, (ii) industrial and technological conditions, (iii) the types of technologies and formal/informal institutions and (iv) social contexts, including networks connecting entrepreneurs and industry (Franco-Leal *et al.*, 2019).

Evidence indicates that Africa's youth have immense entrepreneurial potential, but translating this potential into viable activity has been hindered by dysfunctional entrepreneurship ecosystems in most of countries (Fal, 2013; Devine and Kiggundu, 2016; Guerrero *et al.*, 2020). Nonetheless, support for novice entrepreneurs is available in the form of incubator programmes, but - as Isenberg (2016) points out - the development of entrepreneurship should not merely be reduced to incubator initiatives to support startups. Within the contexts and domains of functional entrepreneurship ecosystems, universities are integral to the institutional context, e.g. in providing technology transfer offices, university research and incubators, and university accelerators and mentorship centres. The entrepreneurship ecosystem also includes other formal institutions, such as government-led science parks (Franco-Leal *et al.*, 2019). These institutions offer training, mentorship, technical advice, linkages with industry and financial services, including pilot funds, as well as an overall platform for developing the competitive advantages of startups and opening avenues for them to access resources (Autio *et al.*, 2014).

As institutional agents, universities are therefore important facilitators of healthy entrepreneurial ecosystems through: (i) their triple functions of teaching, research and outreach, (ii) their ability to lobby, mobilize and influence policy and public opinion and (iii) their ability to successfully pilot activities and generate invaluable lessons. Universities and other higher-education institutions have an important role in anchoring entrepreneurial activities and providing leadership along the path to the attainment of economic and social benefits (Civera *et al.*, 2019). In recent years, policy-makers have increasingly come to perceive universities as vehicles for promoting growth and innovation (Sandström *et al.*, 2018). In turn, universities have responded by promoting academic entrepreneurship and generating spillovers and spinoffs, all of which accelerate innovation-led enterprise development and contribute to economic growth (Lehmann *et al.*, 2020).

12.3 Working Models of Education and Training in Entrepreneurship

Goals and Intended Outcomes

Many African universities have instigated entrepreneurship programmes to increase their social relevance by promoting economic growth and innovation. The incorporation of entrepreneurship modules or programmes within university curricula is also recognized as a means of promoting entrepreneurship as a viable career path (Mbeteh and Pellegrini, 2018). The catalytic role of entrepreneurship education in the emergence of new and innovative business startups has gained attention and support in the wider society and political arena. Universities have responded by developing course content at both undergraduate and graduate levels (Hayes *et al.*, 2020). Entrepreneurial education has generally been observed to increase students' entrepreneurial aspirations (Boldureanu *et al.*, 2020). Early evidence from within SSA indicated that many universities were offering entrepreneurial courses, but with limited specialization and practice (Kabongo and Okpara, 2010). Universities needed to adapt their practices to become more action-oriented, emphasizing learning by doing (Rasmussen and Sørheim, 2006). Collaboration, innovation, adaptability and interpersonal co-operation are some of the essential skills. All this has led to greater specialization, with universities increasingly adapting the training they offer to cover the spectrum of education *about* entrepreneurship, *for* entrepreneurship, *in* entrepreneurship and *through* entrepreneurship (Table 12.2). This response is intended to ensure that students graduate with the capacity to deal with real-world entrepreneurial challenges (Windeløv-Lidzélius, 2018).

Table 12.2. Classifications of instruction in entrepreneurship

Classification	Description
Education *about* entrepreneurship	Following academic tradition, instruction is concerned with explaining and understanding entrepreneurship
Education *for* entrepreneurship	Instruction is designed for individuals interested in starting and running a business; curricula emphasize cognitive competencies
Education *in* entrepreneurship	Instruction stresses practical experience designed to develop entrepreneurial skills in students
Education *through* entrepreneurship	Instruction is designed to equip students with the competencies needed to adopt an entrepreneurial approach to achieving societal goals

Source: Constructed by authors from Pittaway and Edwards (2012); Hoppe *et al.* (2017) and Ndofirepi (2020).

Adapting and Innovating Models of Entrepreneurial Education

The focus of entrepreneurship remains contentious. Early on, Smith and Di Gregorio (2002) and Daily *et al.* (2002), for example, suggested that entrepreneurship focuses on novel forms of products, processes and markets as drivers of wealth creation. Shane and Venkataraman (2000), on the other hand, argued that entrepreneurship involves discovering and exploiting profitable opportunities for wealth creation. Either way, the concept incorporates an element of valorizing - i.e. enhancing the value of - existing opportunities. Accordingly, universities have adapted their programme and course offerings to incorporate innovative training approaches designed to catalyze practical entrepreneurial education.

As an example, two East Africa universities - Egerton University and Gulu University - piloted an entrepreneurship training approach with funding support from the Regional Universities Forum for Capacity Building in Agriculture (RUFORUM). The universities developed the students enterprise scheme (SES) model (Fig. 12.3) as an adaptation of a training model developed by EARTH University in Costa Rica (Osiru *et al.*, 2016). The SES model moves beyond traditional teaching on entrepreneurship (i.e. education *about* entrepreneurship) by: (i) involving both graduate and undergraduate students, (ii) adopting a practical orientation, (iii) identifying and developing business plans to address opportunities within the community and (iv) translating the resulting business plans into pilot enterprises. Supporting infrastructure is provided, including financing (up to US$4,000), mentorship, coaching and linkages with the private sector. In short, SES provides a microcosmic entrepreneurship ecosystem, allowing students to synergize their theoretical knowledge of agriculture, entrepreneurship and agribusiness with the practical realities of identifying, conceptualizing, initiating, managing and sustaining small and medium-sized enterprises (Kalule *et al.*, 2017).

Key lessons learned through this innovative initiative are diverse:

1. Students are adaptive and have the potential to transition with their enterprises once they complete their degrees.
2. Universities must reorient both academic and non-academic staff from education purely *about* entrepreneurship to education *for*, *through* and *in* entrepreneurship, which requires new skills.
3. The private sector has strong interest in providing mentorship and coaching to novice entrepreneurs; universities simply need to create the necessary platforms for entry and engagement.
4. With the intake of each successive cohort of students, ventures gradually evolve from being social enterprises to technology-led enterprises.
5. After their first year of successful operation, on average, student enterprises generate employment for at least one to two people.
6. At the institutional level, university-wide alignment and curriculum re-alignment are required to ensure that the entrepreneurship project is fully integrated into the curriculum.

Building on Lessons Learned

Building on the lessons from the pilot SES programme at Egerton University and Gulu University,

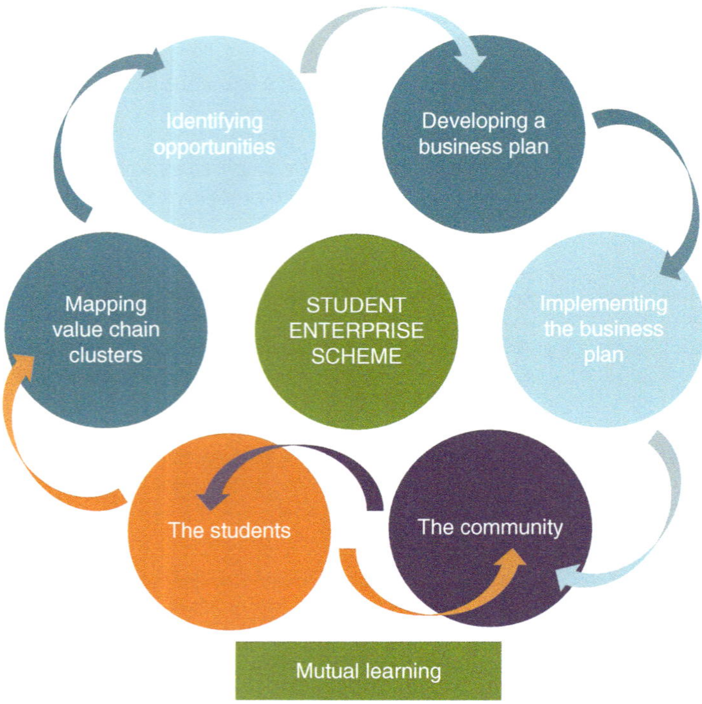

Fig. 12.3. Student enterprise scheme model implemented at Gulu and Egerton universities
Source: Adapted by authors from Kalule *et al.* (2017, Fig. II).

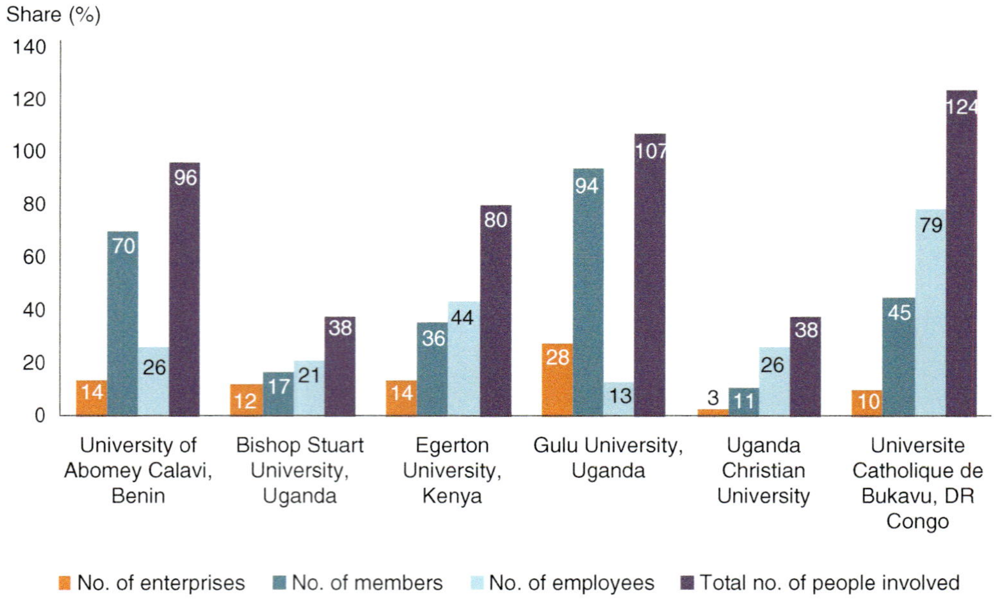

Fig. 12.4. Number of student enterprises and jobs created at six African universities, 2018–2020
Source: Constructed by authors from RUFORUM (2020a).

and with support from the the Bill & Melinda Gates Foundation and the Mastercard Foundation, RUFORUM initiated an eight-year project: Transforming African Agricultural Universities to Meaningfully Contribute to Africa's Growth and Development (TAGDev). This project enabled RUFORUM to scale up its support to universities in developing their entrepreneurship programmes through the RUFORUM Entrepreneurship Challenge Programme (RECAP), which offers universities the opportunity to establish mini-incubation hubs as pilot projects to nurture entrepreneurs. Through these efforts, RUFORUM has facilitated the establishment of 16 agribusiness incubation hubs in Kenya (Egerton University and South Eastern Kenya University), Uganda (Gulu University, Bishop Stuart University, Busitema University, Uganda Christian University, Mbarara University of Science and Technology, Makerere University and Muni University), Eswatini (the University of Eswatini), Ghana (the University of Cape Coast), Benin (the University of Abomey Calavi), South Sudan (the University of Juba), Ethiopia (Haramaya University) and Burundi (the University of Burundi). Based on the success of the RECAP model, AfDB granted RUFORUM additional funding to support two mini-incubation hubs at Mbarara University of Science and Technology (Uganda) and Université du Sine Saloum El-Hâdj Ibrahima NIASS (Senegal). Through these initiatives, at least 702 students gained the opportunity to experience practical entrepreneurship training. The initial six hubs nurtured 81 enterprises engaging 273 students during 2018–2020. By the time of the 2020 COVID-19 outbreak, these student-led enterprises had created 209 jobs.

A recent study on agribusiness incubation and acceleration in SSA (Roulin, 2020) reported that there were more than 300 enterprise incubators/accelerators across SSA. This is a dramatic increase in the past 20 years, mostly with a strong youth focus. While this represents significant regional progress, most of these incubators/accelerators have limited experience in facilitating entrepreneurial processes. Their management capacity is limited and their staff are generally younger compared with businesses in the North, many of which have been in operation since the 1980s. Greater management, facility and infrastructure capacity are needed for incubation programmes in SSA to succeed.

An RUFORUM assessment of 1,187 applicants for entrepreneurship training from 39 countries revealed that the majority of startup enterprises had been in operation for less than two years (Fig. 12.5), and most involved agriculture (Fig. 12.6). An assessment of training needs indicated a requirement for capacity building in business management, financial management, marketing, entrepreneurial mindsets, business innovation, resource mobilization and business growth, among other skills. This information points to the need for continuous training and mentorship of young entrepreneurs.

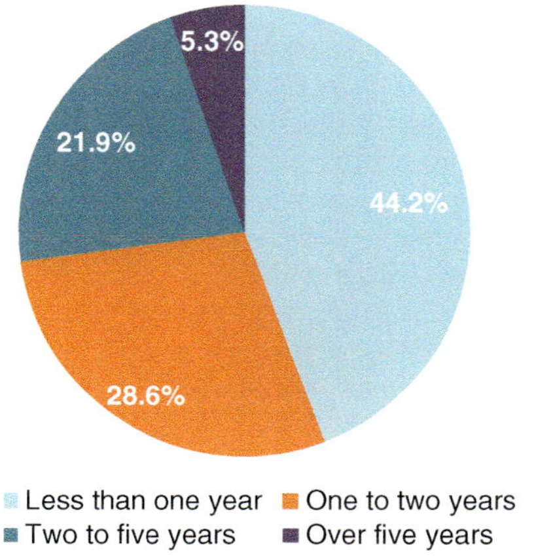

Fig. 12.5. Length of business operation among entrepreneurship training applicants
Source: Constructed by authors from RUFORUM (2020b).

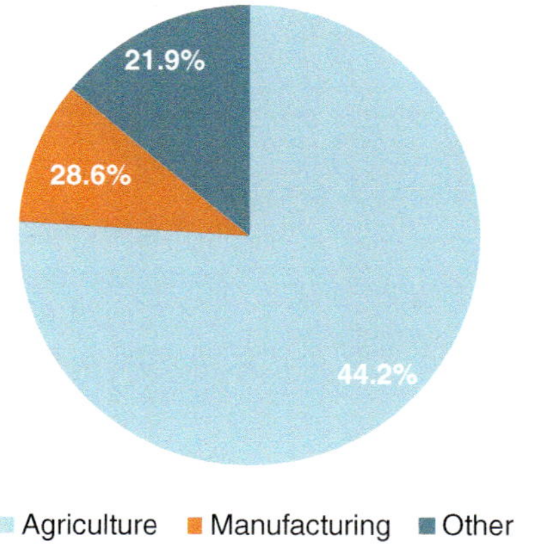

Fig. 12.6. Sectors where the entrepreneurship trainees operate
Source: Constructed by authors from RUFORUM (2020b).
Note: 'Other' includes food processing, technology, tourism, dairy, pharmacy and other retail businesses.

To build management capacity within university-based incubator hubs, RUFORUM runs a support programme which links young startup entrepreneurs with mentors. Accordingly, the University of Nairobi's incubation hub received six months of technical and training support from the African Agribusiness Incubators Network. During this period, the hub developed necessary operating systems and processes, while also supporting young entrepreneurs in innovating new products and services. Examples of enterprises supported include (i) MuMiSET Co. Ltd., which focuses on providing planting material, (ii) Cassava Econet Co., which focuses on providing logistics, bulking and primary processing, (iii) EHENA Foods, which focuses on producing healthy, high-quality consumables from cassava, (iv) MUHOGO TIBA, a service enterprise, which adds value by leveraging and exploiting existing ecosystem gaps using information and communications technology diagnostics to detect bacterial and viral diseases and (v) Duke Enterprises, which produces six guava products, including guava wine.

Responding Innovatively to Entrepreneurs' Needs

As a network organization, RUFORUM is faced with four kinds of challenges in its drive to promote entrepreneurship: (i) ensuring that universities move to adopt and implement practical entrepreneurship training, (ii) mobilizing government and donor funding to support universities in translating their goals to provide practical entrepreneurship training into reality, (iii) maintaining momentum by routinely engaging the network to share innovative programming, including skills development and (iv) developing a pan-African entrepreneurial ecosystem of relevance to African universities. RUFORUM's activities for developing entrepreneurship capacity are structured under two streams (i) entrepreneurship education and (ii) entrepreneurship training programmes (Fig. 12.7). Within this framework, entrepreneurship education is supported at the community level, through secondary and tertiary education, and including both technical and vocational education and training, and the university system. Egerton University and Gulu University have been supported in reviewing and developing their curricula to facilitate the integration of entrepreneurship education across at least ten academic programmes. At the same time, these and other universities have been supported in establishing the aforementioned incubation hubs, and entrepreneurship training has been enhanced through mentorship, business linkages and startup funding.

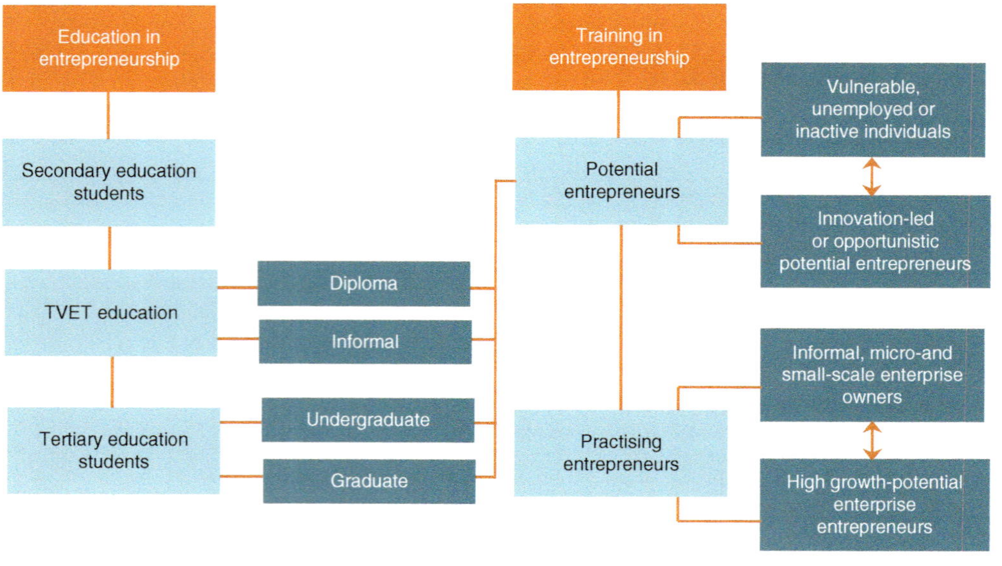

Fig. 12.7. RUFORUM's entrepreneurship programmes
Source: Constructed by authors from Valerio *et al.* (2014).

Based on its capacity needs assessment (as discussed above) RUFORUM undertook in-person and online training to engage the network[1]. All students admitted through the TAGDev programme underwent entrepreneurship training during their orientation to set the stage for their active participation in the incubation hubs during the course of their studies. Online training was also initiated to serve the broader network and maximize the impact of private-sector trainers and experts from within and outside Africa, and this approach could be expanded. The online training was designed to support opportunities for growth in response to the needs of: (i) potential entrepreneurs, (ii) those in the early stages of their entrepreneurial journey and (iii) practising entrepreneurs in informal, micro- or small-scale enterprises. Short training modules were also offered online on the following topics: (i) discovering opportunities and the entrepreneurial mindset, (ii) design thinking, (iii) the art of pitching, (iv) business assessment, (v) the canvas of business models, (vi) sales, marketing and distribution, (vii) managing business operations and (viii) data for business growth. Institutions like RUFORUM can significantly contribute to regional efforts to develop entrepreneurial activities, expertise and resources by attracting guest trainers and providing technical support and mentorship to emerging entrepreneurs across Africa.

Connecting Young Entrepreneurs with the Private Sector

Linkages between universities and the private sector in most of SSA are generally weak and need strengthening. RUFORUM takes a progressive approach to building linkages between universities and the private sector, starting with step-by-step nurturing and growth for young entrepreneurs. The CURAD incubator (Consortium for Enhancing University Responsiveness to Agribusiness Development Limited) is providing a conducive platform to link novice entrepreneurs with the private sector. CURAD is a public–private partnership based in Uganda and promoted by Makerere University, the National Union of Coffee Agribusinesses and Farm Enterprises Limited, the National Agricultural Research Organization, the University of Copenhagen and NIRAS International. CURAD's operating model is founded on strong private-sector leadership, which ensures students gain a solid grounding in practical entrepreneurship. 'Incubatees' have ample opportunity to embark on manufacturing and processing using the incubator's shared facilities, as well as its linkages to financial institutions to access credit.

The second element of RUFORUM's approach to fostering entrepreneurship is connecting young entrepreneurs with individuals from industry who can mentor them as they develop their businesses. The third element is engaging the private sector at national, continental and international levels to invest in university efforts to catalyze growth in entrepreneurship and business development through the incubation processes. Morocco's Mohamed VI Polytechnic University (UM6P) launched a series of initiatives which support both mentorship and greater private-sector engagement through the Impulse accelerator programme. This is a unique programme developed in partnership with MassChallenge and supported by the OCP Group and its subsidiary OCP Africa. This 12-week programme targets innovative startups operating in various fields of science and industry, such as agricultural technology, biotech, mining technology, materials science and nanotechnology. Impulse's main objective is to support the rapid growth of startups and help them forge a profitable and lasting partnership with the OCP Group, Mohammed VI Polytechnic University and their entrepreneurship ecosystem. Startups participating in Impulse benefit from the following resources and opportunities to strengthen their projects and businesses:

- Mentoring and coaching from OCP experts, UM6P professors and MassChallenge's international network of mentors
- Access to business opportunities via OCP Group's, OCP Africa's and UM6P's networks
- A cash prize of US$250,000 shared among the winning startups on 'Demo Day'
- Access to UM6P's infrastructure, including Fablab (an innovation lab for operations working towards developing and managing infrastructure, demonstrators and other test tools in the area of industrial digitization), experiment laboratories and living

laboratories equipped with state-of-the-art equipment enabling startups to develop and test prototypes of their products and services
- Study trips to the innovation ecosystems of Boston and Lausanne
- Access to a 430 square metre working space, equipped to enable startups to develop their businesses
- Access to financing through a set of national and international investment funds and 'business angels'

Benin's University of Abomey-Calavi established the business incubation programme 'UAC Startup Valley'. This incubator builds on the university's strategic focus to establish a model of entrepreneurial and managerial education, responding to labour-market demands through entrepreneurship training. UAC Startup Valley is run as a programme of the University of Abomey-Calavi Foundation. It operates an innovative (three to four year) business incubation process, with a particular focus on innovation, learning by doing and cooperative dynamics. This is realized through a four-phase process of: (i) recruitment (4 months), (ii) pre-incubation (8 months), (iii) incubation (24 months) and (iv) post-incubation (12 months). This four-phased approach has had positive impacts, not only through the successful creation of enterprises by young graduates, but also by improving graduates' competitiveness in the job market (Box 12.1).

A fourth undertaking by universities has responded to the private sector's call for graduates with greater capacity in the area of what is known as 'soft skills', which are considered the cornerstone of successful enterprise development. To this end UAC Startup Valley developed a model of entrepreneurial pedagogy to facilitate the successful integration of entrepreneurship education into Benin's universities and to support the realization of its intended outcomes. These efforts are part of the incubator's focus on the critical factors contributing to the sustainability of enterprises established by its incubatees. These efforts by UAC Startup Valley have previously been recognized in assessment studies conducted by the World Bank on the organizations supporting entrepreneurial endeavours in West Africa. UAC Startup Valley was acknowledged as an incubator with a unique governance and an innovative model capable of being scaled up.

Box 12.1. Tangible outcomes of UAC Startup Valley

An assessment of UAC Startup Valley's incubation process showed that it had increased its students' likelihood of becoming successful entrepreneurs by more than 7 times, and of obtaining a well-paying job within three years of graduating by 13 times (Abihona and Kissezounon, 2019). UAC Startup Valley has been in operation since 2014 and has recruited six cohorts of trainees drawn from 322 groups from various universities in Benin. This has led to the training of 1,059 fellows in entrepreneurial and managerial education, 35% of whom were women. In total, 163 business ideas were transformed into business projects after the pre-incubation phase (50.6%), leading to the creation of 99 agribusiness enterprises, technologies and value-adding services generating over 1,000 direct jobs and an annual turnover of slightly more than one billion CFA Francs.

UAC Startup Valley, which has drawn positive attention for the private sector, has helped incubatees mobilize over US$300,000 in capital financing from various sources for business acceleration. Forty incubatees received US$5,000 each from the Tony Elumelu Entrepreneurship Program, and another incubatee received €25,000 from the Hackathon P-Digitech 2020 Entrepreneurship Development Program in the Pineapple Sector. Thirteen incubatees obtained a subvention of US$8,500 from the Niche Project. One incubatee obtained a subvention from Coopération Technique Belge amounting to US$15,500, and one incubatee received a €3,000 climate prize at the United Nations Climate Conference, among other examples. In 2018, UAC Startup Valley also benefited from a US$50,000 grant from RUFORUM's RECAP, funded by the Mastercard Foundation. Through this grant, the incubator has been able to train and graduate 100 student incubatees. In addition to entrepreneurial and managerial education through learning by doing, these incubatees also received 'proximity coaching' to improve the survival rate of their enterprises. Accordingly, 14 enterprises were created by the student incubatees.

Source: Compiled by authors.

12.4 Conclusion

The university-based incubation hubs described in this chapter provide proof-of-concept that young university graduates can be mentored and supported, not only to become successful entrepreneurs, but also to establish enterprises which create jobs. Students are adaptive and have the potential to transition into their enterprises once they complete their degrees. However, universities must reorient both academic and non-academic staff from education purely *about* entrepreneurship, to education *for*, *through* and *in* entrepreneurship - and this requires new skills. Capacity strengthening is needed, particularly in the area of business management, so universities can more effectively coordinate the entrepreneurship processes they support. The private sector has strong interest in providing mentorship and coaching to novice entrepreneurs; universities simply need to create the necessary platforms for entry and engagement. Institutions such as RUFORUM facilitate universities' endeavours to establish practical entrepreneurship education through networks, partnerships and linkages with the private sector. The very promising results already being demonstrated by university-based incubation hubs will be multiplied with the continued expansion and scaling of these entrepreneurship and training models across Africa. Importantly, adopting and adapting online training, learning and mentorship is a pivotal factor in taking full advantage of the invaluable knowledge, experience and advice of experts from across Africa and the world and integrating it into local entrepreneurial ecosystems.

Endnote

[1] These trainings were intensified during the COVID-19 lockdowns and have played a significant role in engaging students while university campuses were closed. Participants in the online training programmes came from Benin, Burundi, Cameroon, Congo, Chad, Côte d'Ivoire, Djibouti, Mauritius, Mali, Malawi, Lesotho, Kenya, Ghana, Ethiopia, Gambia, Morocco, Mozambique, Namibia, Niger, Nigeria, Rwanda, Somalia, South Africa, South Sudan, Sudan, Tanzania, Togo, Uganda, Zambia and Zimbabwe; trainees from outside African came from China, Costa Rica, France, Germany, India, Saudi Arabia, Switzerland, the United Kingdom and the United States.

References

Abihona, S. and Kissezounon, G. (2019) Effects of the business incubators 'UAC Startup Valley' on the professional insertion of graduates from Abomey-Calavi University. *Revue d'Economie Théorique et Appliquée* 9(2), 107–124. Available at: https://www.retanet.org/lire_article.php?num_art=Vol9-n2-abst2&vol=9&semester=2 (accessed 3 June 2021).

AfDB (African Development Bank) (2011) Entrepreneurship Development. Ch. 6 in *African Development Report, 2011*. Available at: https://www.afdb.org/fileadmin/uploads/afdb/Documents/Publications/African%20Development%20Report%202011%20-%20Chapter%206-Entrepreneurship%20Development.pdf (accessed 3 June 2021).

AfDB (2017) *African Economic Outlook 2017: Entrepreneurship and industrialisation*. Available at: https://www.afdb.org/en/documents/document/african-economic-outlook-aeo-2017-95818 (accessed 28 June 2021).

Autio, E. Kenney, M. Mustar, P. Siegel, D. and Wright, M. (2014) Entrepreneurial innovation: The importance of context. *Research Policy* 43(7), 1097–1108.

Boldureanu, G. Ionescu, A. M. Bercu, A.M. Bedrule-Grigoruţă, M.V. and Boldureanu, D. (2020) Entrepreneurship education through successful entrepreneurial models in higher education institutions. *Sustainability* 12(3), 1267.

Calderon, G. Iacovone, L. and Juarez, L. (2016) Opportunity versus necessity: Understanding the heterogeneity of female micro-entrepreneurs. *World Bank Economic Review 30*(Supplement 1), S86–S96.

Civera, A. Meoli, M. and Vismara, S. (2019) Do academic spinoffs internationalize? *Journal of Technology Transfer* 44(2), 381–403.

Daily, C.M. McDougall, P.P. Covin, J.G. and Dalton, D.R. (2002) Governance and strategic leadership in entrepreneurial firms. *Journal of Management* 28(3), 387–412.
Devine, R.A. and Kiggundu, M.N. (2016) Entrepreneurship in Africa: Identifying the frontier of impactful research. *Africa Journal of Management* 2(3), 349–380.
Devlin, K. (2013) *Reducing youth unemployment in sub-Saharan Africa*. Population Reference Bureau. Available from: https://www.prb.org/resources/reducing-youth-unemployment-in-sub-saharan-africa/ (accessed 3 November 2021).
Fal, M. (2013) Accelerating entrepreneurship in Africa. *Innovations: Technology, Governance, Globalization* 8(3/4), 149–168.
Franco-Leal, N. Camelo-Ordaz, C. Fernandez-Alles, M. and Sousa-Ginel, E. (2019) The entrepreneurial ecosystem: Actors and performance in different stages of evolution of academic spinoffs. *Entrepreneurship Research Journal* 10(2). Available at: https://doi.org/10.1515/erj-2018-0228 (accessed 3 June 2021).
Glassco, J. and Holguin, L. (2016) *Youth and Inequality: Time to Support Youth as Agents of their Own Future*. Oxfam Briefing Paper. Oxfam International. Available at: https://www-cdn.oxfam.org/s3fs-public/file_attachments/bp-youth-inequality-global-120816-en_0.pdf (accessed 3 June 2021).
Guerrero, M. Liñán, F. and Cáceres-Carrasco, F.R. (2020) The influence of ecosystems on the entrepreneurship process: A comparison across developed and developing economies. *Small Business Economics*, 1–27.
Haas, M. (2013) *Necessity entrepreneurship: Individual, environmental and public policy-related factors influencing the process of opportunity exploitation under unfavorable circumstances*. Thesis No. 5830, Doctoral programme in Technology Management, École Polytechnique Fédérale de Lausanne, Switzerland.
Hayes, D. Subhan, Z. and Herzog, L. (2020) Assessing and understanding entrepreneurial profiles of undergraduate students: Implications of heterogeneity for entrepreneurship education. *Entrepreneurship Education* 3, 151–195. Available at: https://doi.org/10.1007/s41959-020-00025-5 (accessed 28 June 2021).
Hoppe, M. Westerberg, M. and Leffler, E. (2017) Educational approaches to entrepreneurship in higher education: A view from the Swedish horizon. *Education + Training* 59(7/8), 751–767. Available at: https://doi.org/10.1108/ET-12-2016-0177 (accessed 27 June 2021).
ILO (International Labour Organization) (2018) World Employment Social Outlook: Trends 2018. Geneva. Available at: https://www.ilo.org/wcmsp5/groups/public/---dgreports/---dcomm/---publ/documents/publication/wcms_615594.pdf (accessed 3 June 2021).
ILO (2019) Work for a Brighter Future. Report of the Global Commission on the Future of Work. Geneva, ILO. Available at: https://www.ilo.org/wcmsp5/groups/public/---dgreports/---cabinet/documents/publication/wcms_662410.pdf (accessed 27 September 2021).
Isenberg, D.J. (2010) How to start an entrepreneurial revolution. *Harvard Business Review*, 88(6), 40–50.
Isenberg, D. (2011a) *How to Foment an Entrepreneurial Revolution*. The Babson Entrepreneurship Ecosystem Project, Babson Global. Available at: https://edisciplinas.usp.br/pluginfile.php/2326986/mod_resource/content/5/A9-How-to-foment-an-entrepreneurial-revolution-IEF-jan-12-20111.pdf (accessed 11 May 2021).
Isenberg, D. (2011b) *The Entrepreneurship Ecosystem Strategy as a New Paradigm for Economic Policy: Principles for Cultivating Entrepreneurship*. The Babson Entrepreneurship Ecosystem Project, Babson Global. Available at: http://www.innovationamerica.us/images/stories/2011/The-entrepreneurship-ecosystem-strategy-for-economic-growth-policy-20110620183915.pdf (accessed 11 May 2021).
Isenberg, D. J. (2016) Applying the ecosystem metaphor to entrepreneurship: Uses and abuses. *The Antitrust Bulletin* 61(4), 564–573.
Kabongo, J.D. and Okpara, J.O. (2010) Entrepreneurship education in sub-Saharan African universities. *International Journal of Entrepreneurial Behavior & Research* 16(4), 296–308.
Kalule, S.W. Mugonola, B. and Ongeng, D. (2017) The student enterprise scheme for agribusiness innovation: A university-based training model for nurturing entrepreneurial mind-sets amongst African youths. *African Journal of Rural Development* 2(1), 55–66. Available at: https://doi.org/10.22004/ag.econ.263300 (accessed 27 June 2021).
Kazeem, Y. (2017) More than half of the world's population growth will be in Africa by 2050. *Quartz Africa* (29 June). Available at: https://qz.com/africa/1016790/more-than-half-of-the-worlds-population-growth-will-bein-africa-by-2050/ (accessed 27 June 2021).
Kew, J. (2015) *Africa's Young Entrepreneurs: Unlocking the Potential for a Brighter Future*. International Development Research Centre. Available at: https://www.idrc.ca/sites/default/files/sp/Images/Articles/africas-young-entrepreneurs-unlocking-the-potential-for-a-brighter-future.pdf (accessed on 3 June 2021).

Lehmann, E.E. Meoli, M. Paleari, S. and Stockinger, S.A. (2020) The role of higher education for the development of entrepreneurial ecosystems. *European Journal of Higher Education* 10(1), 1–9.

Mbeteh, A. and Pellegrini, M.M. (2018) Entrepreneurship education in developing countries: A study of the key challenges in Sierra Leone. In Dana, L.-P. Ratten, V. and Honyenuga, B.Q. (eds.) *African Entrepreneurship*. Palgrave Studies of Entrepreneurship in Africa. Palgrave Macmillan, Cham. Available at: https://doi.org/10.1007/978-3-319-73700-3_5 (accessed 3 June 2021).

McKague, K. Wong, J. and Siddiquee, N. (2017) Social franchising as rural entrepreneurial ecosystem development: The case of Krishi Utsho in Bangladesh. *International Journal of Entrepreneurship and Innovation* 18(1), 47–56.

Mo Ibrahim Foundation (2013) *2013 Ibrahim Forum Facts & Figures. Africa Ahead: The Next 50 Years*. Addis Ababa. Available from: http://static.moibrahimfoundation.org/downloads/publications/2013/2013-facts-&-figures-an-african-conversation-africa-ahead-the-next-50-years.pdf (accessed 3 June 2021).

Ndofirepi, T.M. (2020) Relationship between entrepreneurship education and entrepreneurial goal intentions: Psychological traits as mediators. *Journal of Innovation and Entrepreneurship* 9(2). Available at: https://doi.org/10.1186/s13731-020-0115-x (accessed 27 June 2021).

Ocici, C. (2017) Entrepreneurship is a marathon, not a sprint: A business focus interview. In *Africa on the Move: The Role of Universities in Transforming Africa's Agricultural Sector*. Africa Higher Education Insights 1 (April–June), RUFORUM, Kampala. Available at: https://blog.ruforum.org/2017/07/07/african-higher-education-insights-ruforum-quarterly-newsletter-issue-one/ (accessed 3 June 2021).

Osiru, M. Nampala, P. and Ekwamu, A. (2016) African faculties of agriculture within an expanding university sector. Ch. 9 in Lynam, J. Beintema, N.M. Roseboom, J. and Badiane, O. (eds.) *Agricultural Research in Africa: Investing in Future Harvests*. International Food Policy Research Institute, Washington, DC. Available at: https://www.ifpri.org/publication/agricultural-research-africa-investing-future-harvests (accessed March 24, 2021).

Pittaway, L. and Edwards, C. (2012) Assessment: Examining practice in entrepreneurship education. *Education+ Training*, 54(8/9), 778–800.

Purbasari, R. Wijaya, C. and Rahayu, N. (2019) The entrepreneurial ecosystem as a network-rich system: A systematic mapping study. *Academy of Entrepreneurship Journal* 25(2) 1–17.

Rabana, R. (2019) *How Africa's Entrepreneurs Are Changing the Direction of Globalization* World Economic Forum Available at: https://www.weforum.org/agenda/2019/01/african-entrepreneurs-changing-the-direction-of-globalization/

Rasmussen, E.A. and Sørheim, R. (2006) Action-based entrepreneurship education. *Technovation* 26(2), 185–194.

Roulin, A. (2020) *Study on agri-business incubators and accelerators in sub-Saharan Africa: Entrepreneur support organisations*. West & Central Africa Webinar, 15 October. Agripreneurship Alliance. Available at: https://hub.nourishingafrica.com/documents/1603725709Agribusiness%20Incubation%20-%20West%20and%20Central%20Africa%20webinar%2015-10-2020.pdf (accessed 3 June 2021).

RUFORUM (Regional Universities Forum for Capacity Building in Agriculture) (2020a) *Unpublished RUFORUM monitoring, evaluation and learning data* Kampala, RUFORUM.

RUFORUM (2020b) Unpublished entrepreneurship trainee assessment data. RUFORUM, Kampala.

Sandström, C. Wennberg, K. Wallin, M.W. and Zherlygina, Y. (2018) Public policy for academic entrepreneurship initiatives: A review and critical discussion. *Journal of Technology Transfer* 43(5), 1232–1256.

Shane, S. and Venkataraman, S. (2000) The promise of entrepreneurship as a field of research. *Academy of Management Review* 25(1), 217–226.

Smith, K. and Di Gregorio, D. (2002) Bisociation, discovery and the role of entrepreneurial action. In: Hitt, M., Ireland, R.D., Camp, S.M., Sexton, D., Nixon, R.D. *et al.* (eds.) *Strategic Entrepreneurship: Creating a New Mindset*, Blackwell Publishers, Malden, Massachusetts, USA.

Stam, E. (2015) Entrepreneurial ecosystems and regional policy: A sympathetic critique. *European Planning Studies*, 23(9), 1759–1769.

Stam, E. and van de Ven, A. (2021) Entrepreneurial ecosystem elements. *Small Business Economics* 56, 809–832. Available at: https://doi.org/10.1007/s11187-019-00270-6 (accessed 3 June 2021).

Valerio, A. Parton, B. and Robb, A. (2014) *Entrepreneurship Education and Training Programs around the World: Dimensions for Success*. World Bank, Washington, DC. Available at: https://openknowledge.worldbank.org/bitstream/handle/10986/18031/9781464802027.pdf (accessed 28 November 2020).

Windeløv-Lidzélius, C. (2018) The school as a living lab: The case of Kaospilot. Ch. 5 in Dezuanni, M. Foth, M. Mallan, K. and Hughes, H. (eds.) *Digital Participation through Social Living Labs*. Chandos Publishing. Available at: https://doi.org/10.1016/C2016-0-01911-3 (accessed 28 June 2021).

13 Transformative Technical and Vocational Training in Tertiary Agricultural Education in Africa

Abraham Sarfo[1] and Caroline Mutepfa[2]*
[1]*Independent Consultant, ABSL Consulting, South Africa;* [2]*African Union Development Agency, South Africa*

13.1 Introduction

The ongoing demographic transition of Africa's population, which is expected to more than double by 2050 to reach 1 billion (AfDB/OECD/UNDP, 2015), is probably the region's most consequential challenge. This unprecedented shift will add significant pressure to both rural and urban labour markets. Africa's workforce is expected to expand by 880 million people - almost 800 million of whom will be located in sub-Saharan Africa (SSA), representing two-thirds of growth in the global workforce. Moreover, 370 million youth are projected to join the labour market in SSA by 2030 (Dahlman, 2015). About 50% of the 10 million graduates pouring out of African universities annually do not succeed in securing employment. In contrast, studies show that one of the greatest challenges faced by agro-processing firms and commercial agriculture is lack of a skilled workforce (Acquaye, 2012). Thus, it has become imperative for Africa to prioritize multisectoral rural and location-specific strategies using participatory approaches as a means of empowering rural populations, creating jobs and developing the skills needed to implement the African Union's agenda of integration and structural transformation. It is especially important for Africa to adopt policies in support of youth training to ensure the competitiveness of the workforce and the region's ability to meet current and future challenges.

Among these is the tremendous challenge of generating 'decent'[1] employment and income-earning opportunities, beyond merely raising people out of hunger and poverty. This requires catalyzing long-term sustainable development and inclusive economic growth. In the short to medium term, the agricultural sector has a key role to play in absorbing a substantial share of the rural and urban labour force in agriculture-based economies. At the same time, the sector's productivity must increase, which will ultimately depend on having an African labour force equipped with the appropriate skills. In many African countries, standard education fails to equip rural youth with the knowledge and skills needed to seize the few available productive and decent employment opportunities in the agriculture sector. This limits young people's capacity to pursue viable livelihood alternatives, including starting their own rural enterprises. Meanwhile, formal technical and vocational education and training (TVET) programmes often ignore the specific characteristics of rural labour markets, especially in agriculture, and fail to address rural

*Corresponding author: carolinem@nepad.org

 Transforming Tertiary Agricultural Education in Africa (eds D. Kraybill, J. Lynam and Adipala Ekwamu)
DOI: 10.1079/9781789246544.0013

workers' specific needs for vocational training. Even when TVET institutions do offer relevant education for the agricultural sector and rural areas, it can be difficult for rural youth to access because it is either prohibitively expensive or facilities are located too far away.

The development of agricultural skills is also key to the creation of sustainable livelihoods and ultimately to reducing poverty. It is evident that the broad-based growth necessary to ensure poverty reduction reaches the majority requires skills and training to meet evolving economic demands. Youth employment and job creation depend on both supply and demand factors. The supply side of human-resource development comprises knowledge and skills acquired through training and experience. The demand side requires that businesses have the capacity to absorb newly-trained youth through private-sector-led economic growth and improved access to resources (such as land, finance and markets), supported by enabling policies to facilitate youth and agricultural initiatives. Analyzing successful approaches of the past, it is clear that skills need to be developed in both the formal and the informal sectors, combining technical and social (i.e. 'life') skills. Due consideration of the often-poor background and low level of education of the target group is also pivotal. With limited public- and private-sector employment opportunities, another key factor is building entrepreneurial skills. To address employability, programmes should incorporate the development of the business-oriented skills needed to create jobs and labour-market services for the youth.

Also, the circumstances and needs of rural women and other marginalized groups call for more inclusive approaches. To promote women's economic empowerment, innovative gender-transformative approaches are currently being employed in order to empower women and girls to develop the relevant skills with which to take control over their own lives. This involves raising awareness, building confidence, expanding available choices, increasing access to and control over resources and promoting the transformation of the structures and institutions which perpetuate gender discrimination (European Commission, 2019). Empowering women means more than simply developing their skills; the focus also needs to be on creating the necessary socioeconomic environment to enable them to participate and succeed.

13.2 Agricultural Technical and Vocational Education in Africa

Agenda 2063, the blueprint for Africa's development, recognizes youth as a major constituency, a key driver, an important resource and a critical enabler of Africa's transformation (see Ch. 1, this volume). Attaining the goals set out in Agenda 2063 hinges, in part, on empowering youth and developing their skills. The Continental TVET Strategy (AUC 2014a) aims to support young people in acquiring professional skills through TVET programmes. The mission of the strategy is to position TVET within the education system as a tool for empowering African people, especially youth. According to the strategy, implementation should focus on the following factors:

- Value chains in productive and service sectors, especially in agriculture
- Innovative approaches to developing and delivering curricula, emphasizing entrepreneurship
- The inclusion of all learning pathways - i.e. formal, informal, non-formal (see Box 13.1)
- Research linkages with industry, science and the workplace
- The development of trainers' capacity
- Improvement of infrastructure, including laboratories, experiment fields and equipment
- The promotion of organic linkages with industry through internships, mentorships, innovation hubs and innovation incubators
- Finance and investment in the TVET sector

Box 13.1. Definition of training categories

Formal training. Instruction given in educational and training institutions or designed training sites, including enterprises participating in formal apprenticeship systems. Training is structured and systematic, and it follows pre-defined content and precise learning objectives

Informal training. Unstructured education and training that takes place outside the formal education and training system

Non-formal training. Organized and systematic learning activities conducted outside the educational system emphasizing activities directly associated with work.

Source: Rosas (2006).

The Malabo Declaration (AUC, 2014b) reaffirms the Comprehensive Africa Agriculture Development Programme (CAADP) and identifies nine principles for achieving sustainable and equitable economic development, job creation, poverty reduction and income growth for African people through agricultural productivity. Women and youth are identified as the central focus, with commitments to:

- establish or strengthen inclusive public–private partnerships for at least five priority agricultural value chains with strong linkages to smallholder agriculture in each country;
- create job opportunities for at least 30% of the youth in these agricultural value chains; and
- support and facilitate preferential entry for and participation by women and youth in gainful and attractive agribusiness opportunities.

Among the proposed strategic actions in the Malabo Implementation Strategy and Roadmap (AUC, 2015) is enhancing skills, knowledge and agricultural education. This will be done by developing innovative methods and incentives to increase the enrolment of youth in agriculture-related vocational and tertiary education programmes with especially training programmes focusing on agribusiness and value chains.

Through this and many other initiatives, agricultural technical and vocational education and training (ATVET) is receiving increased regional and national attention, although investment is still low, and the pace of reform remains below what is needed. Progress has been made towards mainstreaming the development of agricultural skills within national development agendas. Mechanisms include: (i) promoting holistic reforms in the TVET sector by bringing agriculture under its umbrella, (ii) promoting youth entrepreneurship and employability and (iii) facilitating policy dialogue and advocacy amongst stakeholders regarding the challenges youth face in attaining skills and becoming ready for the job market. ATVET reforms have also focused on developing competence-based training (CBT) curricula targeting agricultural value chains with training delivered in formal, informal and non-formal systems.

13.3 Agricultural Technical and Vocational Education and Training Challenges

ATVET is fragmented across Africa and in many countries is not integrated into an overall TVET system. Many governments perceive it as a low priority, so it also lacks sufficient funding. Insufficient effort has been invested in identifying private-sector training needs and, in general, ATVET leaders do not consider the private sector to be an important stakeholder. Interaction with public and private employers and service providers would increase ATVET's ability to respond to the needs of a modern agricultural sector and open avenues to expand training programmes. Several stocktaking studies in Africa have analyzed the challenges facing ATVET (NEPAD, 2014, 2015; DAFF, 2015). Key challenges across 12 African countries are summarized below.

Policy challenges. Many African countries have lacked policy coherence on ATVET, and little attention has been paid to the need to develop agricultural skills in national development planning. Most national agricultural investment plans do not adequately address investment in agricultural training and skills development, and the fragmented nature of training does not meet the needs of the rapidly growing agricultural and food sector.

System challenges. Compared with progress in developing TVET systems in general, ATVET has only received marginal attention. ATVET is not usually included in the overall TVET systems, and - although they often mention agriculture - reform agendas and processes do not include specific measures related to agricultural training. Typically, national qualification frameworks (NQFs), and in some cases technical qualification frameworks, do not include agriculture.

Training challenges. Agricultural training colleges deliver technical skills training mostly on a theoretical basis, but with little or no practical learning. Hence, learners do not acquire the competencies needed to become employable or self-employed. Training institutes have low capacity, inadequate and outdated training materials and equipment and a lack of qualified trainers.

Teachers and trainers lack practical, pedagogical and didactic skills, and the knowledge and competence to develop curricula to meet the needs of modern agricultural system. Efforts to rehabilitate training centres and programmes tend to build on old systems instead of promoting more innovative and efficient methods.

Private-sector linkages and support. Compared with the progress made in general TVET systems, ATVET has been only given marginal attention to potential private-sector linkages in developing curricula, planning and delivering training and monitoring trainees.

Poor perception of ATVET. Public perception of the inferiority of vocational training in agriculture is widespread. As a professional career, agriculture is generally viewed negatively compared with technical or administrative/office work, and this has suppressed student interest in ATVET.

13.4 Agricultural Training Centres

Overview of current status

Within the ATVET system, agricultural training centres (ATCs) are public or private centres, colleges or schools offering formal, informal and non-formal training for learners and farmers. Unlike universities, many of which are autonomous and design their own training delivery models, ATCs directly translate government policies and strategies into training delivery mechanisms. A recent paradigm shift towards training for what has been termed 'agripreneurship' aims to transform agricultural trainees and graduates into entrepreneurs possessing the necessary skills and competencies to not only become employable, but also to be employers. ATCs with well-developed programmes of agripreneurship could become change agents in transforming the traditional agricultural knowledge and practice of farmers, extension agents, women, youth and others to commercial agriculture and profitable enterprises which create jobs, generate income and strengthen food security, as is envisioned in the CAADP framework and the Malabo Declaration.

Improving ATVET delivery in Africa requires modernizing ATCs to enable them to respond to the dynamic needs of the agricultural sector. ATCs should be catalysts of ATVET responsiveness to changing labour-market demands, especially those of the private sector. In most African countries, formal vocational education and training systems are geared towards starter education and training for young people at the secondary, post-secondary or tertiary levels. The vast majority of these systems involve full-time, classroom-based training. A small number of countries, such as South Africa, also offer state-regulated apprenticeship training schemes which confer state-recognized qualifications. In the past, ATCs in Africa were mostly configured to train agricultural extension officers and other government officials. In the wake of structural adjustment programmes, lack of investment in ATCs coupled with the collapse of extension services has led to some ATCs being upgraded to universities. In the process, these transformed institutes have mostly lost their mandate to provide lower- and middle-level training, and instead are competing with universities to train graduates for the mass market.

In Benin, Lycée Agricole Mèdji de Sékou (LAMS) is an ATC which has been at the forefront of secondary training for most of the country's agriculture graduates. Like most African institutes, it was designed to provide basic training for agricultural engineers and public service officers. LAMS's mandate has not changed significantly, but curricula now include entrepreneurship and practical training. With the near collapse of the country's agricultural extension services and lack of opportunities for most graduates, it is time for institutions like LAMS to move beyond the old structure to focus on enhancing employability and creating opportunities for entrepreneurship, self-employment and private-sector jobs.

In Kenya, ATCs became prominent in the 1950s after the launch of the Swynnerton Plan, which called for intensification of African agriculture (MOA, 2011). These centres played a major role in the development of livestock and crop production. Until the constitutional change that transferred implementation powers to county governments, ATCs were part of the extension delivery system and operated under the National Agricultural Sector Extension

Policy within the Kenyan Ministry of Agriculture. Following the shift to local implementation, ATCs in different agroecological zones focused on offering in-service training to public-sector employees and farmers, and short, on-demand courses to both the public or the private sectors.

In Ethiopia, policies aimed at smallholder commercialization have led to the modernization of training and extension services. The focus of agricultural extension delivery shifted to sustainably intensifying productivity, adding value and developing smallholder competitiveness for both domestic and export markets (Tefera *et al.*, 2015). In addition to the existing ATCs, thousands of farmer training centres (FTCs) were established as part of a strategic shift to knowledge-based training and development of smallholders. ATCs concentrated on training extension officers in order to support the delivery of training services to farmers at the FTCs. As of 2014, the government had established close to 11,000 FTCs, as well as 25 ATCs to train development agents in various specializations to support the FTCs. Ethiopia therefore presents one of Africa's best examples of reforming ATCs to meet national strategic priorities needs for agricultural development.

Challenges facing agricultural training centres in implementing agricultural technical and vocational education and training

In a self-assessment, 12 African ATCs - in Ghana, Kenya, Namibia, Cameroon, Tunisia, Sierra Leone, Togo, Benin, Burkina Faso, Rwanda, Uganda and South Africa - identified their key challenges to include management, trainer capacity, linkages with the private sector and resource mobilization (AUDA, 2019). Other studies and reports have identified similar challenges in ATCs across Africa (Hartl, 2009; NEPAD, 2014; Jones, 2016). Further details are provided below.

Policy, governance and management. On the side of ATVET supply, weak governance structures hinder efficient and effective institutional management. Many training centres and colleges have shut down, and for many of those which have survived, inadequate financing, poor management and ill-adapted organizational structures lead to poor results. Many of the governance problems emanate from the nature of ownership and control of ATCs. Most are controlled and managed by ministries of agriculture, while the general TVET programmes and institutions delivering them are managed and controlled by ministries of education or, in francophone West Africa, by ministries of TVET. The governance of ATCs has important implications for national regulation of ATVET curricula through NQFs, which are instruments for developing and classifying qualifications according to a set of criteria for various levels of learning. Most countries do not recognize or accredit ATCs offering ATVET courses. At the school-management level, most of the principals do not have training or experience in the ATVET-management system. In Ghana, for example, ATCs are managed by the Human Resources Department of the Ministry of Food and Agriculture. Heads and principals are posted to ATCs without the requisite qualifications for school management. Consequently, many heads have not been trained as educators and hence are not able to guide their staff adequately to deliver training.

Training delivery. Many ATVET training institutions have low capacity, deficient materials and equipment and a shortage of qualified trainers. Curricula are outdated and do not correspond to the needs of target groups, nor are they oriented to value chains or developing 'agripreneurial' skills. Many teachers and trainers lack technological knowledge and practical skills, pedagogy and methodology, and advanced teaching and learning materials. Furthermore, they lack the competence to create curricula and develop a system of instruction - i.e. didactics, media, course design, evaluation methods and tracer studies. Modern competence-based training approaches recommend at least 60% practical training delivery, but most ATCs do not have the requisite laboratories, practical training fields or workplace learning experiences.

Engagement and linkages with the private sector. Forums and frameworks to facilitate interaction between ATCs and the private sector are lacking. Interaction is hindered by differences in the culture and concepts used by the two parties and, more importantly, their expectations of benefits from the engagement.

13.5 Reform Of Agricultural Technical and Vocational Education and Training

Comprehensive efforts will be needed to develop ATVET into demand-driven systems - combining education, training, knowledge-development and skill-enhancing techniques - which comprise both public and private actors and measures of integration into national TVET systems. New and innovative models are needed to incorporate current and emerging agricultural conditions into existing ATVET curricula and to generate new institutions for agriculture. The need to develop professional and vocational capacity must be recognized as a cross-cutting strategy under national development processes and at the regional and Africa-wide levels. In addition, best-practice examples should be shared within existing continental initiatives and mechanisms.

Agricultural education and capacity-building programmes should be developed for women and youth using participatory approaches which focus on best practices and knowledge sharing. Rural youth and young farmer's platforms and councils must be created to determine training and capacity-building needs. Training on 'soft' skills, leadership and lobbying should be implemented for rural youth, making use of information and communication technology (ICT). Women and men should have equal access to training and education, and gender aspects must be taken into consideration when determining training themes and setting the timing of training.

Ten goals have been identified for reforming ATVET in Africa (NEPAD, 2014, pp. 14–15):

1. *Demand orientation.* The ATVET system is geared towards improving the competitiveness of economic sectors through integrated, demand-driven and competence-based ATVET systems, producing a qualified workforce responding to labour-market demands.

2. *Outcome-based approach.* The system instils trainees with the necessary knowledge, information and skills to enable them to perform according to occupational standards and hence receive certification.

3. *Harmonization with existing national qualification frameworks.* NQFs are intended to improve access to work and further learning by ensuring that qualifications are relevant to employment and learning.

4. *Quality and relevance.* Effective quality management continuously monitors the relevance of programmes to policies and development strategies, and provides support and guidance to private and public ATVET providers.

5. *Open access with equal opportunities/ gender sensitivity.* The system promotes social inclusion by increasing overall access to ATVET training opportunities by all target groups needing to develop skills. ATVET is accessible, irrespective of education level, sex, ethnicity or religious affiliation. Policies are developed to ensure that women and girls are not discriminated against.

6. *Permeability.* The system provides vertical and horizontal mobility across different occupations and qualification levels.

7. *Flexible mode of delivery and modularization.* Outcome-based ATVET provides high flexibility in organizing and delivering training according to the needs of target groups and occupational requirements. Training modules are accessible for graduates unable to attend classes continuously.

8. *Integration of formal, informal and non-formal vocational education and training.* ATVET recognizes and certifies competencies obtained through formal, informal or non-formal training. It integrates good practices and training on value chains into formal training programmes.

9. *Combination of theoretical and practical training.* Curricula provide a higher ratio of practical training compared with theoretical training (the recommended breakdown being 70% to 30%). Offerings include internship programmes and cooperative training - a mode of ATVET provided in partnership between the private sector and training institutions.

10. *Stakeholder involvement and private-sector participation.* ATVET operates at the interface of various sectors (i.e. education, the labour market, medium- and small-scale enterprises, farmers, farmer and trade associations and public and private training providers). Involving all stakeholders from the early stage of policy discussions is key to building trust and ensuring continued participation in the process of change.

In developing ATCs, interventions and capacity building should focus on training students to become employees, entrepreneurs and (ultimately) employers. These outcomes should be the top priorities in developing curricula and determining technology, knowledge-transfer methodologies and value-chain targets. Consequently, the content of training should focus on technical and vocational skills, and also entrepreneurship. In building ATC capacity, initiatives should focus on technological and practical knowledge and skills, business models and cases (support by industry) and the design and implementation of workplace-based learning, industry attachments, business and technology incubation, mentorships and student projects. Technical skills should focus on hands-on and practical vocational and entrepreneurial approaches. New fields of training should be included for an holistic approach to making trainees meet current global value-chain markets, including food safety, quality systems, Global G.A.P.[2], modern processing technologies, aggregation and distribution (logistics management), marketing, business management, use of ICT and agri-vocational skills development.

ATCs need pre-service and in-service teacher training through train-the-trainer and master-trainer programmes, along with other approaches to improve teachers' capacity to deliver high-quality, demand-driven training. Teacher training is also needed to implement value-chain and agripreneurial training approaches, in which many ATC trainers are currently not well-versed. Capacity development should include training ATC administrators and teachers to engage the private sector through such strategies as cooperative training, workplace-based learning and job placements. Organizational development work is also important to improve the quality of ATVET service delivery. Areas of organizational development requiring attention include the following:

1. ***Inter-organizational linkages:*** establishing and maintaining close work relationships with farmers, agribusinesses, universities and research centres
2. ***Effective and strategic governance:*** establishing a board with representation from the private sector, community and farmers' associations
3. ***Strategic management:*** implementing efficient and effective management systems, including strategies to manage change
4. ***Monitoring:*** initiating, motivating, planning, implementing and monitoring institutional development processes
5. ***Evaluation:*** establishing a learning culture to promote ongoing staff development and the formation of innovative departmental teams

ATC institutions require ongoing capacity development to include curricula, teaching methods and industry linkages supporting the implementation of demand-driven training. External resources, investments and industry knowledge are all required to strategically and continuously guide ATCs.

13.6 Improving Teaching and Learning Systems

Innovative approaches to demand-driven delivery of formal, informal and non-formal ATVET training and business development are needed as part of the process of reforming and improving ATVET services in Africa. Four such approaches are described in turn below.

Competence-based training

As previously indicated, many existing curricula for agricultural knowledge and skills are outdated and fail to meet employers' needs. To address this weakness, the CBT approach is used worldwide in TVET institutions and has been adopted in some African countries to improve ATVET. Curricula inspired by this approach are designed and developed to meet private-sector needs for skills and training, thereby preparing the trained workforce to produce the inputs or services demanded throughout the value chain and hence enable them to either find formal employment or become self-employed. CBT operates on the concept that people can learn skills and that most training is transferable. A person may already have the skills and knowledge to gain a unit of competency, or even a full qualification, without taking part in a full training programme. In contrast, ATVET needs a more inclusive and innovative approach to designing and implementing relevant curricula for the African context. Unlike the traditional CBT approach, the ATVET approach should focus its curricula along agricultural value chains, taking specific skills gaps into consideration.

Successfully implementing CBT requires investment, and an agreement on funding based on the documented needs and expectations of stakeholders. The investment is principally in the capacity of teachers and instructors, and in a pool of experts who can deliver CBT training through innovative approaches. Employers, industry and trade associations and professional bodies must be involved in identifying the occupations requiring skills development, and act as partners in delivering training and offering work-experience. Investment should include adequate resources for quality assurance to ensure its integration into all planning, development and implementation processes.

The process of implementing CBT, based on the value-chain approach from CAADP's ATVET Project in 12 African countries (BMZ, no date), comprises 12 general steps which can be modified to suit individual countries' certification and accreditation systems (Fig. 13.1).

Informal Training

Informal training differs from formal learning to the extent that it takes place outside the institutional education and training system. It does not require such structures as defined learning sites, curricula, teachers and trainers. Informal training often includes training undertaken by civil society or community-based organizations in the course of their programming or advocacy work. Informal training meets an important need because formal vocational education and training systems are generally not geared to the needs of the informal economy. The informal economy has specific needs and challenges, including people's general inability to spend long periods of time away from work. Such people

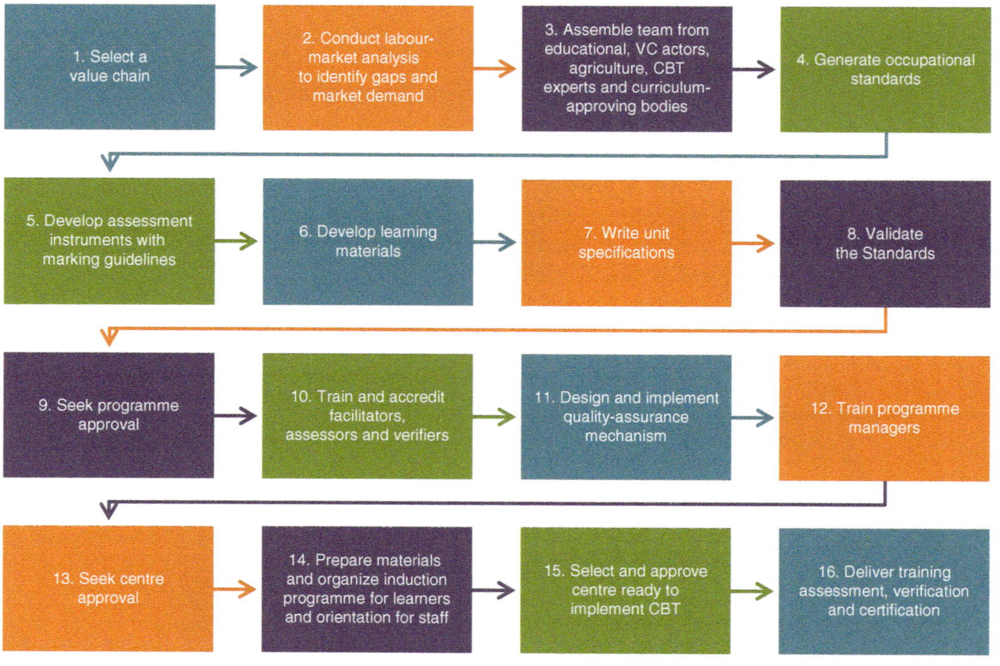

Fig. 13.1. Overview of the competency-based training process for agricultural technical and vocational education and training in Ghana
Source: AUDA (2017, p. 35).
Notes: CBT = competency-based training; VC = value chain.

rarely have a formal basic education to allow them to qualify for formal training, and this limits their access to more structured forms of training.

To address this barrier, informal training activities should be integrated into formal and non-formal ATVET in order to recognize prior learning and open alternative pathways for youth, especially in rural areas, who have limited access to formal training. Informal training can be developed and implemented through outcome-based, demand-driven training modules focusing on value chains and delivered through cooperation of public–private partnerships, e.g. with farmers, processing companies and marketing firms.

Providing informal training through ATVET is a challenge, but the successful example of Koberside Cooperative in Benin[3] shows how a standardized approach to non-formal training can improve the conditions of those working in the informal economy. This can be achieved, for example, by establishing public training centres for informal training. Centres could be oriented around (i) identifying and providing relevant information to particular target groups (e.g. women and youth), (ii) collaborating with actors at different levels to ensure the participation of target groups, (iii) tailoring vocational education and training interventions, materials and approaches to suit the needs of target groups, (iv) providing certification to bridge the gap to formal education or training opportunities where appropriate and (v) embedding training in social movements, local organizations and the broader educational system so as to create institutional structures and linkages.

Cooperative skills training and creating 'agripreneurs'

Cooperative ATVET is a mode of training delivery provided as a partnership between ATVET institutions and private actors (e.g. farms, processing companies and marketing firms). The core feature of this model is *direct* private-sector participation in the provision of practical training on farms, in workshops and on production lines. In some countries and regions, cooperative ATVET schemes have been organized as formal apprenticeship-style training, implying a contract between the trainee and the company. Practical training, however, can be organized in two different formats: (i) as attachments during school holiday breaks and dedicated training sessions or (ii) as a dual model, whereby private companies deliver practical training and ATVET institutions teach the theoretical aspects. The two major advantages of cooperative training models are: (i) introducing trainees to the workplace and (ii) the implications of cost-sharing by both public and private sectors.

Agripreneurial and entrepreneurial training aims to improve the commercial performance of target groups. It is a comprehensive set of training instruments which makes use of an action-oriented approach and experiential learning methods to develop and enhance the business management and personal competencies of the learner. Agripreneurial and entrepreneurial training is now implemented in Africa and other parts of the world. Various concepts and approaches have been applied to create sustainable employment strategies and programmes to develop youth as agripreneurs. In each country, extension education and training institutions must respond to their constituencies' needs.

The Innovation-Centre Business Model

Innovation centres offer a broad range of services for different target groups, such as farmers, agribusiness workers, processors, agripreneurs and startups (Fig. 13.2). Some innovation centres develop non-academic ATVET programmes for specific target groups, such as farm managers or supervisors in processing companies. Centres may provide regular diploma and certificate programmes for rural youth and extension officers. Curricula are designed according to defined occupational standards, and should be designed to open pathways to higher levels of education. Innovation-centre training emphasizes practical skills development combined with an appropriate theoretical grounding. On completion of training, graduates should be qualified to become farm labourers (employees) or farmers (employers or self-employed), or to become self-employed in agriculture-related enterprises. Training programmes for extension officers provide comprehensive training on technology and extension methods. Some innovation centres offer long-term

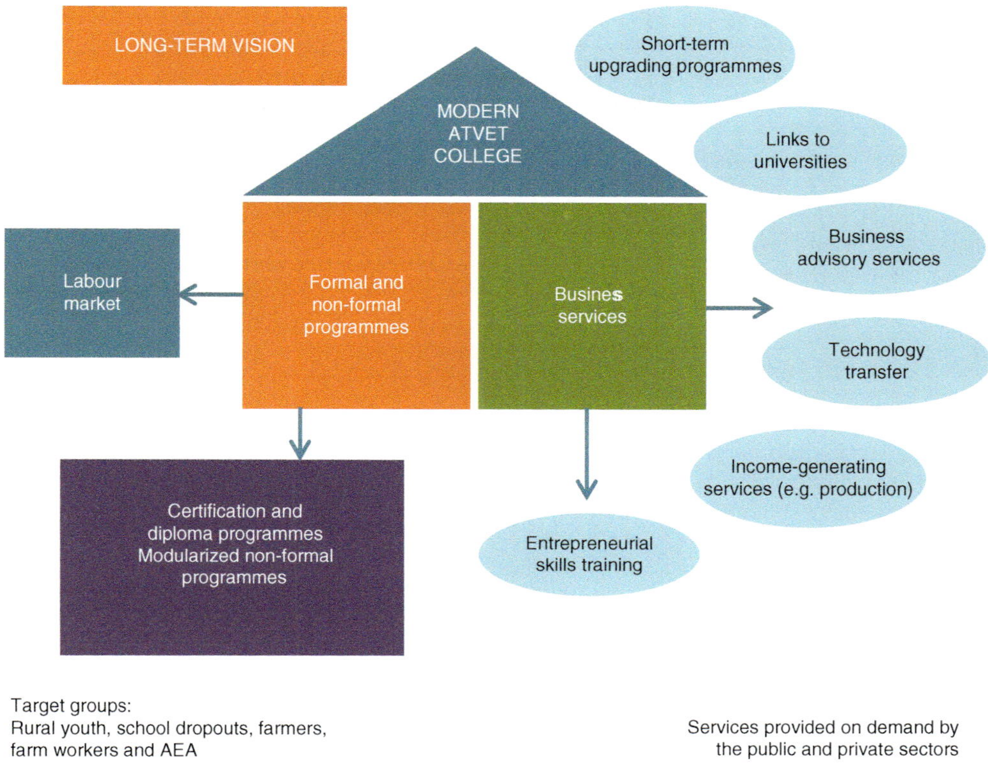

Fig. 13.2. Model for a modern ATVET institution as an innovation centre
Source: Constructed by authors.
Notes: Non-formal training includes systematic learning activities conducted outside the educational system. AEA = agricultural extension agents.

training programmes, often using flexible training modes, such as modularized training delivery, in addition to full-time delivery. Students enrolled in formal programmes are typically certified according to national regulations.

To promote business development, innovation centres may provide practical and applied technology transfer as a means of generating their own income. Such services include (i) business-development for farmers and micro-, small and medium-sized enterprises, (ii) value-chain advisory services, (iii) ATVET training consultancies for farm businesses, processing industries, the public sector and other ATVET institutions and (iv) research on process development and action research, tracer studies or feasibility studies. Innovation centres (and other ATVET institutions) should develop a marketing plan. They should also develop close links with the public and private sectors to be visible to clients and to ensure their products and services are well known. A monitoring and evaluation instrument is also needed to encourage feedback from clients on service quality and acceptability. Innovation-centre staff working in business advisory services and technology transfer may need to be trained in marketing, external collaboration and monitoring and evaluation.

13.7 The Private Sector as a Key Stakeholder

The private sector plays a critical role in the successful development of ATVET. In most cases, the private sector serves as the primary end user of ATVET outputs by providing jobs opportunities for graduates. Moreover, it is through the private sector that networking and cooperation can be fostered along the value chain. Under CAADP's ATVET model, the private sector is a key stakeholder, so strategies are specifically designed to engage private enterprise. The private sector's role within the ATVET system is described below.

Provision of workplace facilities for training. Within the training ecosystem, private organizations can offer their premises for practical and hands-on training. Where training institutions do not have adequate infrastructure, the private sector can be an invaluable partner. In addition to offering facilities, the private sector can often provide technical experts from the workplace to support student training.

Support curriculum development. In Kenya, the National Industrial Union plays a critical role in developing occupational standards for curriculum development. In Ghana, industry experts participate in selecting value chains, identifying skills gaps and providing information needed for skills training. Industry experts also lead the development of occupational standards, which serve as the basis for CBT curriculum development.

Provision of resources. The private sector can provide resources for training, including student scholarships, funding for teacher training, equipment, opportunities for work-based learning (i.e. attachments), learning materials, laboratories and equipment. The private sector can also provide financing for applied research relevant to their production and delivery requirements.

Strategic and industry information. The private sector generally has an in-depth understanding of market trends, growth patterns, labour-market requirements, future dynamics and threats and opportunities in particular industries. This information is relevant to determining the kind of training ATVET should offer.

Policy advocacy. A critical private-sector role is helping to shape policy. Industry groups - such as chambers of commerce, commodity value-chain councils, private enterprise federations or foundations, export associations, agro-processing firms and producers - can influence government policies on reforming skills-development programmes and determining resource allocation.

13.8 Linkages Between Agricultural Technical and Vocational Education and Training, and Tertiary Agricultural Education

Linkages between the different agricultural education and training streams are important to ensure an adequate and appropriately trained workforce is available to support the modernization of African agriculture. Effective commercialization of smallholder farming systems requires national linkages among education, research and extension (Swanepoel *et al.*, 2014) and an integrated agricultural innovation system to ensure productivity gains (FAO, 2014).

A key weakness of the agricultural education and training (AET) system is a disconnect between ATVET and tertiary agricultural education (TAE). This weakness is compounded by a negative perception of ATVET, and lack of vertical and horizontal mobility between the two systems. When CAADP's ATVET project began, many AET providers did not see the value of linkages between ATVET and TAE. However, as tertiary education has become more vocation-oriented, the partnership between the two education sectors has gained acceptance. Transformation of the agricultural sector requires that AET address the gamut of research, extension services, capacity building, skills improvement, knowledge transfer and feedback and monitoring in order to improve smallholder agriculture, youth employability and women's economic empowerment.

Table 13.1. Examples of partnerships and projects

Programme	Focus
KULIMA FRONTLINE	Agricultural extension services
CELIB Incubar	Agribusiness incubator for the livestock value chain
DAIRYTRAIN	Building capacity to deliver competent graduates to enhance competitiveness in Kenya's dairy value chain
USINKENYA	Trilateral partnerships for food security

Source: Compiled by authors.

Case study of linkages between universities and agricultural technical and vocational education and training

The lack of coordination between research (the production of knowledge and technology) and extension (the transmission of knowledge and information to farmers and other users) and the vocational and technical training needs of farmers and youth has limited the growth of the agricultural sector in Africa. The result is lack of the much-needed skilled workers to apply research recommendations and technologies throughout agricultural value chains.

To improve coordination among training levels, the Regional Universities Forum for Capacity Building in Agriculture (RUFORUM), Gulu University, Egerton University and the MasterCard Foundation conceptualized and implemented the project, Transforming African Agricultural Universities to Meaningfully Contribute to Africa's Growth and Development (TAGDev). TAGDev seeks to: (i) pilot connections between TAE and rural communities, especially smallholder farmers, (ii) strengthen agribusiness and entrepreneurship at the two participating universities and selected TVET institutions and (iii) upscale the newly-developed model for agricultural education to other universities and TVET institutions. The aim is to increase collaboration and mutual learning among institutions and agencies in order to implement and influence innovative TAE in support of rural transformation in Africa. In implementing the project, a consultant was tasked to evaluate and suggest best approaches to prioritize employability as a goal of higher education and to foster partnerships and other linkages between TAE and ATVET institutions.

Under RUFORUM's MasterCard capacity-building programme – which has built a new training model worthy of replication - Egerton University in Kenya is partnering with several ATVET institutions, including the Dairy Training Institute, Bukura College of Agriculture and the Animal Health and Industrial Training Institute. The programme's unique characteristics are highlighted below.

The training model and modules. Egerton University has adopted the value-chain and CBT approach, and has developed comprehensive curricula using industry and academic experts. The curriculum will be reviewed every five years with strong private-sector engagement to identify the skills required by particular value chains in the near future. Occupational opportunities were identified within the commodity value chains, and practical work constitutes 60% of the students' examination evaluation. For practical work, groups of students are assigned supervisors who monitor their reporting time, individual contributions to team work, level of skills acquired, outputs and performance. This model starts in Year 1 and continues until the programme's completion in Year 4. In addition to bringing new partnerships to the university, this has made Egerton graduates highly sought-after by employers in all fields.

Training infrastructure. Egerton has some of the requisite training infrastructure needed for practical training. Key infrastructure includes a feed mill, greenhouse, piggery, poultry houses, livestock-rearing buildings, milking facilities, irrigated land for crop production and state-of-the-art computer lab for monitoring and modelling. Egerton is implementing a number of projects to improve infrastructure, expand resources and gain support for student's practical work.

Backward linkages with agricultural technical and vocational education and training institutes. Egerton has established formal partnerships with three ATVET institutions, providing such services as developing curricula, training teaching staff, providing facilities for their students' training, collaborating on research and providing oversight and quality assurance. This partnership facilitates academic progression for ATVET graduates to be admitted to the university.

Strong industry networks and strategic alliances. Egerton has established strong partnerships with private-sector companies, especially in the dairy industry. These companies engage Egerton to conduct research and provide training and technical support. Egerton benefits from industry knowhow and from opportunities for students to be hired by the companies. Egerton also places some of its master's degree students in companies to conduct their thesis research.

Making agriculture attractive to youth. Egerton's agricultural students take pride in the course they are undertaking, and showcase their activities to other young people. The students have instituted a reality show dubbed 'Miss and Mr. Agriculture' in which they compete for this title. The winners become the face and voice of the faculty for a year, visiting private companies and high schools to promote Egerton's Faculty of Agriculture.

Pan-African ATVET Project

The African Union Development Agency implemented an ATVET project as a model for improving agricultural education and skills within the context of Agenda 2063 and CAADP. The project is supported by the German Development Cooperation (GIZ) through the GIZ–CAADP Support Programme commissioned by Germany's Federal Ministry for Economic Cooperation and Development. In 2017, the project was commissioned to scale-up to six additional countries, with an additional module, 'Women in ATVET', focusing on developing skills to promote gender transformative approaches within the agricultural sector. Both the original and the expanded versions of CAADP's ATVET project focus on developing the capacity of farmers, women and youth to access labour-market and income-enhancing entrepreneurial opportunities in the agrifood sector. These interventions help support African Union member states to design approaches, tools and implementation instruments to meet the Malabo Commitment of halving poverty by 2025 through inclusive agricultural growth and transformation. The CAADP indicators associated with this goal are 'Creating job opportunities for at least 30% of the youth in agricultural value chains' and 'Creating preferential entry and participation by women and youth in gainful and attractive agribusiness' (AUC, 2015, p. 4).

The ATVET project develops curricula on agricultural value chains using CBT, with training delivered through formal, informal and non-formal systems. The project also supports mainstreaming agriculture in the overall TVET system in the initial six participating countries and is developing tools and instruments to guide African Union member states and other key stakeholders in scaling up the model. The blueprint of implementation in African Union countries involves supporting capacity building of actors and stakeholders, including: (i) ministries of agriculture, education and labour, (ii) TVET authorities, ATCs, private-sector employers and trainers, (iii) farmers and other value-chain actors and (iv) women and youth. This is implemented by integrating ATVET into national policies and strategies, mainstreaming agriculture in national TVET systems, building the capacity of ATCs to design and implement CBT curricula and linking trainers and trainees to the private sector through industrial placements, attachments and other practical approaches to impart hands-on training. The training strongly emphasizes agripreneurship.

As of 2019, 36 competency-based and gender-sensitive curricula targeting specific value chains had been made available in Africa, and 20 of the selected ATCs had adapted their training modules to the ATVET approach for the value chains included in the project. A key example focusing on gender-transformative approaches is how a cohort of 122 women graduated as farm and tractor managers and operators in Ghana. The Africa Union Development Agency and participating countries have published approximately 79 examples of ATVET practice.

Table 13.2. First countries to implement agricultural technical and vocational education and training, and the value chains developed

Country	Value chains developed
Benin	Meat and rice
Burkina Faso	Rice, sesame seed and cashews
Ghana	Pineapples and citrus fruit
Kenya	Dairy produce, horticulture and aquaculture
Malawi	Mangoes, pineapples and aquaculture
Togo	Rice and aquaculture

Source: Adapted by authors from AUDA (2017, Fig. 1).

13.9 Conclusion

Recommitment from Africa Head of States to the CAADP process through the Malabo Declaration and its implementation strategy has laid emphasis on the importance of youth engagement in agriculture and agribusiness and ATVET is now being prioritized in many agriculture investment plans in African countries. The innovative value chains and competence-based training delivery approach employed in curriculum development and delivery has ensured the adoption of an occupation-based modular-training approach, as opposed to the conventional training modules that focus on crop and animal science.

With Africa's youth posing a significant opportunity for economic growth, their participation in agriculture requires a modern approach that caters to their technologically-advanced mind set. Therefore, if agriculture skills and training is to contribute meaningfully to Africa's development challenges (food security and youth unemployment) barriers to agricultural TVET need to be addressed. This TVET system reform must lean on international best practices while ensuring that solutions are based on local contexts and national realities. Private-sector engagement remains the biggest opportunity to ensure, not only that skills provided by public institutions are demand-driven, but also that co-operative training delivery includes efficient internships and apprenticeships to ensure graduates are work ready as soon as they complete their training.

The revitalization of agricultural training centres to become functional entities cannot be underestimated. Transforming theoretical agricultural training is a necessity and can only be fully achieved through university partnerships ensuring that research is practically converted into practice. The adoption of value chain-based training approaches to agriculture with an ecosystem perspective will go a long way towards ensuring that more young people can take up the entrepreneurial opportunities offered within the sector. In conclusion, policy reform must spearhead the successful transformation of agricultural TVET systems and institutions.

Endnotes

[1] According to the International Labour Organization (ILO, no date), 'decent work sums up the aspirations of people in their working lives. It involves opportunities for work that is productive and delivers a fair income, security in the workplace and social protection for families, better prospects for personal development and social integration, freedom for people to express their concerns, organize and participate in the decisions that affect their lives and equality of opportunity and treatment for all women and men'.

[2] Global G.A.P. is a worldwide standard initiated by food-system retailers for good agricultural practices to address product safety, environmental impact and the health, safety and welfare of workers and animals.

[3] The Koberside Farmers Organization supports smallholder farmers in acquiring standardized skills which benefit smallholder farmers with contracts to supply their products to the biggest supermarket chain in Cotonou, Benin (https://www.facebook.com/koberside).

References

Acquaye, D. supported by Frimpong-Manso, A. (2012) *The Roles and Opportunities for the Private Sector in Africa's Agro-Food Industry.* UNDP African Facility for Inclusive Markets. Available at: https://www.enterprise-development.org/wp-content/uploads/UNDP_AFIM_Agro-food-industry.pdf (accessed 29 March 2021).

AfDB/OECD/UNDP (African Development Bank, Organisation for Economic Co-operation and Development, and United Nations Development Programme) (2015) *Africa Economic Outlook 2015: Regional Development and Spatial Inclusion*. OECD Publishing, Paris. Available at: https://www.oecd-ilibrary.org/development/african-economic-outlook-2015_aeo-2015-en (accessed 29 March 2021).

AUC (African Union Commission) (2014a) *Continental Strategy for Technical and Vocational Education and Training (TVET) to Foster Youth Employment*. Available at: https://au.int/en/documents/20181022/continental-strategy-technical-and-vocational-educational-and-training-tvet (accessed 29 March 2021).

AUC (2014b) *Malabo Declaration On Accelerated Agricultural Growth and Transformation for Shared* Prosperity and Improved Livelihoods. Available at: https://www.au.int/web/sites/default/files/documents/31247-doc-malabo_declaration_2014_11_26.pdf (accessed 22 March 2021).

AUC (2015) *Implementing Strategy and Roadmap to Achieve the 2025 Vision on CAADP: Operationalizing the 2014 Malabo Declaration on Accelerated Africa Agricultural Growth and Transformation for Shared Prosperity and Improved Livelihood*. Available at: https://www.esaff.org/wp-content/uploads/2017/07/AU-MALABO-Implementation-Strategy-Final-version.pdf (accessed 20 March 2021).

AUDA (African Union Development Agency) (2017) *Toolkit on Agricultural Technical Vocational Education and Training (ATVET). Series 1: Stakeholder Engagement & Curriculum Development*. Available at: https://www.nepad.org/file-download/download/public/117609 (accessed 29 March 2021).

AUDA (2019) *Toolkit on Agricultural Technical Vocational Education and Training (ATVET). Series 2: Roles and Responsibilities of Agricultural Training Centers (ATCs)*. Available at: https://nepad.org/publication/toolkit-agricultural-technical-vocational-education-and-training-roles-and (accessed 29 March 2021).

BMZ (German Federal Ministry for Economic Cooperation and Development) (no date) *CAADP: Supporting Agricultural Technical Vocational Education and Training (ATVET) (Completed)*. Available at: https://www.giz.de/en/worldwide/15974.html (accessed 29 March 2021).

DAFF (Department of Agriculture, Forestry and Fisheries) (2015) *Agricultural Education and Training Strategy for Agriculture and Rural Development in South Africa*. Pretoria, South Africa. Available at: https://www.nda.agric.za/doaDev/sideMenu/educationAndTraining/AET_%20executive_summary.pdf (accessed 29 March 2021).

Dahlman, C. (2015) A new rural development paradigm for developing countries in the 21st century. OECD Development Center, New York.

European Commission (2019) *Gender Transformative Approaches in a Rural World: Tackling Root Causes of Discrimination against Rural Women and Girls*. Directorate General, International Cooperation and Development, EuropeAid, Directorate Sustainable Growth and Development, Brussels. Available at: https://ec.europa.eu/international-partnerships/system/files/gender-transformative-approaches-in-a-rural-world-brief-20190412_en.pdf (accessed 13 June 2021).

FAO (Food and Agriculture Organization of the United Nations) (2014) *Building a Common Vision for Sustainable Food and Agriculture Principles and Approaches*. Available at: http://www.fao.org/3/i3940e/i3940e.pdf (accessed 29 March 2021).

Hartl, M. (2009) Technical and Vocational Education and Training (TVET) *and Skills Development for Poverty Reduction: Do Rural Women Benefit?* International Fund for Agricultural Development, Rome. Available at: http://ddugky.info/dms/documents/1529234406_Hartl-formatted_01.pdf (accessed 29 March 2021).

ILO (International Labour Organization) (no date) *Decent work*. Available at: https://www.ilo.org/global/topics/decent-work/lang--en/index.htm (accessed 1 April 2021).

Jones, C.(2016). *The Role of Agricultural Technical and Vocational Education and Training in Developing Countries: A Review of Literature, Issues and Recommendations for Action* Pennsylvania State University.

MOA (Ministry of Agriculture, Kenya) (2011) *Demonstration Farms Fund (DFF) Operations Manual for Agricultural Training Centers*. MOA, Nairobi.

NEPAD (New Partnership for Africa's Development) (2014) *Review of Agricultural Technical Vocational Education and Training (ATVET) in Africa: Best Practices from Benin*, Ethiopia, Namibia and Sierra Leone. Available at: https://nepad.org/file-download/download/public/15927 (accessed 29 March 2021).

NEPAD (2015) *Agricultural Education and Skills Improvement Framework (AESIF) 2015–2025*. Available at: https://nepad.org/file-download/download/public/15864 (accessed 29 March 2021).

Rosas, G. (2006) *Glossary of Key Terms on Learning and Training for Work*. International Training Centre, International Labour Office, Geneva.

Swanepoel, F. Ofir, Z. and Stroebel, A. (eds.) (2014) *Towards Impact and Resilience: Transformative Change in and through Agricultural Education and Training in Sub-Saharan Africa*. Cambridge Scholars Publishing, Newcastle upon Tyne. Available at: https://www.cambridgescholars.com/product/978-1-4438-5680-5 (accessed 29 March 2021).

Tefera, L.T. Sehai, E. and Hoekstra, D. (2015) *Status and Capacity of Farmer Training Centers (FTCs) in the Improving Productivity and Market Success (IPMS) Pilot Learning Woredas (PLWs)*. International Livestock Research Institute, Addis Ababa. Available at: https://core.ac.uk/download/pdf/132636837.pdf (accessed 29 March 2021).

14 Network Approaches to Transforming Tertiary Agricultural Education in Africa

Florence Nakayiwa[1]* and John Lynam[2]
[1]*Regional Universities Forum for Capacity Building in Agriculture, Uganda;*
[2]*Independent scholar, based in Kenya*

14.1 Introduction

Africa continues to be challenged by human-resource constraints. Moreover, its capacity to build capacity is itself a conundrum, in terms of how best to strengthen teaching and research capacity both in Africa's universities and in agricultural and technical vocational education and training (ATVET) institutions. Despite the growing number of universities in Africa - and the small number of centres of excellence, usually focusing on specific disciplines - few institutions have the capacity to offer high-quality graduate (i.e. master's and doctoral) programmes. Africa's higher-education institutions have adopted different approaches to increase their internal capacities and mitigate the challenges associated with their mandate of providing teaching, conducting research and engaging with the community. Given the obstacles of distance and lack of reliable communications services, crossborder networks are an invaluable means of harnessing economies of scale in order to catalyze and support growth in Africa's tertiary agricultural education (TAE) sector. The network approach is a more formal version of what is known as 'communities of practice' in many industries, from manufacturing to research. Such associations can be based on similar or divergent capacities, but with common functions or objectives. Variations in functions and configurations do exist, however (Verburg and Andriessen, 2011). For African universities, a plethora of networks have emerged to provide individual universities with resources they would not be able to access independently. The network approach has enabled universities to leverage their comparative advantages and maximize economies of scale in order to access resources for research and outreach, while also establishing nodes of specialization with viable regional academic programmes.

This chapter examines the network approach from the perspective of the Regional Universities Forum for Capacity Building in Agriculture (RUFORUM), which was created to foster TAE institutions, increase effective research outputs and elevate community-based action research. The broader goal is to support national agricultural development plans by promoting African agricultural research and thereby the development of technologies to meet the unique needs and challenges of Africa's smallholder farmers. Note that the intention of the chapter is not to provide a compendium of African university networks, but rather to explore the potential of networking to strengthen TAE, especially at the graduate level. The chapter offers an assessment of RUFORUM's approaches to

*Corresponding author: f.nakayiwa@ruforum.org

 Transforming Tertiary Agricultural Education in Africa
(eds D. Kraybill, J. Lynam and Adipala Ekwamu)
DOI: 10.1079/9781789246544.0014

developing capacity within its member universities, outlining the challenges and benefits of interventions within the agricultural sector and providing lessons to inform future capacity-building programmes.

14.2 The Typology and Popularity of Networks in Africa

In the context of the higher-education sector, networks involve a collection of universities which have chosen to cooperate in order to leverage their comparative advantages across disciplinary, institutional and national boundaries (Moock, 2016). Other scholars have defined networks as durable sets of social arrangements directed to coordinate a set of common purposes (Cloete *et al.*, 2018). University networks exist in different formats across Africa. Several are discipline-specific, involving departments across institutions. Others are clustered around a research project, mainly in response to a multi-institutional call for proposals. Generic institutional networks which provide a framework for variants of groupings within and across institutions are also common. These generate smaller, time- or output-based linkages which can be considered as 'networks within networks'. Yet another format is cross-institutional networks, whereby universities consider themselves to have reached a level of development which precipitates a need to distinguish themselves on the basis of prestige. Formal arrangements for all forms of networks take the form either of (generic) memoranda of understanding or contracts with specific deliverables. Functionally, networks can be categorized as simple or complex, with gatekeepers, rules and regulations controling entry and exit, as well as expected contributions to the network and benefits gained from the association (Scott, 1987).

Simple networks are clustered around a single node acting as a fulcrum for activities. Complex networks, on the other hand, have multiple nodes which drive processes based on potential risks and benefits. These networks are robust. They have different points of contact, some of which are outside the framework of the network's focus. Gateways to complex networks are fluid, and members easily move in and out depending on the expected benefits. The RUFORUM network would fall into the complex category of networks, especially because its member universities extend beyond just agriculture to other disciplines. Similarly, the RUFORUM network includes both well-established and emerging universities, as well as public and private universities. RUFORUM has also expanded into post-conflict and fragile state universities. Nevertheless, most member universities are tightly clustered around TAE and agricultural research focusing on Africa, and include only limited coverage of other disciplines.

The primary intention of adopting the network approach is to generate the necessary collective capacity to advance scientific knowledge and disseminate research findings. For sub-Saharan Africa (SSA), the network approach is particularly important for the following reasons:

1. Most higher-education systems at national and institutional levels are too small to stand alone or benefit from economies of scale.
2. Capacity constraints resulting from under-investment remain.
3. Country-level diversification and the potential for focusing efforts on target objectives remain limited.
4. Collaborations offer the potential to focus on problems uniquely relevant to SSA.

The main downside of the network approach is that it limits independent engagement and learning because the assigned activities are attached to specialized nodes, which newer or emerging units may yet to have established.

Over time, new networks have emerged, and others have mutated beyond their initial focus and areas of operation. The focus of the majority of networks has been discipline-specific - e.g. the Directory of Selected African Research Networks has listed close to 50 discipline-specific networks. Other networks based on geographical scope have moved beyond regions to serve the whole continent. Examples of existing regional university networks include the Southern African Regional Universities Association, a membership-based association of vice chancellors of public and private universities in the Southern African Development Community. The African Research Universities Alliance was inaugurated in Dakar in March 2015, bringing together 16 of the region's leading universities. Other examples are the Association of African Universities, which is the African forum for consultation, exchange of information and cooperation among institutions of higher education. The African Higher Education Centres of Excellence is another configuration of networks based on the creation of a hub of universities which promote regional specialization and differentiation among participating universities.

It is, however, worth noting that the organizational framework for RUFORUM and for many research networks in Africa is neither hierarchical nor market-based. Rather, it is configured around a process designed to serve a specific target or purpose at any given point in time. Such network processes are centred around a common area of focus and available resources, which are shared within the network for the benefit of all members (Adelle *et al.*, 2018). For RUFORUM, the network is based on the need to produce high-quality graduates who will improve the performance of agriculture across the value chain. The incentive was universities coming together to influence production, marketing, policy and development of new agricultural technologies within the African context. Since membership in the network is voluntary, RUFORUM endeavours to sustain commitment and active enthusiasm among members for purposes of continuity. These attributes are continually shifting depending on members' expectations and the internal and external environments within which the universities operate - which may be the reason why universities join several networks. Members of any network must both derive benefits and add value (Cloete *et al.*, 2018). At the same time, a tension exists between individual and institutional benefits resulting from affiliation with the network. The network is more likely to survive if individuals from member institutions receive direct benefits and drive the process of their institution's participation.

Within Africa, many universities - especially those categorized as 'flagship' universities - belong to numerous networks. The rationale for joining one network or another varies even in cases where the networks appear to duplicate roles and functions. While some can be perceived as exclusive clubs, it can be argued that, from a global perspective, partnerships and networking have become essential arrangements for universities throughout the world to leverage their resources for greater impact. Furthermore, Kotecha (2011, p. 18) states the following:

> Networks are considered a fundamental source of innovation, allowing researchers to pursue complex research; provide fast and widespread technology transfer to society and industry and unlock the potential of theoretical research to produce both social benefits and commercial applications.

This assertion is applicable to all forms of networks, especially as they apply to the higher-education sector.

The motivation for joining networks varies from network to network as much as it varies from university to university. In some cases, these networks can be seen as a form of social identity, where universities consider membership to the network as a form of legitimacy. From the perspective of regionalization, Knight (2012) contends that the network approach is a process whereby connections and relationships are intentionally nurtured among higher-education actors and systems. The approach could also be considered a coping mechanism for African universities. This particularly applies to the changing configuration of public and private resource provision. Research networks across Africa have stood out as facilitators of advocacy for scientific knowledge in policymaking by increasing researchers' capacity and by enhancing the visibility and dissemination of the research produced. Moreover, RUFORUM was established almost at the same time as the Higher Education Councils within the East African region. From another perspective, RUFORUM is a mechanism through which universities derive prestige as they compete through the quality of their programmes against peers within the sector. Institutions such as the University of Cape Town (South Africa), Makerere University (Uganda), the University of Nairobi (Kenya) and the University of Rwanda have joined to four of the five networks scoped for this chapter (Table 14.1). Invariably, all networks have a capacity-development component, mainly for graduate training and research. Apart from academic cooperation and knowledge sharing, these networks can be perceived as promoting social identity among the universities. Allegiance to several networks may indicate that a single network cannot meet all of its members' requirements for prestige, research and resources.

Table 14.1. Participation in African university networks by selected universities

University/network	ARUA	RUFORUM	AAU	ACE	SARUA
University of Lagos, Nigeria	✓		✓		
University of Ibadan, Nigeria	✓		✓	✓	
Obafemi Awolowo University, Nigeria	✓			✓	
University of Ghana	✓	✓	✓	✓	
University of Dar es Salaam, Tanzania	✓	✓	✓		
University of Nairobi, Kenya	✓	✓	✓	✓	
University of Cape Town, South Africa	✓	✓	✓		✓
University of the Witwatersrand, South Africa	✓	✓	✓		
University of Rwanda	✓	✓	✓	✓	
University Cheikh Anta Diop, Senegal	✓	✓		✓	
Makerere University, Uganda	✓	✓	✓	✓	
University of Stellenbosch, South Africa	✓	✓			✓
University of Pretoria, South Africa	✓	✓			
Rhodes University, South Africa	✓	✓			✓
University of Kwa-Zulu Natal, South Africa	✓	✓			
Addis Ababa University, Ethiopia	✓	✓		✓	
Total number of universities in network	*16*	*129*	*374*	*na*	*66*

Source: Compiled by authors.
Notes: ARUA = African Research Universities Alliance (http://arua.org.za); AAU = Association of African Universities (https://www.aau.org); ACE = African Higher Education Centers of Excellence Project; SARUA = Southern Africa Regional Universities Association (https://www.sarua.org); RUFORUM = Regional Universities Forum for Capacity Building in Agriculture (www.ruforum.org); na = not available.

14.3 Implications of the Status Quo for Graduate Training

Africa's higher-education sector has expanded significantly over time, from less than 400,000 enrolments in 1970 to 7.2 million as of 2013 (Darvas *et al.*, 2017). Equally impressive is growth in the number of tertiary education institutions at an annual rate of 15% over the past three decades. By 2017, the number was estimated to be 1,650. Indeed, the literature shows that, although absolute numbers are still low, Africa's growth in tertiary gross enrolment ratio (GER) over the past decades averaged 4.3% per annum compared with a global average of 2.8% (Darvas *et al.*, 2017). Conversely, growth in the number of academic staff has not kept pace with growth in either enrolment levels or the number of higher-education institutions. Moreover, public institutions now face substantial competition from private institutions both for staff and other resources, as well as for students.

Growth in the population and in secondary and primary school enrolments have led to significantly increased demand for tertiary education. Given the importance of agriculture to Africa's livelihoods and its contribution to GDP in most countries, the TAE sector should be at the forefront of the expansion, but interest in careers in agriculture and related degree programmes has actually waned, as Africa's youth consider agriculture to be unrewarding compared with more lucrative academic programmes (Dramé-Yayé *et al.*, 2011). This is exacerbated by outdated pedagogical practices, lack of role models and increased youth unemployment. Most universities are inadequately resourced, which has limited staff research capacity due to poor remuneration, heavy teaching loads, lack of mentorship programmes and inadequate research infrastructure. Government subventions are the main source of revenue for most public universities in Africa[1]. Funding levels, however, cannot be sustained at the levels needed to support high-quality learning environments, research and extension throughout the value chain. These constraints are magnified by the fact that several public universities still support board and lodging for students, as well as municipal services for the university community (Akilagpa, 2004).

African universities continue to experience academic disruptions from staff and student strikes. Several of the universities face challenges in their human-resource capacity, with marked shortage in science, technology, engineering and mathematics disciplines, as well as agriculture- and health-related disciplines (Montenegro and Patrinos, 2013). The average share of PhD degree-qualified faculty in public higher-education institutions in Africa is estimated to be less than 20%, and in many universities these faculty members are over 50 years of age and hence are approaching retirement (see Ch. 5, this volume). Several departments have only one or two professors, which severely limits mentorship opportunities for younger faculty members. Female scientists are not adequately represented, and most staff, institutional development and research programmes are externally financed. In a review of eight of Africa's flagship universities, data revealed the share of faculty with PhD degrees in 2015 ranged from as high as 67% at Cape Town University, South Africa, to as low as 21% at Eduardo Mondlane, Mozambique (Fig. 14.1). These figures correspond with a study across all disciplines by the British Council and German Academic Exchange Service (2018), which revealed national shares of PhD-qualified faculty members of 8% in Ethiopia, 31% in Ghana, 33% in Kenya and 43% in Nigeria. Other studies have indicated the overall share of permanent instructional and research scientists with PhD degrees in South Africa to be 44%. That country produces 2,500 PhD-degree graduates per year, compared with a target in the National Development Plan of 5,000 per annum by 2030 (Gower, 2018). These levels obviously limit the universities' capacity to provide the required tertiary and research training for the next generation of educators and students.

Structural adjustment programmes in the 1980s and 1990s - which promoted market liberalization coupled with reduced government spending - led to regression of Africa's higher-education sector. Thereafter, growth in the number of public and private universities occurred while existing universities were still relatively fragile. Mamdani (2007) reported that the

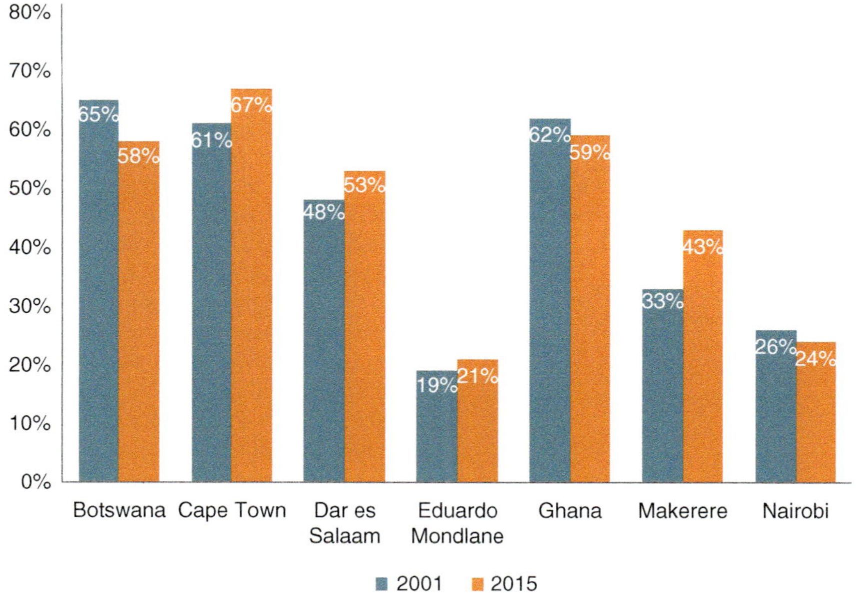

Fig. 14.1. Share of PhD-qualified academic staff employed at selected African universities, 2001 versus 2015
Source: Adapted by authors from Bunting *et al.* (2017).
Note: The full names of the universities are the University of Botswana; the University of Cape Town (South Africa); the University of Dar es Salaam (Tanzania); Eduardo Mondlane University (Mozambique); the University of Ghana; Makerere University (Uganda) and the University of Nairobi (Kenya).

sector was a low priority and continued to be inadequately funded by both the private and public sectors. More recently, UNESCO (2015) argued that with 148 researchers per 1 million people and less than 1% of the world's research output, Africa would need a critical mass of PhD-qualified scientists to harness global knowledge prospects. The mismatch between expanding student enrolments and numbers of appropriately-qualified staff required universities to evolve mechanisms to increase their human-resource capacity. Key avenues for addressing the multiple challenges confronting the higher-education sector in Africa have included the establishment of stronger collaboration and partnerships between and among universities, between industry and academic institutions and between the public and private sectors.

The inability to increase the supply of PhD-qualified agricultural scientists to match growing demand has been coupled with persistent 'brain drain' - a phenomenon which has beleaguered the higher-education sector. The loss of highly-trained and specialized academic staff departing Africa for better conditions and facilities in Europe and the United States has influenced the interventions RUFORUM has championed to its member universities. These interventions target relevant research and staff capacity development in agriculture, as well as broader areas of science technology and innovation, with the added advantage of being context-specific to Africa. RUFORUM evolved from a Rockefeller Foundation programme, which was designed with the following mandate (RUFORUM, 2018, p. 2):

> ...to provide experiential learning for students, retain the best faculty and ensure adequate working environments and career prospects; and promote closer and better cooperation between higher education institutions across borders.

RUFORUM facilitates Africa's universities in forging linkages with each other, with farming communities and industry and with other entities in both the Global South and the Global North. In this way, RUFORUM's activities stimulate international integration of African universities' activities (Knight, 2012), while also promoting the relevance of TAE in Africa.

In the context of the Comprehensive Africa Agriculture Development Programme (CAADP), RUFORUM was designed to link universities with the rural communities they serve and facilitate their ability to respond to relevant agricultural issues. These interventions align with World Bank (2007) recommendations that, in order to boost economic growth in Africa, significant investments should be made in human-resource capacities through tertiary education focusing on agriculture, science and technology. Collaborative regional MSc programmes and aggressive efforts to expand local PhD programmes are highlighted as avenues for addressing local staffing shortages and stimulating local research in agriculture and other disciplines. World Bank (2007, p. 10) also argues that:

> The quality of tertiary agricultural education is critical because it determines the expertise and competencies of scientists, professionals, technicians, teachers, and civil service and business leaders in all aspects of agriculture and related industries.

TAE also contributes directly to research and advisory services.

14.4 The Ruforum Model of Network Approaches

RUFORUM was established by 10 African vice chancellors with the vision of fostering a vibrant agricultural sector interlinked with African universities capable of producing high-performing graduates and high-quality research and innovations in response to the demands of Africa's farmers. The ultimate goal was to generate sustainable livelihoods and support national economic development. The design's central focus was (i) regionally-based MSc- and PhD-level training programmes, (ii) competitively-selected research in the area of agricultural science and technology, (iii) innovation activities which could be addressed through collaborative partnerships and (iv) institutional strengthening. The partnerships, however, feature both cooperation and competition, because institutions compete for resources on the one hand, but collaborate to develop programmes and attract resources on the other.

The theory of change underlying RUFORUM is that appropriately-skilled graduates and more relevant research could be produced through the combination of (i) competitively-awarded graduate research grants, (ii) scholarships for MSc and PhD training and (iii) direct support for new or upgraded/reformed training programmes. It was further assumed that grants given to university researchers for institutional strengthening and community-action research would promote collaborative research to solve issues of mutual interest. RUFORUM programmes would (i) enhance training and mutual learning, (ii) allow industry and communities to utilize research and graduates to enhance agricultural productivity and (iii) increase universities' visibility locally, regionally and globally. The RUFORUM model adopts programmatic initiatives which address some of the challenges of providing TAE. Several of the programmes directly originate from within the university system, building on member universities' comparative strengths. The model supported pooling the use of resources, strengthening graduate study, providing mentorship for young faculty members and improving the management of research (Whitworth *et al.*, 2008). The model also supports the provision of common products and services to member universities to improve their relevance and productivity, at the same time taking advantage of collective responsibility and economies of scale and scope to mitigate risks and challenges (Adelle *et al.*, 2018). Member universities work collaboratively, both individually and with actors within and outside Africa, to build the required human and institutional capacity needed to drive Africa's development agenda, especially as it relates to ensuring food and nutrition security.

At the member university level, agricultural departments are paired with other units - including those focusing on sociology, engineering, geography, information and communications technology and medicine - to harness the multidisciplinary capacity which distinguishes RUFORUM graduates. The underlying premise was that universities needed to focus on providing experiential learning for their students. Community-based participatory research approaches were adopted by universities to integrate teaching and research by linking universities both with industry, through agribusiness and entrepreneurship schemes, and with communities, through outreach and student-centred extension and advisory services.

Beyond its member universities, the RUFORUM network extends to include research institutes, government ministries and other agencies, non-governmental organizations and private enterprises. Researchers from diverse institutions received graduate research grants to address key issues affecting communities. The scale of RUFORUM's impact is evident from the breadth of individuals participating in its research programmes. Grants were designed to support smallholders and build capacity through MSc-level training, thereby improving the relevance of academic programmes to Africa's agricultural sector. Another distinguishing aspect of the RUFORUM model is bringing academia together with agricultural research practitioners - from research institutes, farmers' associations, government ministries, policy institutions and so on - to address common challenges in African agriculture. These individuals open up networks, internships and research sites for MSc students beyond the traditional university boundaries. Similarly, the participation of key researchers from the different agencies and national agricultural research systems demonstrates RUFORUM's capacity for outreach and community engagement beyond the education system. Ownership and leadership of initiatives occurs across multiple levels, with farmers not only benefiting from the work, but also directly contributing to it.

The Community Action Research Program, an initiative focusing on specific commodity value chains in particular communities, is an exceptional example of comprehensive engagement by the different actors, placing farmers at the centre of the intervention (Rukarwa *et al.*, 2018). These experiences have shown that partnership approaches stimulate fruitful relationships with industry, local business and local communities, enabling universities to be more responsive to local needs, not only in terms of research-based solutions, but also in terms of providing relevant education programmes, curricula and pedagogy. Moreover, the focus on graduate students enables agricultural faculties to explore new methods of operations, while stimulating new ideas and enthusiasm. As a result, instead of duplicating generic centres of excellence, anchor universities provide capacity in specific disciplines or competencies. Key examples under the RUFORUM network are the Plant Breeding Centre at Makerere University (Uganda) and the

Fish and Aquaculture Center at Lilongwe University of Agriculture and Natural Resources (Malawi). These initiatives acted as the fulcrum for RUFORUM's regional PhD and MSc programmes. Collectively, the seven PhD and three MSc regional training programmes have trained close to 500 specialists at the MSc-degree level (40% of whom were female) and 420 specialists at the PhD-degree level (30% of whom were female). These programmes serve the entire continent and have played a pivotal role to developing specialized human-resource capacity in Africa. Four of the programmes have since transitioned to become centres of excellence in higher education supported by the World Bank[2].

The critical mass generated by the network enables farmers to access extension services across different agricultural disciplines. In this way, the network facilitates and motivates the collective engagement needed for universities to meet their outreach missions. Under the graduate training assistant programme, for example, network universities send staff to member universities under a cost-share, tuition-waiver arrangement, which has been lauded as a means of leveraging collective resources for capacity development across the network. The potential drawback is the limit of the influence of the network on the broader operations of the member universities beyond the faculties/colleges of agriculture. The question still remains as to whether the lessons learned under the RUFORUM model can be scaled out at the institutional level to harness even greater impact. For most universities, RUFORUM support contributed to improving the functional and technical competencies of their staff and students. In addition, the research grants strengthened the wider systems and structures for delivering effective TAE and agricultural research. Overall, the universities are stronger, more credible and better equipped to fulfil their missions (RUFORUM, 2018). An assessment of membership outcomes revealed that most deans and principals perceived the greatest benefits of participation in the network to be (i) networking and sharing experiences, (ii) funding for research and (iii) exposure of staff and students to good agricultural practices (Fig. 14.2).

By offering high-quality, cost-effective TAE options and increasing the potential for relevant, context-specific research, RUFORUM's regional training programmes are shifting the dynamic of Africa's graduate agricultural training from dominantly South–North collaborations to South–South collaborations. Several research outputs have been applied across the region. *Maksoy*, for example, is an improved soybean variety developed under Makerere University's plant-breeding programme and adopted across East and Southern Africa. Collaboration with other education and research institutions in the North helped raise the standards of member universities by exposing Africa's young researchers and educators to different teaching, research,

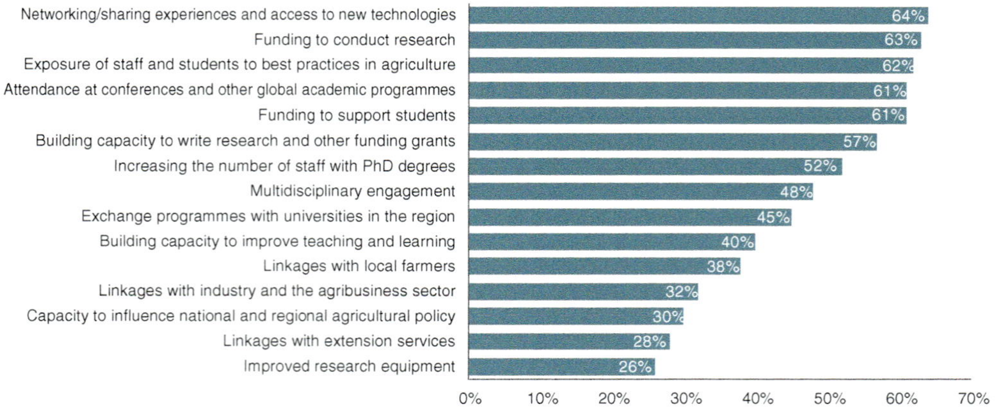

Fig. 14.2. Most significant benefits of RUFORUM membership from the perspectives of deans and principals, 2018
Source: RUFORUM (2018, Fig. 5).
Note: Data are based on a sample of 46 responses from 20 RUFORUM member universities.

services and management methods. International cooperation, experience and exposure enabled these researchers and educators to expand their work, publish it in recognized journals and present it at professional meetings.

In responding to the challenge of globalization, universities recognize that their capacities can be enhanced through collaboration. For RUFORUM, the network expands beyond the confines of the lecture room and laboratory to incorporate national agricultural research organizations, ATVET institutions, smallholders, the private sector and students as active participants in developing technologies and practices which have impact on agriculture in their communities. This is done through a number of different mechanisms:

1. the competitive grant system, as adopted by RUFORUM, which includes graduate research grants to enable principal investigators to create research teams comprising both seasoned and emerging researchers;
2. the Community Action Research Program, focusing on community-based commodity value chains;
3. RUFORUM's Entrepreneurship Challenge Program, which enables universities to explore the business potential of the technologies developed;
4. field-attachment programme awards;
5. institutional strengthening/nurturing grants and
6. community-engagement projects and doctoral grants managed as scholarships.

Member universities can take advantage of one or more of these mechanisms, and the network expands its impact on Africa's national agricultural research systems to a degree that no single university could achieve on its own. Importantly, resources are leveraged, maximized and strengthened.

The network has facilitated established universities in forging valuable international connections. For emerging universities, it has enhanced staff exposure and developed their capacity by providing access to research funding, projects and teams. Centres of leadership have arisen, recognizing that most African universities excel in specific disciplines: even if they are considered 'flagships', they cannot claim excellence in all disciplines (Cloete *et al.*, 2018). RUFORUM's strategy has been to identify and support specialization hubs which promote complementarity and synergy across the network. In this way, RUFORUM provides catalytic support functions. These include the facilitation of: (i) advocacy through Africa and international platforms, (ii) linkages to African and regional strategies and plans, (iii) systemic capacity strengthening at national levels and (iv) the capacity for collaborative sharing and learning among participating institutions. RUFORUM's effectiveness lies in its capacity to develop a complex system of networks within networks, both vertically and horizontally, which has spurred different aspects of capacity building (Box 14.1).

RUFORUM's network and partnerships have been highly beneficial and influential in attracting grant funding to member universities from such

Box 14.1. Networks within networks

RUFORUM does not work alone. Progress over the years has been made possible through strategic partnerships at national, regional and international levels. RUFORUM, for example, works closely with the African Union Commission, Association of African Universities, Forum for Agricultural Research in Africa, sub-regional agricultural research organizations, Pan-African Farmers' Organization, African Women in Agricultural Research for Development, African Doctoral Academy and other higher-education networks, such as the European Network of Universities and Agricultural Research Institutes and the American Public and Land Grant Universities. In recent years, RUFORUM has brokered partnerships with universities in East, Central, Southern and West Africa and is taking a phased approach to bringing more universities into its network. RUFORUM is currently in the process of strengthening its linkages with international universities and CGIAR Centers, and has signed a five-year partnership with the International Crops Research Institute for the Semi-Arid Tropics.

Source: Authors.
Notes: The four sub-regional research organizations are the Association for Strengthening Agricultural Research in Eastern and Central Africa, the Centre for Coordination of Agricultural Research and Development for Southern Africa, the West and Central African Council for Agricultural Research and Development and the Northern Africa sub-Regional Research Organisation.

sources as industry, foundations and international organizations. The number of projects directly implemented under the RUFORUM network grew from one in 2004 to 91 in 2018, with funding supported by multiple donors. Two key development partners - the Bill & Melinda Gates Foundation (two grants in 2009 and 2014) and the MasterCard Foundation (one grant in 2016) - provided long-term funding amounting to 71% of the resources raised (RUFORUM, 2018). Linkages to pan-African programmes, such as the African Union Commission's Science, Technology and Innovation Strategy for Africa (Priority Area 1 on reducing poverty and ensuring food and nutrition security) increased the credibility of the network to champion African TAE. Similarly, such initiatives as the Committee of Ten Heads of State and Government Championing Education, Science and Technology elevated RUFORUM's profile as an advocate for science and technology innovation in African higher education.

14.5 Implications for the Evolution of Network Governance and Ownership

RUFORUM's governance structure is a hybrid between the committee-based university system and the private sector's practice of annual general meetings. This fosters collective responsibility in a more flexible decision-making process which resonates with familiar institutional processes (Fig. 14.3). All vice chancellors of member universities act as board members who meet with other stakeholders annually to set the network's agenda for the coming year, including defining policies and admitting new members (RUFORUM, 2017). This stakeholder-led governance and management structure creates visibility for both the network and its individual member institutions. It also promotes ownership, which has attracted universities to join the network. Nesting the annual general meeting within a set of dialogues on future academic and policy directions has been beneficial in building linkages with pan-African and international agencies and policy-makers at the highest levels of government, while also attracting participation by academics, researchers, students, the private sector and civil society. These partnerships and networks have contributed to the sustainability of the network and to driving activities and processes. The large and growing size of the board has, however, presented challenges in terms of organization, continuity and consensus. To mitigate these challenges, a 10-member Executive Committee, with regional and non-university representation nominated by members, ensures

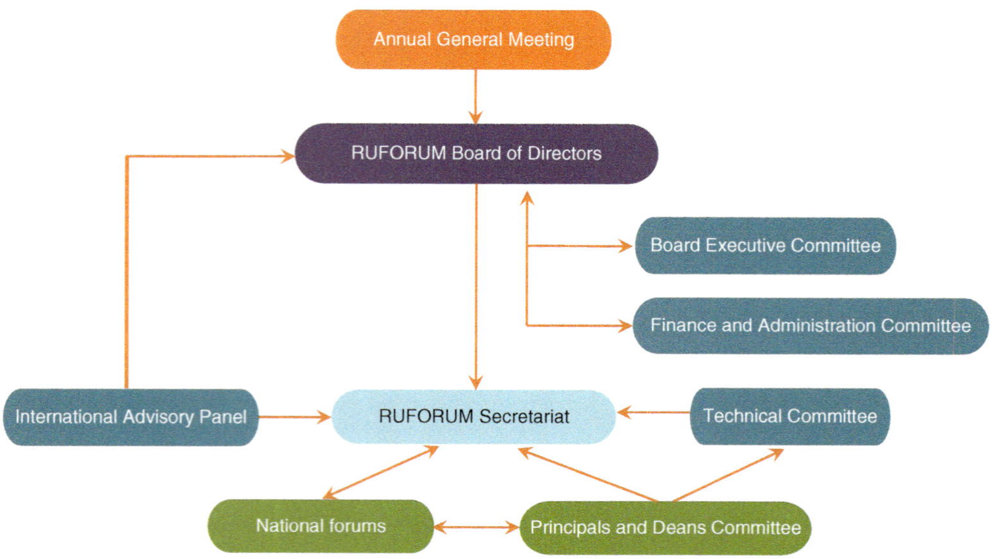

Fig. 14.3. RUFORUM governance structure
Source: RUFORUM (2017).

implementation of resolutions and takes decisions on behalf of the board for urgent and emerging issues. The Finance and Administration Committee supports the Executive Committee by providing oversight for financial operations, including budgets and audits. An International Advisory Panel ensures the network is responsive to global financial and TAE trends (RUFORUM, 2017). The panel oversees strategy and links the network with resource providers to facilitate the network's sustainability. Through this structure, RUFORUM gains credibility and legitimacy from member universities and conforms to international best practices for governance (Abbott and Snidal, 2001).

At the technical level, the principals and deans of participating faculties act as technical experts, which facilitates the direct application of the decisions, practices and leadership capacity generated through network. Deans act as nodes for network activities within member universities. Even when they are not the direct beneficiaries of awards, the deans are held accountable for activities related to curricula development, research, supervision and community engagement at their universities. The Deans and Principals Committee keeps the RUFORUM Secretariat informed of the progress of activities and of issues emerging at university and country levels. The Deans and Principals Committee also synthesizes issues from the national forums and regional thematic groups, and provides feedback to the Secretariat and Technical Committee (RUFORUM, 2017). The Technical Committee - which draws 40% of its membership from the universities and 60% from the non-university system - provides content-specific support in the preparation of research proposals and other grants awarded to member universities or principal investigators. The Technical Committee also provides independent input into grant proposals for support by the network, drawing on lessons from the private sector and farmer communities, and advises the Secretariat and Board on training needs and research priorities (RUFORUM, 2017). This, combined with the national forums - which contextualize outputs and other network activities within the national framework - ensures that the RUFORUM's activities are both relevant and context-specific to local and national needs. While the Technical Committee is fully functional, the national forums primarily operate in the few countries from which the network's initial members were derived. The RUFORUM Secretariat is the network's knowledge and coordination hub, and the primary node through which the network mobilizes and channels its resources. The Secretariat is the custodian of the network's technical and financial accountability frameworks (and processes and documents), which it channels to the appropriate actors to advance the network and its member universities. Ultimately, the coordination undertaken by Secretariat ensures the network's knowledge-production function and success in generating economies of scale. The network's joint activities, knowledge-sharing and scaling of agricultural technologies all manifest themselves at this level.

African networks, such as the ARUA and RUFORUM, were initially established by the leadership of originating member universities. The vice chancellors saw the need to address capacity gaps not being met by existing networks and to enhance their universities' ability to meet internal and external stakeholder expectations[3]. The networks derive legitimacy from this foundation, stemming from the inherent need for collective engagement and marshalling resources. Networks support individual institutions in overcoming their vulnerabilities. They symbolize new forms of engagement within and across universities, and have evolved to encourage fresh thinking about developing agricultural systems. They also provide a channel and an incentive for development partners to support universities' capacity and other needs. Such networks encourage systems thinking and provide access to a spectrum of knowledge (Moock, 2016).

The associated benefits of the network are the broader visibility of African TAE and the capacity not only to bring forward issues affecting higher education, but also to provide a forum for TAE actors to connect. Additional key initiatives are also emerging, including: (i) Agricultural Higher Education Science Technology and Innovation (ii) RUFORUM's 10-year Vision 2030 and (iii) Strengthening Higher Agricultural Education in Africa, a variant of African agricultural centres of excellence. These events have further galvanized RUFORUM's membership, which has grown from 10 universities in 5 countries in 2004 to reach 129 in 38 countries in

2021. Whether the current format of the governance structure will meet stakeholders' varying expectations is yet to be determined. Membership has expanded rapidly from its original focus in East and Southern Africa, so the potential exists to modify both the fundamental nature of the network and its governance structure. The general shift in the focus of tertiary education toward science and technology may actually have a negative impact on Africa's TAE sector. Faculties of agriculture already represent a declining share of universities' resources and enrolments. As a network focusing on building capacity, other components may need to be explored in response to broader issues, while at the same time securing the necessary resources to fund programmes. Moreover, increased competition is expected from other networks addressing higher education, science and technology and innovation.

14.6 Conclusion

RUFORUM has played an important role in improving the quality of education in Africa's rapidly changing TAE landscape. The network is founded on: (i) resource-sharing across the network, (ii) infrastructure or the internal systems and structures to support the network and (iii) the network's capacity for jointly creating and adding value. The motivation for joining the network varies among member universities. The flagship universities have taken leadership in several academic programmes, including regional MSc- and PhD-degree programmes, which have been instrumental in building Africa's TAE capacity and contributing to the development of newly-established and emerging universities. Established universities equally benefit from the network by forging international linkages, improving curricula, establishing centres of excellence and attracting resources from development partners and national governments. The RUFORUM model demonstrates how network approaches can catalyze the transformation of Africa's TAE sector. Several important milestones could not have been reached without the leverage and collaborative synergies the network facilitates. It remains to be determined, however, whether RUFORUM's existing structures and programmes will continue to be applicable as its membership growth and its geographical scope expands.

Endnotes

[1] The government provides the majority of funding for most public universities, largely supporting salary-related and development costs. For most institutions, tuition fees are below the recurrent costs of operations.
[2] For more information, see https://projects.worldbank.org/en/projects-operations/project-detail/P126974.
[3] Both ARUA and RUFORUM were born out of prior initiatives. ARUA stemmed from the Higher Education Research and Advocacy Network in Africa, which highlighted deficiencies in the research capacity of flagship universities as engines for development. RUFORUM grew out of FORUM a capacity-building initiative to support agricultural faculties, develop local capacity, reduce brain drain and strengthen African universities.

References

Abbott, K.W. and Snidal, D. (2001) International 'standards' and international governance. *Journal of European Public Policy* 8(3/Special Issue), 345–370.

Adelle, C. Elema, N. Chakauya, E. and Benson, D. (2018) Evaluating 'homegrown' research networks in Africa. *South African Journal of Science* 114 (3/4). Available at: http://dx.doi.org/10.17159/sajs.2018/20170070 (accessed 15 May 2021).

Akilagpa, S. (2004) African universities and the challenge of research capacity development. *Journal of Higher Education in Africa* 2(2), 213–242.

British Council and German Academic Exchange Service (2018) Building PhD capacity in sub-Saharan Africa. *International Higher Education*. Available at: https://www.britishcouncil.org/sites/default/files/h233_07_synthesis_report_final_web.pdf (accessed 15 May 2021).

Bunting, I., Cloete, N. and vanSchalkwyk, F. (2017) *An Empirical Overview of Emerging Research Universities in Africa*. African Minds, Cape Town, South Africa.
Cloete, N., Bunting, I. and vanSchalkwyk, F. (2018) *Research Universities in Africa*. African Minds, Cape Town, South Africa.
Darvas, P. Gao, S. Shen, Y. and Bawany, B. (2017) *Sharing Higher Education's Promise beyond the Few in Sub-Saharan Africa*. Directions in Development, World Bank, Washington, DC. Available at: https://doi.org/10.1596/978-1-4648-1050-3 (accessed 15 May 2021).
Dramé-Yayé, A. Chakeredza, S. and Temu, A.B. (2011) *Why Do Agricultural Faculties Fail to Attract the Best Students? Background paper prepared for the ASTI–IFPRI/FARA Conference, Agricultural R&D: Investing in Africa's Future Analyzing Trends, Challenges, and Opportunities* Accra, Ghana, 5–7 December.
Gower, P. (2018) Global network to boost HE capacity and PhD numbers. *University World News* 24 August. Available at: https://www.universityworldnews.com/post.php?story=20180824114059890 (accessed 15 May 2021).
Knight, J. (2012) *A Conceptual Framework for the Regionalization of Higher Education in Asia*. Routledge, London.
Kotecha, P. (2011) The value of research networks in Africa. *International Higher Education* 65(fall), 17–19.
Mamdani, M. (2007) *Scholars in the Market Place: The Dilemmas of Neo-Liberal Reform at Makerere University 1989–2005*. Fountain Publishers, Kampala.
Montenegro, C.E. and Patrinos, H.A. (2013) *Returns to Schooling Around the World. Background Paper to the World Development Report*. World Bank, Washington, DC.
Moock, J.L. (2016) Network innovations: Building the next generation of agricultural scientists in Africa. Ch. 10 in Lynam, J. Beintema, N.M. Roseboom, J. and Badiane, O. (eds.) *Agricultural Research in Africa: Investing in Future Harvests*. International Food Policy Research Institute, Washington, DC. Available at: https://www.ifpri.org/publication/agricultural-research-africa-investing-future-harvests (accessed 24 March 2021).
RUFORUM (Regional Universities Forum for Capacity Building in Agriculture) (2017) *Governance and Structure*. Available at: https://ruforum.org/governance-and-structure (accessed 17 June 2021).
RUFORUM (2018) *Building Africa's Capacity to Train for Africa: RUFORUM@15 Summative Outcome Evaluation* RUFORUM, Kampala. Available at: https://ruforum.org/sites/default/files/RUFORUM%20Outcome%20Evaluation_Final%20Report_03%20December%202018-short.pdf (accessed 15 May 2021).
Rukarwa, R.J., Mensah, S. and Egeru, A. (2018) *Universities Responding to Advancing Community Transformation through Action Research in Africa*. RUFORUM Working document series, Kampala.
Scott, J. (1987) *Social Network Analysis: A Handbook*. Sage Publications, London.
UNESCO (United Nations Educational, Scientific and Cultural Organization) (2015) *UNESCO Science Report: Towards 2030. Executive Summary*. UNESCO, Paris.
Verburg, R.M. and Andriessen, E.J.H. (2011) A typology of knowledge sharing networks in practice. *Knowledge and Process Management* 18(1), 34–44.
Whitworth, A.G. Kokwaro, G. Kinyanjui, S. Snewin, V.A. Tanner, M. *et al.* (2008) Strengthening capacity for health research in Africa. *The Lancet* 372(1), 1590–1593. Available at: https://www.thelancet.com/pdfs/journals/lancet/PIIS0140-6736(08)61660-8.pdf (accessed 15 May 2021).
World Bank (2007) *Cultivating Knowledge and Skills to Grow African Agriculture: A Synthesis of an Institutional, Regional, and International Review*. World Bank, Washington, DC. Available at: https://www.agrilinks.org/sites/default/files/resource/files/World%20Bank_cultivatingknowledge.pdf (accessed 15 May 2021).

15 A Case Study of Transformation in Four African Universities

David Kraybill[1]* and Moses Osiru[2]
[1]*Ohio State University, United States;* [2]*International Centre of Insect Physiology and Ecology, Kenya*

15.1 Introduction

Rapid growth of tertiary enrolments, particularly at the undergraduate level, have strained the capacities of African tertiary education systems and institutions to maintain academic quality (see Chs. 4 and 5, this volume). Graduate (i.e. MSc- and PhD-level) training has equally been affected, with research output representing only a 2.2% share of global research output, despite Africa's contribution of over 17% of the world's population (Osiru *et al.*, 2016). As of 2014, tertiary enrolments had increased from 400,000 in 1970 to 7.2 million in 2013 (Darvas *et al.*, 2017). And although private universities currently outnumber public universities, public universities are still larger, on average, and account for the bulk of enrolments. Observers cite decreasing ratios of faculty to students, inadequate facilities for teaching and research and an alleged gap between the training delivered by universities and the competencies required for employment and self-employment (see Ch. 4, this volume; Shabani *et al.*, 2014). Many of these portrayals are accurate but, at the same time, have the effect of masking the successes of progressive African tertiary education institutions. Real life examples can be powerful motivators. In that spirit, this chapter presents selected analytical narratives of successful institutional transformation of tertiary agricultural education (TAE) from four African universities. The aim of the case studies presented is to inspire other African universities to pursue transformation in their own institutions with the aim of increasing the relevance and excellence of their academic programmes (see Ch. 6, this volume). The case studies are also intended to inform the public of advances being made by African universities and the resulting impacts on the populations they serve.

Africa's tertiary education institutions face enormous challenges in attempting to deliver high-quality and relevant higher education. Key challenges include: (i) inadequate financing of higher education, which limits access - particularly for poor and disadvantaged students; (ii) a mismatch between the supply of skills and the needs of employers, particularly in the emerging private sector; (iii) weak structures, support systems and incentives for knowledge generation, which constrain scientific output; (iv) inadequate staff capacity, particularly at PhD level, and difficulty retaining staff; (v) inadequate resources to meet the high cost of constructing and maintaining buildings, information and communication technology (ICT) and other infrastructure and (vi) overall inadequate investments in the education

*Corresponding author: kraybill.1@gmail.com

 Transforming Tertiary Agricultural Education in Africa (eds D. Kraybill, J. Lynam and Adipala Ekwamu)
DOI: 10.1079/9781789246544.0015

sector (Teferra and Altbach, 2004; USAID, 2014). These challenges stretch the capacity of universities to respond and adapt to the emerging needs of African countries. African universities have struggled to find their fit within global higher education and national development contexts since independence from colonial powers (Yesufu, 1973). This struggle for relevance and quality continues (Nakayiwa *et al.*, 2016). In practice, the goals of quality and relevance are in conflict when tertiary education institutions emphasize one to the exclusion of the other. This conflict is exacerbated by the tendency of the global higher education community to judge university status by rankings which emphasize publications and citations in peer-reviewed scientific journals. However, most global university ranking systems are based on a narrow definition of quality which tends to reward scientific rigor without attention to economic and social relevance.

The four universities analysed are Jomo Kenyatta University of Agriculture and Technology, Kenya (JKUAT), Gulu University, Uganda (GU), University of Abomey Calavi, Benin (UAC) and the University of Venda, South Africa (UNIVEN). The next section outlines the study's methodology and profiles the four case study institutions. Subsequent sections examine *what* each of the universities changed, and *how* they went about effecting and supporting those changes. The discussion focuses on transformation of both 'soft changes' in organizational culture and strategies, and 'hard changes' in enrolments, staffing, programmes, departments, faculties/colleges and campuses. Outcomes of the transformation process are described, along with lessons learned.

15.2 Study Methodology and Institutional Profiles

Analytical approach

The approach used in this study to analyze the universities' transformation process is derived from appreciate inquiry (AI), a method of organizational development and research (Cooperrider and Whitney, 2007; Bushe, 2012). AI emphasizes the positive ways in which organizations strive to transform. It focuses on: (i) processes by which an organization's members collectively envision the future, (ii) how they work together to build both morale and capacity and (iii) processes by which the organization continues to evolve. AI is used both to bring about change in organizations and to study how organizational change occurred, after the fact. Data for AI research is typically based on interviews and analyses of organizational documents.

Data

Quantitative and qualitative data on the four universities were gathered for a ten-year period beginning in 2007/08. Interviews and a questionnaire were used to gather data on the historical and current status of institutions, and key changes. The interviews were standardized and open-ended, such that the same questions were asked across all four institutions, but interviewees had the freedom to answer in their own ways, without having to choose among pre-determined answers. The individuals interviewed were vice chancellors, deans or principals, department heads, and students. The questionnaire gathered data on enrolment and staffing levels, organizational structure and strategies at the beginning and end of the ten-year period. The questionnaire was completed by one or more individuals as designated by the vice chancellor of each university. Additional information was gathered from the universities' websites and official publications.

Institutional profiles

Jomo Kenyatta University of Agriculture and Technology. JKUAT is a public university in Kenya established in 1981 as a middle-level college - Jomo Kenyatta College of Agriculture and Technology - with funding from the Kenyan and Japanese governments. In 1988, it became a constituent college of Kenyatta University and, in 1994, a fully-fledged university. The main campus of JKUAT is located in Juja, 36 kilometres northeast of Nairobi. Between 2002 and 2015, branch campuses of JKUAT were established at ten other Kenyan locations, as well as one in Rwanda (Kigali) and one in Tanzania (Arusha).[1] JKUAT comprises five colleges: (i) health sciences,

(ii) engineering and technology, (iii) pure and applied sciences, (iv) agriculture and natural resources (v) human resources, and (vi) development; two institutes: (i) biotechnology and (ii) energy and environment; and one centre: urban studies. In addition, JKUAT hosts the Pan African University Institute for Basic Sciences, Technology and Innovation, one of the Pan African University's five campuses.

Gulu University. GU opened as a public university in Uganda in 2002 under the name of Gulu University of Agriculture and Environmental Science. Initially, the university focused on agriculture, but it soon expanded to include non-agricultural disciplines, and the name was changed to Gulu University. Today, GU has three campuses, all in post-conflict northern Uganda, with the main campus located in the city of Gulu. GU has six faculties: (i) agriculture and environment, (ii) medicine, (iii) science, (vi) education and humanities, (v) business and development studies and (vi) law and two institutes: (i) peace studies and (ii) research, graduate studies and staff development.

University of Abomey-Calavi. UAC is a public university in Benin.[2] It is the largest campus of the National University of Benin, which was founded in 1970 as the University of Dahomey. UAC operates its main campus in Abomey-Calavi, 18 kilometres northwest of the city centre of Cotonou, and has two branch campuses. UAC comprises six faculties: (i) agronomic sciences, (ii) arts and humanities, (iii) health sciences, (iv) economic sciences, (v) law and political science and (vi) science and technology; four schools: (i) polytechnic, (ii) applied economics and management, (iii) national administration and (iv) Ecole Normale Supérieure; and five institutes: (i) Arabic languages and Islamic cultures, (ii) technology and teacher training, (iii) public health, (iv) mathematics and physical sciences and (iv) physical education and sports.

University of Venda. UNIVEN is a public university in South Africa, established in 1981 for students from the Venda Bantustan on the campus of the former Dimani Agricultural School. Upon the country's transition to democracy in 1994, UNIVEN shifted its focus to science and technology, and in 2002 was authorized by the Department of Education to become a comprehensive university. UNIVEN has a single campus located in the town of Thohoyandou in Limpopo Province in the northern part of the country. Today, the university comprises eight schools: (i) agriculture, (ii) education, (iii) environmental science, (iv) health sciences, (v) human and social sciences, (vi) management sciences, (vii) mathematical and natural sciences and (viii) law.

15.3 The Four Universities' Transformations

For each of the four universities, the following sections describe the opportunities and challenges presented by the external environment, the ways in which the universities responded over the 2008–2018 period, changes in enrolment levels and academic staffing, the universities' inferred priorities, changes in mindsets within university communities and the extent of the resulting transformations. Opportunities facing the universities as a result of changes in national and international conditions include: (i) burgeoning demand for higher education as a result of rapid growth in the youth population and expansion of secondary education, (ii) new training and research opportunities as new technologies and emerging economic sectors open up new occupations, (iii) freedom to open branch campuses as national agencies regulating tertiary education in many African countries embraced competitive models of higher education, including privatization and (iv) the spread of ICT, allowing educational activities to be cost-effectively coordinated across branch campuses. Challenges facing the universities include: (i) declining per-student operating subsidies from national governments; (ii) inadequate infrastructure and a relative decline in government funding for capital investments; (iii) weak pre-tertiary educational systems which inadequately prepare prospective university students for further study; (iv) a weak research culture and inadequate research facilities, leading to low research output; (v) the perception of many stakeholders, especially employers, that universities are providing students with insufficient 'soft skills' and practical technical skills for employment; (vi) weak quality-assurance mechanisms and incentives and (vii) lack of leadership training for university leaders and limited exposure to best practices in leadership and management in tertiary education.

Jomo Kenyatta University of Agriculture and Technology

External environment and internal response. During the period 2008–2018, JKUAT faced major external challenges and opportunities. Competition for students and public funding increased as the national number of public universities grew from 7 to 31. As the number of universities seeking financial support increased, JKUAT's inflation-adjusted operating subsidy from the Kenyan government declined in 2017/18 to just 20% of what it was in 2007/08, and inflation-adjusted capital funding from the government was just 33% of the 2007/08 level. The university compensated for the loss of government revenue, in part, by increasing enrolments through branch campus expansion.

The Kenyan government requested the university's assistance in addressing problems of food security, affordable housing, manufacturing and healthcare. Businesses sought improved skills training for workers and help in developing improved technologies. In response to these challenges, Vice Chancellor Mabel Imbuga created a Resource Mobilization Committee (RMC), gave it an operating budget and appointed ten members of the university who had previously revealed themselves to be champions of change. The RMC was charged with finding additional financial resources and addressing stakeholder needs through new instructional, research and outreach programmes, and new campuses.

During this period of change, JKUAT built strategic planning capacity and sought and adapted ideas from universities in Europe, North America and South Africa. The university's capacity to establish and maintain relationships with industry was strengthened through frequent meetings with stakeholders, the establishment of a Public–Private Partnership Committee, periodic review of curricula and the creation of adjunct professorships to bring experts from industry and government. To improve the quality of programmes and activities, JKUAT put incentives in place which aligned the annual performance evaluations of administrative and academic units and individuals to annual workplans based on the university's strategic plan. To improve the quality of administrative services, the university adopted service charters which reduce wait times and enhance efficiency.

Student enrolment and academic staffing levels. Over the decade from 2007/08 to 2017/18, JKUAT expanded at a rapid rate (JKUAT, 2018). Undergraduate enrolment levels increased 285% during this period (from 10,650 to 41,000 students). The number of MSc and PhD students graduating per year increased by 475% (from 207 to 1,190). The number of academic programmes grew 188% (from 138 to 397). The number of academic staff members - including professors, associate professors, senior lecturers, lecturers, assistant/tutorial fellows and teaching assistants - increased 66% (from 523 to 870).

Inferred priorities. Interviews and document reviews indicate that JKUAT seeks to balance the developmental relevance and scientific excellence of its programmes, monitoring its impact on local and national stakeholders, as well as its reputational standing among national, regional and pan-African universities. Budgetary allocations to the university from the Kenyan government are based on the extent to which the university meets performance goals set out in its workplan, which is prepared each budget cycle. Responsibility for implementation of the workplan begins with the vice chancellor and, from there, cascades down to faculty deans, department heads and individuals. Employee evaluations are based on contributions to the university and unit-level workplans.

Changes in mindset. JKUAT's transformation initiative was accompanied by an intense effort of the vice chancellor and her management team to communicate the reasons for and nature of the changes taking place. Open communication created empowerment, brought a widespread sense of ownership of the changes and improved morale. The vice chancellor created a collective ethos in which members of the university came to understand that decisions were to be made together, not by the top administration alone. Dialogue with communities and businesses led to willingness on the part of university administrators and academic staff to alter curricula for greater relevance. Service charters and ISO 9001 certification (which provides a framework for systematically improving an organization) encouraged students and staff to expect and demand high-quality, rapid services from administrative offices.

Extent of the transformation. The changes at JKUAT transformed the university from an educational institution with relatively little impact in the country, and a meagre reputation outside the country, into a beacon of local relevance and academic quality, both nationally, and internationally. Its visibility has increased; the academic quality of degree programmes has improved; the quality of in-coming students has improved; external linkages with businesses, communities and government have increased; involvement in national, regional and pan-African programmes and networks has expanded; the volume and quality of research have increased; the management of research has improved; outreach has improved and national and local (county) governments increasingly look to JKUAT for training, research and problem-solving. As a result, staff retention has also improved.

Gulu University

External environment and internal response. Gulu was launched during the final stage of a long period of civil unrest and violence in the surrounding northern region of Uganda. The university was intended by its founders and the Government of Uganda to provide human resources and practical solutions for rebuilding the northern region. From this environment emerged the idea of work–study programmes, in which students live and work in local villages for extended periods of time during their studies. Initially, local leaders were reluctant to allow university students to live within their communities. However, the outcomes produced by this university–community collaboration had a large, positive developmental impact in the initial host communities, and soon communities throughout the region sought to participate. In response to requests for assistance from communities, GU opened up two branch campuses in the northern region.

During the early part of the past decade, GU was in the enviable position of receiving more in government subsidies than most of the country's peer institutions, measured in both total and per capita terms, due to special treatment by the Ugandan government following cessation of conflict in the northern region. However, by 2017/18, these subsidies were reduced substantially.

Student enrolment and academic staffing levels. From 2007/08 to 2017/18, undergraduate enrolment at GU grew 18% (from 3,507 to 4,146 students). The number of MSc and PhD students graduating per year increased by 60% (from 55 to 88). The number of degree programmes grew by 33% at the undergraduate level (from 21 to 28) and by 267% at the graduate level (from 3 to 11). The academic staff grew by 41% (from 170 to 240), while the non-academic staff grew by 43% (from 134 to 192).

Inferred priorities. GU appears to place a very high priority on economic and social relevance in its teaching and community engagement. Both students and local community members are treated as important partners in the ecosystem of learning. Students are encouraged to bring field knowledge back from their village work–study experiences and to update professors on the biological, economic and social conditions they observed, further improving the relevance of the knowledge imparted. Assessment of student performance is undertaken jointly by the faculty and community, an important innovation which ensures the relevance of education at GU to local needs. Research at GU tends to be applied and practical, so the number of publications in international academic journal per faculty member is lower than at many other African universities.

Changes in mindset. In the early days of the university, when facilities were scarce, the vice chancellor and academic staff shared a single office, an arrangement which helped top administration and the academic staff develop a unified vision of the university's mission. The practical mindset at GU is summarized by the current vice chancellor, who told the authors, 'When you're surrounded by problems, you cannot run an elitist university'. GU has developed a community-based mentality instead of the teacher-based mentality found in most universities.

Extent of the transformation. Gulu is outward-looking and developmental in its orientation to a much greater extent than most African universities. It connects with the region's problems and concerns through its work–study programme, which embeds students in the surrounding region for extended periods of time. Prior to the start of students' work–study experiences, GU promotes critical thinking and soft-skills development, preparing students for their dual learning–teaching role. Students come to understand that villagers are often valuable sources of knowledge on agriculture and other matters, and that helping to solve the region's problems requires listening to and learning from local residents. At the same time, students learn that they, themselves, can help improve the quality of life in local communities by teaching and demonstrating scientific-based methods of agriculture, animal husbandry and healthcare. When students return to campus, they share with their professors their newly-acquired knowledge of conditions and challenges faced by the communities. Professors rely on the students' reconnaissance to revise courses with new topics and updated information. Professors and graduate students also rely on the knowledge brought back by returning work–study students to identify high-priority topics for research. GU has also partnered successfully with a number of foreign universities, such as Earth University in Costa Rica, ICRA in the Netherlands and many other institutions in a two-way process of sharing best practices among similar-minded centres of excellence.

University of Abomey-Calavi

External environment and internal response. In 2006, for the first time, Benin universities were allowed to elect their top administrators (rectors, vice-rectors, deans and chairs) democratically - a step which set the stage for subsequent changes at UAC. Employers' complaints that university graduates lacked the skills and knowledge needed for employment acted as an important stimulus. In 2011, the second democratically-elected rector launched a modernization process with the goals of increasing the relevance of the university's degree programmes for employment and increasing the quality and quantity of research conducted. To gain a better understanding of the country's economic and social needs, UAC created a council which meets regularly with private-sector and governmental stakeholders. In response to their requests, UAC began reforming its curricula, targeting defined competencies, and developed a business incubator open to students, graduates and non-student innovators. UAC also established relationships with many universities throughout West Africa and the world for the purpose of sharing and learning from other academic institutions.

Like many other African universities, UAC has faced declining government support in recent years as the number of universities vying for subvention has increased. The university's inflation-adjusted per-student subsidy from the Benin government in 2017/18 was only 55% of the 2007/08 level. UAC received no capital-budget allocations during this period, although it is allowed to divert the government subsidy to capital spending. These financial stresses led UAC's administration to engage in cost-cutting measures, such as reducing the number of centrally-owned and maintained vehicles. UAC greatly increased its enrolments during the ten-year study period, generating revenue from fees, which partly offset the decline in the government's per-student subsidy.

Student enrolment and academic staffing levels. From 2007/08 to 2017/18, undergraduate enrolments at UAC grew by 71% (from 38,350 to 65,712). The number of MSc and PhD students graduating per year increased by 480% (from 582 to 3,420). The number of degree programmes grew by 10% at the undergraduate level (from 30 to 33) and by 279% at the graduate level (from 38 to 144). Teaching staff (i.e. lecturers and professors) grew by 7% (from 829 to 887) while non-academic staff grew by 43% (from 609 to 870).

Inferred priorities. Interviews and document reviews indicate that UAC seeks to balance the socioeconomic relevance and scientific rigor of its academic programmes. The university values engagement with external stakeholders to discover the skills and knowledge needs of the country's private and public sectors. UAC also places a high priority on published research by its academic staff, and takes its international academic reputation seriously. It values inter-university collaboration in research and teaching, and has agreements with many other African and international universities and institutes. UAC also values the ability of its graduates to succeed in employment or self-employment.

Changes in mindset. Under the leadership of the vice chancellor elected in 2011, the university administration adopted an open and transparent style of communication within its community. Even administrative budgets were openly shared. Through dialogue and collaboration, the administration promoted an organizational culture of inclusive decision-making on campus and openness to external stakeholders, building trust and creating a sense of joint ownership. Among the staff, they promoted the possibility of the university becoming more relevant to national development, while upholding and improving scientific rigor in teaching and research. Those who excelled were rewarded through annual ceremonies, and the notion of public service was promoted among students, creating a programme through which graduates could work as volunteers on campus for a year after they graduated and prior to seeking employment outside the university.

Extent of the transformation. The transformative changes at UAC are evident in the quality of administration and academic programmes. The shift to selecting top administrators by vote rather than by political or administrative appointment improved both the quality of leadership and staff morale. Under visionary leaders, UAC has become proactive in its future planning. In 2016, the university created both a planning and an evaluation unit to ensure the proper implementation of its strategic plan and associated workplan. External stakeholders are now consulted to a much greater extent than in the past. Curricula and pedagogy have been revised to enhance the employability of graduates by supplementing classroom-based training with attachments to businesses and training in soft skills. The university's new business incubator is now helping to prepare students for innovation and job creation.

University of Venda

External environment and internal response. When it was established by the planners of apartheid leading into the final decade of that system, UNIVEN was intended to be a Bantustan institution primarily serving Venda speakers from the surrounding region. With the advent of democracy in South Africa in 1994, the university's leadership changed, but facilities and funding were insufficient to rectify the poor employability of graduates, low research output and lack of national and international reputation. In 2005, UNIVEN opened an institutional development office, and in 2007 a new vice chancellor was hired with a mandate to make the university competitive with other South African and foreign universities. The government invested heavily in UNIVEN's buildings and other infrastructure from 2008 onward. To facilitate efficient management, the new vice chancellor initiated strategic planning for the first time, beginning with mission and vision statements to define and motivate change. The aim was to become an international leader in southern Africa. In 2010, South Africa's Council of Higher Education conducted a quality audit which identified many deficiencies in UNIVEN's programmes and operations, leading to introspection and preparation of a plan for improvement. To improve leadership and administration, the vice chancellor launched a staff exchange programme which sent campus administrators to other universities to observe good management practices first-hand. Steps were taken to improve governance, broaden enrolment beyond the northern region, make curricula more relevant to the needs of the economy, increase research outputs and foster engagement with communities and businesses. UNIVEN began offering a mentoring programme and life-skills training for incoming students, targeting particularly those from rural areas, where many primary and secondary schools are deficient. A Transformation Charter was developed in 2013 to support further transformation in the areas of diversity, institutional culture, people management, student focus and governance.

Student enrolment and academic staffing levels. Over the decade from 2007/08 to 2017/18, undergraduate enrolment at UNIVEN grew by 1214% (from 818 to 10,752). The number of MSc and PhD students graduating per year increased by 166% (from 262 to 696). Teaching staff (i.e. lecturers and professors) grew in number by 27% (from 365 to 465), while the non-academic staff grew by 28% (from 478 to 612).

Inferred priorities. Based on interviews with UNIVEN leaders and students and a review of university documents, the university places a high priority on both stakeholder engagement and research. UNIVEN aims to meet the needs of the people of the northern region of South Africa while, at the same time, striving to make its programmes more relevant internationally and attract students from other regions and countries.

Changes in mindset. UNIVEN re-oriented itself for greater diversity, relevance and excellence as it emerged from the period of being a provincial institution serving only the Venda Bantustan. Changes include a new understanding of the importance of open communication and inclusive decision-making involving top administration, academic staff and non-academic staff working in consultation with students, the government and the surrounding community.

Extent of the transformation. UNIVEN implemented regular processes of strategic planning over the ten-year study timeframe. This process has since been further altered with the creation of a transformation plan focusing on inclusion. University governance has become more open with the creation of a 'Joint Structures Body', which is a group representing top administrators, academic staff and non-academic staff. Many new buildings have been constructed and ICT infrastructure has been installed. UNIVEN's research output per researcher increased 783% from 2008 to 2015, based on research output units defined by South Africa's National Research Foundation. A staff performance and recognition framework was established, including annual awards for teaching and learning, research and community engagement. The Higher Education and Leadership Programme, a staff training programme begun at UNIVEN, was so successful that it is now being replicated in other South African universities.

15.4 Lessons Learned

The preceding sections described *what* changed within the four universities in response to external challenges and opportunities. This section addresses *how* the changes were instigated and nurtured. Two aspects of managing change are highlighted: preparation and building momentum.

Preparing for change

Transformation of a university for greater relevance and academic excellence requires preparation, including changes in mindsets, consultation with stakeholders, mobilizing agents of change and establishing priorities and strategies for change.

Shaping organizational culture. An important step in bringing about the transformation of universities is changing the mindsets of administrators, staff and students (see Ch. 6, this volume). The leaders of the four universities brought about mindset changes through an open and collaborative style of leadership. They established and continue to promote channels and forums for listening to diverse internal and external stakeholders and for sharing leadership responsibilities internally. The leaders, themselves, changed their own mindsets and were willing to forego some of the hierarchical privileges of power which would otherwise have sheltered them from feedback. For the universities in this study, change in organizational culture was not limited to administrators and staff but also extended to students. To varying degrees across the four universities, the creativity and energy of students was harnessed through student-centred pedagogy, emphasizing active learning in which students become partners with their teachers in organizing and transmitting knowledge through active learning methods as opposed to lecturing as the sole method of teaching.

Engaging with stakeholders. Before a strategic plan fully takes shape, leaders typically begin with a vision emphasizing high-level goals for their institution. Top leaders of all four universities in the study told us that stakeholder needs were the primary source of inspiration for their institutions' visions. All four universities conduct forums, ranging from formal to informal, for listening to stakeholders and learning about their needs and expectations. The four universities place a high priority on relevance to national development and to particular stakeholder groups. For example, at GU, the vice chancellor sponsors academic staff members to undertake 'staff attachments' in communities and companies for periods of time to ensure that the university remains up-to-date and responsive to ever-changing national and local realities.

Creating a guiding coalition. To be effective, university transformation initiatives must be steered by university leadership, yet they also require the knowledge, perspectives and wisdom of a wider group who can think and act rapidly. One way of acquiring this kind of input is to create a 'guiding coalition' to advise, plan and even implement change (Kotter, 2012). As utilized by JKUAT, a resource mobilization committee can provide a brainstorming forum focusing on the changes needed and how they will be implemented. JKUAT's vice chancellor even authorized their RMC to communicate directly with the university council. At UAC, the rector created a team to implement the university's new strategic plan. And at GU the vice chancellor's daily interaction with academic staff occurred organically due to lack of office space, but it led to a strong guiding coalition for the university's team.

Developing strategies for action. All four universities in the study developed a strategic plan internally as a living document. In each university, the vice chancellor accepts the development and implementation of the strategic plan as his or her primary responsibility. At JKUAT, the strategic plan is actually a contract with the Ministry of Education. Responsibility for implementing the strategic plan is given to the chair of the university council, who delegates it to the vice chancellor. From there, responsibility for the various parts of the plan flows down to successive administrative levels and finally to the individual level. Each unit of the university develops an annual workplan on which the unit's performance is evaluated by a planning office within the institution. Each university employee develops a personal workplan tied to their relevant unit's workplan. UNIVEN has both a strategic plan and a transformation charter to ensure that institutional change is truly transformative. UNIVEN's charter focuses on particular strategic goals, especially internationalization.

Building momentum

Maintaining frequent and effective communication. Frequent and informative communication is essential for initiating and nurturing transformative institutional change. All four study universities used formal meetings as a key communication tool during their change process. At JKUAT, workshops were held to train deans and chairs in strategic planning and to prepare them for developing their units' portion of the university's website - another important communication tool. An important part of the communications strategy at JKUAT was the creation of a corporate communications department, housed within the office of the vice chancellor and headed by a PhD-qualified staff member. The department prepared and disseminated information for JKUAT's resource mobilization committee. Three of the four universities also communicate to internal and external audiences using newsletters, published monthly by UAC and UNIVEN and bimonthly by JKUAT. A final communications tool mentioned by some of the interviewees was informal impromptu meetings, hallway conversations, email messages and text messages. Informal communication helps fill gaps in formal communications channels and can be used to reinforce key messages, gather feedback, explore new directions and resolve potential conflicts.

Identifying change agents. The leaders of the four universities recognized the importance of identifying individuals willing to champion change early in the transformation process. Within their own ranks, they sought out change agents, engaged them in dialogue and empowered them to begin implementing transformative changes. They understood that 'small wins' are important at the beginning of the transformation process and that each success paves the way for the next (Kotter, 2012). At GU, the new vice chancellor indicated that, in the upcoming phase of the university's transformation, he will look to community members and local officials, among others, for guidance and assistance in implementation. UAC's rector indicated that change agents from all levels of the university were enlisted to help implement the new strategic plan.

Creating early wins. Visible positive outcomes appearing early in a transformation initiative attract attention and help shift the opinions of doubters. One of the quickest and least expensive ways of obtaining a quick win and to signal that transformation is under way is to beautify the physical environment of the campus. At UAC, the campus environment appeared neglected when a new rector began his first term in 2011. The management team promptly launched a campus beautification campaign involving rubbish removal, tree and foliage planting and prohibition of the use of plastic bags.

Incentivizing change.The four universities also created momentum through the use of unit- and individual-level incentives. JKUAT holds a periodic competition to inspire academic and administrative units to improve their portion of the university website, a task which was initially resisted by some units. The award is a lump sum of additional funding which can be used at the unit's discretion. An individual incentive provided by UAC is twice-yearly recognition for high performers through 'model employee' awards.

Initiating innovation. Projects or programmes which emphasize innovative solutions to economic and social problems help consolidate internal and external support for universities' transformation initiatives. JKUAT has launched multiple innovation initiatives since 2017/18, including the Sino-African Joint Research Centre for biodiversity research in Africa; the JKUAT Industrial and Technology Park, where a factory produces laptops to support implementation of the government's literacy programme; the JKUAT Instant Noodles factory and the Innovation and Prototyping Integrated Centre. UAC created an incubator park called 'UAC Startup Valley' to promote and support university graduates in creating and developing innovative businesses. UNIVEN, UAC and JKUAT have aggressively pursued international cooperation as a means of bringing innovation to their programmes.

Coping with resistance. In most organizations, not everyone agrees with proposed changes. The leaders of all four universities reported encountering a certain degree of resistance, the sources of which included: (i) fear of loss of position or power, (ii) incomplete understanding of the benefits of change, (iii) inability to change or (iv) inability to benefit from the proposed change due to imminent retirement. Leaders also reported that they were not surprised by resistance and viewed it as a natural part of change processes. When they became aware of resistance, they pursued dialogue with the individuals in questions, but didn't attempt to silence resistance if it persisted. Instead, they worked with agents of change and attempted to convince through successive small and intermediate wins and to let the evidence 'speak for itself'. Furthermore, leaders understood that organizational change does not require the active participation of everyone involved. As long as most people did not actively oppose the proposed changes, the leaders were able to begin implementation with the help of the change agents they had identified. As results became evident and success built upon success, many of those who were slow to change eventually joined in.

15.5 Conclusion

The four African universities studied for this chapter are unusual in terms of the extent to which they succeeded in effecting positive change during a comparatively short timeframe. Organizational transformation led these universities to create a culture of greater openness to the needs of their constituents, including students, employers and the surrounding communities. In the process of expanding and improving their academic programmes, the universities developed various organizational capacities which shifted their future trajectories and will allow them to continue to respond and adapt to their countries' and regions' emerging needs. All four universities prepared for transformation by shaping their organizational culture, changing outmoded mindsets, engaging and consulting with their stakeholders, mobilizing agents of change and establishing priorities and strategies for change. They built momentum in innovative ways, including engaging in frequent and effective formal and informal communication, empowering champions of change, creating early wins, incentivizing change and initiating and facilitating innovation.

Endnotes

* Gratitude is expressed to persons who responded to interview questions and provided documents at the four universities. In addition, the authors wish to thank the campus coordinators of the university visits: Dr. Duncan Ongeng and Mr. Sam Elolu (GU); Profs. Daniel Sila, Losenge Turoop, and John Wesonga (JKUAT); Prof. Achille Assogbadjo (UAC); and Dr. Segun Obadire (UNIVEN). Dr. Isaac Minde provided useful suggestions on an earlier draft of the chapter.

[1] JKUAT established a total of 12 campuses but in 2018 one of these closed and three others were in the process of closing because of changes in the Kenyan government's regulations.

[2] The official French names of these universities are Université d'Abomey-Calavi, Université Nationale du Bénin and Université du Dahomey.

References

Bushe, G. (2012) Feature choice by Gervase Bushe: Foundations of appreciative inquiry: History, criticism and potential. *AI Practitioner* 14(1), 8–20. Available at: http://www.gervasebushe.ca/Foundations_AI.pdf (accessed 25 May 2021)

Cooperrider, D.L. and Whitney, D. (2007) Introduction. *Appreciative Inquiry: A Positive Revolution in Change*, 1–36.

Darvas, P. Gao, S. Shen, Y. and Bawany, B. (2017) *Sharing Higher Education's Promise beyond the Few in Sub-Saharan Africa*. World Bank, Washington, DC. Available at: https://openknowledge.worldbank.org/handle/10986/27617 (accessed 25 May 2021).

JKUAT (Jomo Kenyatta University of Agriculture and Technology)(2018). *A Decade of Academic Excellence, Research and Innovation for Development* JKUAT, Nairobi.

Kotter, J.P. (2012) *Leading Change*. Harvard Business Press, Boston.

Nakayiwa, F. Osiru, M. Shibru, A. Ochuodho, J. and Assogbadjo, A.E. (2016) Developing higher education systems in Africa: Selected country views. *African Journal of Rural Development* 1(2), 205–217.

Osiru, M. Nampala, P. and Ekwamu, A. (2016). African faculties of agriculture within an expanding university sector. In Lynam, J. Beintema, N. Roseboom, J. and Badiane, O. (eds.) *Agricultural Research in Africa: Investing in Future Harvests*. International Food Policy Research Institute, Washington, DC. Available at: https://doi.org/10.2499/9780896292123_09 (accessed 25 May 2021).

Shabani, J. Okebukola, P. and Oyewole, O. (2014) Quality assurance in Africa: Towards a continental higher education and research space. *International Journal of African Higher Education* 1(1), 139–171. Available at: https://doi.org/10.6017/IJAHE.V1I1.5646 (accessed 25 May 2021).

Teferra, D. and Altbach, P.G. (2004) African higher education: Challenges for the 21st century. *Higher Education* 47, 21–50. Available at: https://doi.org/10.1023/B:HIGH.0000009822.49980.30 (accessed 25 May 2021).

USAID (United States Agency for International Development (2014) *African Higher Education: Opportunities for Transformative Change for Sustainable Development*. USAID, Washington, DC. Available at: https://www.agrilinks.org/sites/default/files/resource/files/African%20Higher%20Education%20Opportunities%20for%20Transformative%20Change%20for%20Sustainable%20Development.pdf (accessed 1 June 2021).

Yesufu, T.M. (1973) *Creating the African University: Emerging Issues in the 1970's*. Oxford University Press, Ibadan, Nigeria.

16 Transnational Partnerships for Tertiary Agricultural Education in Africa

Peter Koehn[1]*, David Kraybill[2] and Isaac J. Minde[3]
[1]*University of Montana, United States;* [2]*Ohio State University, United States;* [3]*Michigan State University, United States*

16.1 Introduction

Collaboration provides an important avenue for tertiary agricultural education (TAE) institutions to share ideas and resources, thereby improving the scope and quality of their academic programmes. Higher-education collaborations differ based on the size of memberships, intended objectives, frequency of interactions and overall duration. On a continuum of collaboration types, partnerships lie at one end, while networks function at the other (Koehn and Obamba, 2014). Compared with networks (see Ch. 14, this volume), partnerships are typically smaller in membership size, tend to have more specific objectives, involve a higher frequency of interactions and are often of shorter duration.

Partnerships are not unique to higher education, but are widely-used vehicles for collaboration in economic and social development. Promoting the building of transnational partnerships is an important function of the African Union (2004). Recently, the African Union called for refocusing the continent's partnerships to respond more strategically to African priorities for growth and transformation in its Agenda 2063 (AUC, 2015). Globally, the Millennium Development Goals (MDGs), agreed upon by the United Nations in 2000, specifically devoted one of the eight goals to 'global partnerships for development'. None of the goals, however, specifically mentioned higher education (Warden, 2015). While it was obvious to some observers that higher education could contribute to the MDGs, the successor set of goals - the Sustainable Development Goals (SDGs), released in 2015 - went further, explicitly identifying a role for higher education. The SDGs devote an entire goal (SDG 4) to 'quality education', emphasizing the role of higher education in ensuring equal access for women and men to affordable and high-quality technical, vocational and tertiary education (United Nations, 2015). Another goal, SDG 17, calls for multi-stakeholder partnerships which mobilize and share knowledge, expertise and technology - areas in which TAE institutions can make enormous contributions. As providers of training, knowledge and technology for agricultural and food systems, TAE institutions have a potentially impactful role to play in poverty alleviation (SDG 1), health and wellbeing (SDG 3), gender equality (SDG 5), decent work and economic growth (SDG 8), responsible consumption and production (SDG 12) and climate change (SDG 13).

Symmetrical partnerships, in contrast to asymmetric donor–recipient relationships rooted in colonial legacies, are motivated by a desire to

*Corresponding author: peter.koehn@umontana.edu

DOI: 10.1079/9781789246544.0016

put international and inter-organizational relations on a more equal footing (Koehn and Obamba, 2014). Within higher education, unequal roles in ownership, decision-making, funding, planning and evaluation undermine the effectiveness of partnerships between African and non-African institutions (Bailey and Dolan, 2011; Ishengoma, 2016; Minde and Monson, 2019). Bilateral and multilateral aid agencies fund most of the partnerships in which Africa's TAE institutions are involved, which (knowingly or unknowingly) contributes to imbalances in their design and implementation. However, the imbalance can be lessened and its deleterious effect on partnership success mitigated by essential 'pre-partnership strategic planning'. Planning to facilitate balanced co-creation involves the collaborative formulation of rules and modalities for engagement, agreement on indicators of co-creation, mutual respect and trust, transparency and periodic monitoring and adjustments to ensure the partnerships goals are being met in an ethically-responsible manner (Minde and Monson, 2019). Several international agreements have addressed the issue of power imbalances in North–South partnerships. Both the Paris Declaration (OECD, 2005) and Accra Agenda for Action (2008) call for greater ownership and control by Southern partners in international development cooperation (OECD, 2012). Building upon these agreements, some foreign donor agencies now support higher education partnerships led and managed by Southern partners (Damtew Teferra, 2016).

The objective of this chapter is to examine ways in which an equal balance of power can be established in designing and implementing higher-education partnerships. The partnerships featured here are transnational - both within and outside Africa - so they typically have different academic structures, traditions and resources. This chapter presents an approach to partnership design and implementation with the intention of helping African TAE institutions strategically pursue arrangements which build enduring institutional and human-resource capacity through joint and equitable engagement in their core mandates: educational instructional, research and outreach. The range of activities undertaken by higher-education partners includes: (i) research, (ii) curriculum development and teaching methods, (iii) degree-programme development and strengthening, (iv) administrative and managerial capacity building, (v) community outreach, (vi) faculty and student exchanges and (vii) seminars, conferences and symposia. Few partnerships encompass all these types of activities, but many engage in more than one.

16.2 Symmetrical and Asymmetrical Partnerships

Material and structural asymmetries are embedded in the political economy of transnational academic cooperation. For instance, Northern universities often provide limited time for faculty to be away from campus engaged in research in the South. Many also place less value on collaborative research published in another language which addresses other country's development problems or generates little, if any, net research revenue for the institution. In many Northern universities, moreover, tenure and promotion are still anchored primarily on individual accomplishments, although some university administrators and funding agencies now recognize the value of collaboration and are attempting to incentivize it (Defazio *et al.*, 2009). Concomitantly, the scholarly context within which most African researchers operate is characterized by lack of access to high-quality information and discussion, scarcity of scientific equipment and materials, inadequate time to conduct research, few (if any) financial resources and dependence on 'invisible' local publication outlets (Damtew Teferra and Altbach, 2003, p. 39).

The asymmetrical Northern-inspired model of development assistance and transnational collaboration reigned supreme until as recently as the 1990s, and 'donor–recipient' attitudes still persist (Utting, 2006). In Northern-dominated divisions of higher-education labour, for instance, 'research agenda setting, activity planning, fund management, data interpretation, results dissemination and basic research components are taken on by Northern researchers, while their Southern counterparts are in charge of data gathering and the more applied research components' (Gutierrez, 2008, p. 21). Holm and Malete (2010, p. 3) posit that one of the underlying assumptions in many North–South collaborations is that the 'developed country academics know what needs to be researched in Africa. The primary role of African colleagues is to organize and carry out data collection and lobby for research permits'. As a consequence of peripheral participation, asymmetrical transnational research partnerships typically result in few collaborative publications by African and Northern researchers (Holm and Malete, 2010).

Symmetrical higher-education partnerships are grounded in mutual respect, trust, influence and responsibility (Minde and Monson, 2019). They encompass jointly-determined and conceptualized project objectives pursued 'through a shared understanding of the most rational division of labour based on the respective comparative advantages of each partner' (Brinkerhoff, 2002, pp. 14–18). They embrace transparency and shared accountability for results (UNCTAD, 2008, p. 96). The envisioned symmetrical North–South partnership incorporates genuinely-collaborative participation in project decision-making, implementation activities and resource contribution. All partners differentially share in the benefits of the symmetrical partnership based on the very diversity which drives the collaboration. Thus, partnering to achieve a shared goal does not mean that each and every aspect of the collaborative project has to be carried out together or equally by all partners (KFPE, 2012). As Samoff and Bidemi (2004, p. 115) have captured the concept, 'a partnership must involve a collaboration that can reasonably be expected to have mutual (though not necessarily identical) benefits. . ..'

16.3 Preparing for Partnerships

Transnational higher-education partnerships are not naturally occurring phenomena. Their crucial formative processes involve initiation, planning and structural integration. Each formative process of a partnership brings trust-building challenges, which are especially important to address in the initial stages.

Partnership Initiation

The impetus for a partnership is the first place to look for asymmetry or symmetry. In most cases, the principal instigator for transnational research and development collaborations is a Northern-based faculty member. Prior institutional or individual faculty member involvement in Africa plays a major part in inspiring many collaborative projects. Northern researchers are in a more advantaged position than are African researchers to identify and select partners, win external funding and pursue unfettered initiatives which are meaningful to them (Jones and Oleksiyenko, 2011). Donor programmes and prospective funding give connected and experienced Northern-based institutions of higher learning influence over projects' agenda-setting processes. Thus, even when extensive subsequent discussions involve the African partners, asymmetry typically is introduced when the Northern partner extends the initial overture.

To inspire initiatives which are strategically driven, members of African higher-education institutions interested in partnering with Northern or other Southern institutions of higher learning should deliberate internally and identify priority needs which are consistent with their own research and development mission and goals (Eddy, 2010, p. x; Wanni *et al.*, 2010, p. 19). This step should be accompanied by the articulation of an institutional and human-resource map (Mihyo, 2008, p. 114). The identified priority needs and resource map should be incorporated into a strategic plan for building research and development capacity. The formative plan should include 'policies that clearly articulate the scope, goals, strategies, priorities, and organizations that the universities intend to emphasize in their partnership efforts' (Obamba *et al.*, 2011, p. 4). Then, transnational partners should be chosen whose own needs, academic initiatives, long- and short-term research and development objectives, distinctive features, organizational commitments and strengths are consistent with the strategic plan (Trani and Holsworth, 2010, pp. 228–229; Francisconi *et al.*, 2011, p. 19).

For TAE institutions seeking partnership opportunities, symmetry is facilitated by open and respectful relationships with scholars at multiple institutions of higher learning. As interpersonal trust develops across boundaries and over time through social interaction, some relationships will lead to joint problem and vision identification and solidified interest in cross-institutional collaboration. Fruitful conversations among faculty and researchers often generate the decisive impetus for exploring the potential for viable strategic partnership based on common values, visions and mutual gains. Then, partnership champions 'use their influence in their home institutions to obtain commitment of organizational capital' (Eddy, 2010, pp. x, 21, 28–29). High-level administrators and research staff, along with all potentially-involved faculty members and departments, need to be engaged in early strategic-partnership discussions and decisions (van de Water *et al.*, 2008, pp. 11, 34). Prior to designing, or even suggesting, a joint project among potential partners, the primary motivations for interpersonal and inter-institutional interactions need to be agreed upon. 'If we are to have symmetry', Holm and Malete (2010, p. 5) astutely observe, 'there needs to be a meeting where both sides discuss a mutually agreeable set of project topics, or African universities need to seek assistance only when the topic fits in with their unique strategic plans.' Considerable evidence shows that a collegial initiation process, where the African university plays 'a major role in identifying the problem to be addressed', contributes to partnership success (Morfit and Gore, 2009, p. 18; Koehn and Obamba, 2014).

An early exchange of institutional visits provides opportunity to explore first-hand the degree of match and depth of commitment among prospective partners (van de Water *et al.*, 2008; Gedde, 2009). Although face-to-face visits are preferable, teleconferences are an alternative means of communication for such discussions. Potential partner institutions will need to contend with complexity in transnational research and development contexts, including diversity in: (ii) legal and research systems, (ii) bureaucratic oversight and publication regulations, (iii) visa, intellectual-property and export-control rules, (iv) insurance, health and emergency-evacuation considerations and (v) research norms (Nebeker, 2011). Gulfs in resources, culture and power need to be bridged through: (i) frequent dialogue, (ii) identification of common understandings regarding the meanings of development, research and partnership, (iii) the exchange of concept notes, (iv) iterative decision-making and (v) other trust-building activities (Brinkerhoff, 2002, pp. 34, 37; Morfit and Gore, 2009, p. 10). Key issues for participants to consider at the initial consultative meetings include (Wanni *et al.*, 2010, pp. 20, 51, 55): 'Does the partnership fit in with institutional plans or national strategic frameworks? Can the institution make a long-term commitment to the partnership? Is the institution able to invest the necessary human and financial capacity and resources involved in running a partnership? Are there enough staff from the organization ready, willing, and able to get involved with the partnership? Will the institution be able to help monitor and evaluate the outcomes of the work to ensure it is meeting the agreed objectives?' Finally, will the partnership 'provide a greater benefit than the time and resources required running it?'

Operational Planning

Once partners have been identified and participants have jointly selected the topical theme(s) or challenge(s) to be addressed, operational planning commences. The principal issues which require operational planning include partnership and project conceptualization, research design, resource mobilization and scheduling.

Partnership and project conceptualization. Symmetry can arise in conceptualizing TAE partnerships and projects for a variety of reasons. One of these is the decision regarding who drafts the partnership proposal. For symmetry to be possible, all partners must be deeply and directly involved in the drafting process. The draft agreement, which articulates a mutually-determined and symmetrical governance process, should be circulated for review and comment by key stakeholders affiliated with all partner institutions before it is signed (Wanni *et al.*, 2010, pp. 46–47).

Research design. Symmetry in research partnerships is based on mutual respect for the priorities and knowledge systems on which the partner institutions' explorations are based. Too often, agenda-setting in partnerships is governed by visible and invisible structural inequalities, including asymmetry in research tools, knowledge-generating systems, information, process commitment and access to grant funds. In asymmetrical research partnerships in Africa, outside scholars determine and control scholarly priorities, thematic commitments, methods of inquiry and the theoretical and conceptual paradigms deployed (Crossley and Holmes, 2001, p. 400). This pattern was established in a period when Africa had relatively few researchers; today, the human-resource situation has improved in many parts of the continent. Numerous African universities and research institutes have scholars possessing the requisite knowledge and skills to co-create the design of research partnerships. Involving African scholars in research design is important for productive research outcomes, as well as for symmetry in partnerships; when research and development agendas are blind to local context, they are unlikely to result in innovations which meet partner objectives (Crossley and Holmes, 2001, p. 396).

Scheduling. The scheduling of partnership interactions is another planning process which is often asymmetrical. One frequently-arising issue involves the limited length of, and at times inconvenience of, visits by the outside partners. 'Too often', as King (2008, p. 6) explains, 'the typical stay in a so-called partner country, often not in the partner institution, is a week to ten days . . .. There is often no time to do research together'. The output emphasis of outside partners may clash with the process or relationship orientation of local partners. Crossley and Holmes (2001, p. 400) observe that many Northern partners 'accord great priority to time schedules, tight deadlines and the meeting of targets, [whereas] Southern partners often emphasize other process and human resource development priorities'. When outside partners fail to spend sufficient time learning about work context and relationships at the African university, one common result is the slowing or scaling back of projects due to 'the overestimation of capacity or underestimation of time' (Wanni *et al.* 2010, p. 38).

Structure and governance. In symmetrical planning processes, partners formulate a *collaborative protocol* which specifies egalitarian roles and principles to guide partnership activity. Walsh and Kahn (2010) suggest that the range of processes covered by a collaborative protocol could include provisions for regularly- scheduled review points, where concerns and new ideas can be articulated and needs for additional training and capability building identified and addressed. Prospects for partnership success are enhanced when clear understanding concerning respective roles and responsibilities exists from the outset and when negotiated arrangements receive the 'full backing' of the chief executives, top academic officers and affected deans at all collaborating higher-education institutions (van de Water *et al.*, 2008, pp. 12, 36).

Transnational partnerships often either establish a separate and parallel administrative unit or focus on 'one element or another of the institution, with most of the rest of the institution uninvolved. . ..' (Kinser and Green, 2009, p. 12). Common consequences of organizational fragmentation and enclaves are 'one hand not knowing what the other is doing' within the same institution (Oleksiyenko and Sa, 2010, p. 378) and among administrators of the partner institutions. This fragmentation can be minimized by co-creating and implementing a strategic plan consistent with the goals and administrative policies and procedures of partner institutions. Integrating the partnership within the host university is facilitated by establishing an agreed-upon organizational framework (for instance, steering committees at each participating institution), assigning staff members individual and team responsibilities, designating suitable space and equipment, identifying specific deliverables for each project objective and setting feasible timetables for different stages of the partnership (Wanni *et al.*, 2010, pp. 33, 40–41, 45–47).

An often-voiced complaint about transnational partnership governance concerns fiscal management. Disproportionate influence of outside partners over the management of project budgets is a source of conflict, contestation and unequal resource distribution in many tertiary-level partnerships involving African universities. Factors contributing to structural inequalities in fund management include Northern-donor distribution practices and legal requirements, inadequate accounting mechanisms at inexperienced universities and much larger cost-share contributions by the Northern partner(s). Yusuf *et al.* (2009, p. 90) confirm that at many African tertiary-level educational institutions, 'weak accounting and reporting systems make budget monitoring difficult'. At universities with limited delegation of expenditure authority, senior officers are required to approve even routine and minor payments (Ashcroft and Rayner, 2011, p. 250). In this situation, excessive financial controls slow the processing of expenditures and unnecessarily occupy university managers' time (Yusuf *et al.*, 2009, p. 91). Faced with such financial management constraints, Northern-university project staff are inclined to maintain strict control over monies and budgeting. The resulting asymmetry in partnership governance carries hidden transaction costs because 'where controlling funds is not a shared responsibility, accountability, too, is likely to be one-sided' (Samoff and Bidemi, 2004, p. 150). Partnerships can minimize asymmetry in financial decision-making by conducting a needs assessment before partnership activities begin to identify and fortify areas of weakness in financial administration. If weaknesses are revealed, training for finance staff and improvements in administrative processes and systems can be launched.

Partners should reach early agreement on a multi-method communication process. Frequent communication is particularly vital in the initial and middle stages of a new partnership. Project teams should possess opportunities to consult - in person or virtually - at least monthly (Toure, 2008, p. 56; Francisconi *et al.*, 2011, p. 20). Face-to-face meetings deepen understanding and empathy for the different working conditions that one's partners face, help build enthusiasm and momentum for joint research and development projects and promote the development of transnational competence (Koehn and Rosenau, 2010). Individual and institutional trust in the partner's commitments to project undertakings and ability to meet deadlines is enhanced by spending 'significant amounts of time in each other's institutions, countries and company' (King, 2008, p. 6).

Asymmetry in the structural design of research partnerships occurs when only outside scholars are principal investigators (PIs) and managers of research funds. Asymmetry also occurs in situations where donors insist that local scholars serve as exclusive PIs on projects. The most symmetrical arrangement calls for co-PIs, one from each participating higher-education institution. Alternatively, when multiple collaborative projects are involved in a single partnership, a balanced distribution of PIs can be arranged. Across all dimensions of institutional interconnection, partners must exercise constant vigilance to maintain symmetry as structural integration is perpetually shifting, rather than stable, throughout the duration of the partnership. Transnational partnership arrangements are dynamic, always becoming something else. At the beginning of the arrangement, therefore, mechanisms need to be put in place for addressing and resolving the confusion, tensions and disagreements which inevitably arise. Successful collaborators learn to channel unanticipated developments and new structural movements creatively in mutually-enabling directions. Although staying focused is encouraged, some level of flexibility along the path has to be accommodated (Minde and Jayne, 2017).

16.4 Managing Symmetry in Partnerships

In symmetrical transnational partnerships, diverse perspectives on management are expressed and embraced (Ashcroft and Rayner, 2011, p. 127), lessons about various management techniques are exchanged and cross-institutional 'best-practices sharing [occurs] amongst senior administrators and project managers' (Maassen *et al.*, 2007, pp. 45, 51, 59–60). The symmetrical management approach 'struggles to balance different forms of engagement rather than a settled view of engagement driven by or effectively serving only the most powerful stakeholders' (Singh, 2007, p. 65, also pp. 75, 77). What often matters most in partnership management are transparency and receptivity to mutual learning processes, where 'interdependencies are explicitly acknowledged' and addressed by reliance on each partner's comparative advantage (Brinkerhoff, 2002, p. 22). Further, symmetrical partnerships among TAE institutions support learning from inevitable mistakes and miscalculations and application of the lessons learned to newly-arising challenges.

Research Management

The symmetrical research partnership pays attention to complementarity in knowledge management. In activities related to agricultural development, the partnership incorporates insights from social sciences, natural (and health) sciences and management studies, and directs attention to the complex interactions and interdependencies among natural and socioecological processes. Local and traditional knowledge offer 'a rich source of lessons for social-ecological adaptations' (Berkes *et al.*, 2003, p. 13). It is crucial for adaptive learning that innovation-seekers avoid homogenizing, diluting or diminishing the diversity of indigenous insights (Folke *et al.*, 2003). Mechanisms must also be in place which bring local knowledge 'back to the university so that it can inform the curriculum' (Lindow, 2009, p. A24). At Makerere University in Uganda, for instance, the AGSHARE project serves as a two-way knowledge conduit between farmers and the university (Kaddu and Haumba, 2016).

The designation of joint project coordinators, who interact on a regular basis and respect arrangements for feedback and information exchange, facilitates local leadership and sustained community impact. Symmetry is advanced by identifying, training and reassigning releasing senior African scholars who are committed to serving as (co-)PIs (Holm and Malete, 2010, p. 9). The Northern team preferably should be led by a faculty member who has successfully worked and lived in the South. To ensure the institutionalization of symmetrical research processes, it is vital that attention be devoted to academic-incentive systems. Highly-motivating incentives include: (i) reduced teaching loads - even in the face of the massification of tertiary education, (ii) seed money to administer collaborative research, (iii) national recognition and (iv) opportunities to co-author with Northern scholars. Symmetry in research partnerships encourages participants to raise new ideas and propose alternative approaches, but does not necessitate unanimity of interpretation. Indeed, collaborative experience among researchers from a UK and a Ghanaian university shows that disagreements and lack of consensus open up 'a discursive and transgressive space' which can generate 'greater awareness of how we are caught in the intricate web of our cultural assumptions' and empower 'insiders and outsiders to engage more critically with each other', as well as with the subjects of their research (Pryor *et al.*, 2009, p. 781).

Building Trust

Trust is a pivotal ingredient in symmetrical transnational relationship-building and mutual learning. When all partners trust that management issues are subject to mutual decision-making, the processes and structures of engagement and undertaking innovative research and curriculum development will be smoother, and the associated transaction costs will be lower. 'Expressions of genuine humility, respect, and eagerness to learn as well as to share are necessary ingredients' in trust-building collaborations (Austin and Foxcroft, 2011, pp. 127, 125). Respect includes being ready to listen to and comprehend the views of others (Minde and Jayne, 2017). Symmetrical transnational partnerships ensure that ample opportunities exist for egalitarian face-to-face interactions (Eckel and Hartley, 2011). Partners often underestimate the importance of spending time socially with counterparts and getting to know their families, interests and aspirations. Eventually, 'if it passes the tests of reliability, delivering on promises, not being underhanded and being honest, the relationship moves on to a possibly more enduring state of "deep trust"' (Williams, 2002, p. 116). Deep trust is built over years of mutually-beneficial collaborations. Deep trust also paves the way for expanded possibilities for collaboration (Austin and Foxcroft, 2011, p. 125).

Egalitarian Interactions

A symmetrical research partnership addressing sustainable agricultural development equitably combines two pairs of actors: a 'driving' African co-PI with a senior international official of the African university on the one hand, and a 'driving' Northern co-PI with a senior international university official on the other. Both sets of actors need to interact strategically and frequently with top academic leaders (Austin and Foxcroft, 2011, p. 124). Symmetry does not require each actor to be involved to the same extent in each management process, or that rigid and inflexible organizational procedures govern partnership operations. Indeed, contextually-appropriate adaptations and trade-offs are consciously implemented. It is essential, however, that the negotiated distribution of roles and responsibilities roughly reflect equal opportunities to influence decision-making across the full spectrum of management functions. These include: scheduling visits, securing project resources, assigning personnel, leading research initiatives, developing curricula, setting and managing budgets, allocating resources, managing grants, attending conferences, making presentations, reporting and monitoring results and assuming accountability (Yusuf *et al.*, 2009, pp. 90–91; Austin and Foxcroft, 2011, p. 124). All partners must possess ample opportunities to influence the 'entire outcome' of the transnational collaboration (Walsh and Kahn, 2010, p. 36).

Navigating Conflicts and Tensions

Navigating among the inevitable and often unpredictable tensions and conflicts which will arise in TAE collaborations requires management processes and actions which facilitate 'adroit steering and constant negotiation' (Singh, 2007, p. 54, also p. 74). Periodic assurance to team members by project leaders from both sides about the project's common purpose and mutually-beneficial outcomes often helps to minimize tensions. For instance, patience often needs to be cultivated in the face of project delays attributable to constraints associated with partner conditions or donor-imposed procedures. Delays are particularly likely to be encountered at the start of new partnerships. When necessary, partners will need to negotiate additional time for activities or a contraction in project objectives (Wanni *et al.*, 2010, pp. 57, 61). Trust in social relationships and persistence in the face of challenging situations are needed to surmount the troublesome and time-consuming hurdles managers and participants confront in symmetrical partnerships.

Project Exit Strategy

In symmetrically and successfully-managed transnational partnerships, it is likely that participants will embark on new and additional grant proposals and projects which incorporate fruitful operating principles based on institutional learning, adjustment and adaptation. Mutually-beneficial *long-term* engagements are a preferred vehicle for TAE institutions committed to sustainable agricultural development (see Koehn and Obamba, 2014, Ch.7). Nevertheless, a viable *project exit strategy* is an essential component in these arrangements. All transnational partnerships need to devote early and persistent attention to the transfer of responsibilities for project management to the participating African institution(s) as its leadership capacity develops. Prior to the end of the agreed-upon project interval, and before undertaking any new collaborative activities, the African partner must be prepared to assume full, long-term management and leadership responsibility for the operation of the initial project. This is another one reason why the development of both institutional and human-resource capacity is fundamental to achieving TAE partnership objectives.

16.5 Building Capacity Through Partnerships

The symmetrical initiative incorporates the partners' objectives for capacity development. Symmetrical partnerships leverage diversity in ways which promote synergy and sustain innovation. Partnership capacity building 'entails action at three levels: *human-resource* development of each partner organization's staff; *partner organizational* capacity building; and capacity building of the *partnership as a whole*' (Brinkerhoff, 2002, p. 92 (emphases added); UNCTAD, 2008, pp. 95–96).

Organizational Capacity Building

Many higher-education institutions in low-income countries confront organizational challenges and limited resources to improve teaching, research and outreach. In the higher-education context, institutional capacity building includes enhancing administrative and personnel management, financial and grants management and resource/infrastructure management (Yusuf *et al.*, 2009, pp. 69, 87, 90–91). In addition to developing institutional management capacity, a defining criterion of a successful transnational academic collaboration is 'building research capacity within collaborating institutions in the South' (Obamba and Mwena, 2009, p. 356). The enhancement of African higher-education research capacity promises mutual North–South gain related to agricultural development and to addressing transnational challenges.

One way to advance this outcome is by building institutional capacity to offer high-quality academic training in higher-education and the management of research projects (Yusuf *et al.*, 2009, p. 86; Hervy *et al.*, 2014, p. viii). Internal capacity building also involves the renovation and extension of infrastructure - particularly libraries, equipment, laboratories and information and communication technology (ICT). The coronavirus pandemic exposed the importance of ICT and remote interactions in African higher education, as well as infrastructural and human-resource gaps.

Mutual enhancement of institutional capacity results from an iterative and sustained process of sharing and learning in Northern and Southern university contexts. In the genuine near-symmetrical partnership, African educational leaders 'progressively take the lead while external partners back their efforts to assume greater responsibility for their own development' (DAC, 1996, p. 13). Symmetrical partnerships place priority on developing the African partners' own capacity-building mechanisms in terms of needs assessment, academic programmes, research undertakings, management training and mentoring, public engagement and quality assurance (Ashcroft and Rayner, 2011, pp. 88–89). The local institution's ability to sustain programmes, benefits and organizational improvements beyond the formal close of an externally-supported project, including absorbing recurrent costs, are all indicators of the partnership's success in enhancing organizational capacity (Brinkerhoff, 2002, p. 6).

Human Capability Building for Enhanced Symmetry

A human-development strategy for enhancing North–South symmetry (i) focuses on the educational and leadership skills of faculty members and managerial staff, (ii) builds expertise in monitoring and evaluating project and programmes (Crossley *et al.*, 2005, pp. 37, 86) and (iii) expands educators' capacity to train students in priority disciplines. The first step is compiling reliable data regarding 'the nature and extent of skills shortages and needs' (Tikly, 2005, p. 308). Symmetrical initiatives designed to build human-resource capacity focus on competencies which unmask and unlock obstacles to local development. Across sub-Saharan Africa, shortages of university graduates are particularly serious in the fields of agriculture, health, engineering and the natural sciences. By building human-resource capacity and providing quality-assurance training, transnational collaborations help to offset the eroding capacity of Southern universities to produce PhD graduates who can contribute to future symmetrical projects and address observed academic weaknesses. As an example, a partnership of universities across East and Southern Africa initiated new doctoral level programmes through the Regional Universities Forum for Capacity Building in Agriculture (RUFORUM), with support from the Bill & Melinda Gates Foundation, the United States Agency for International Development (USAID) and other donors (Lindow, 2009, p. A24).

Local institutions can also encourage and facilitate members of the African diaspora to contribute insights and expertise, either in-country through short-term assignments or virtually from wherever they are situated. Transnational partnerships can be designed to involve diasporic professors in building human-resource capacity 'using their summers or sabbatical times to teach in Africa' and 'building research relationships with African professors or graduate students in African institutions' (Kellogg and Hervy, 2009, p. 10; also Damtew Teferra, 2008, p. 64). Success in tapping this valuable diasporic resource requires a conducive environment, including welcoming attitudes, transnational connectivity, adequate research facilities, reasoned and transparent decision-making and bureaucratic support (Oanda Ogachi, 2016).

Partnership Capacity Building

The success of the symmetrical approach to partnerships depends on the extent to which participants expand and enhance the 'capacity to build capacity' (Johanson and Saint, 2007, p. 51), including the capacity to be an equal partner. Symmetrical partnership-building initiatives ensure that opportunity is provided (i) for long-term mutual reinforcement of collaborative-research capacity (Crossley *et al.*, 2005), (ii) for individual and institutional commitment to contribute to collaborations intended to build knowledge, promote equity and reduce poverty and (iii) for moving partnerships in new project directions (Holm and Malete, 2010, p. 11). Enabling partners to conduct collaborative research might require increasing infrastructure, upgrading equipment and updating technical and human-resource training. Capacity-building efforts should include the provision of state-of-the-art technology, accompanied by training in its uses and application; efforts should also emphasize initiatives which promote symmetry and develop institutional capacity, especially in the areas of writing and managing grants, identifying partners, devising budgets, designing and implementing programmes, dealing with donors and developing curricula and programmes which link the development of professional and research skills with the needs of labour markets.

16.6 Evaluating Partnerships

Evaluation is an essential partnership-management function for enhancing symmetry (see Koehn and Uitto, 2017). Evaluation procedures and tools should be co-created by the partners before any partnership activities are launched, and should then be used throughout the partnership's duration. The co-determined plan should include initial training for partners on how to use the evaluation procedures and tools on how to adjust and adapt the partnership on the basis of evaluation findings.

Key Dimensions of Comprehensive Evaluations

According to the Evaluation Office of the United Nations Development Programme (Naidoo *et al.*, 2013), comprehensive evaluations assess inputs, objectives, processes, outputs, outcomes and impacts. An *output* is a 'tangible product (including services) of an intervention that is directly attributable to the initiative'; *process* refers to the means by which activities are conducted; and an *outcome* is the 'actual or intended changes in development conditions that an intervention seeks to support' (Naidoo *et al.*, 2013, p. 25). Finally, in the context of sustainable development and agriculture-related undertakings, *impact* involves effects on human well-being, agricultural systems and the state of the natural environment (Naidoo *et al.*, 2013).

Asymmetries in Northern-Inspired Evaluations

Most evaluations are asymmetrical in at least two respects. Firstly, they are predominantly Northern-designed and directed. Secondly, they are preoccupied with donor interests. Asymmetrical evaluations arise from exclusive donor determinations of project indicators and baselines against which achievements are measured (Crossley *et al.*, 2005, p. 37). Donors are prone to limit their perceptions of success to timeliness, cost containment and requisite specifications for short-term deliverables (Smith, 2000, pp. 214–215; Neave, 2012, p. 207). For instance, Esther Mebrahtu (2002, p. 510) found that 'the mechanical use' by non-governmental organizations (NGOs) in Ethiopia of externally-required monitoring and evaluation systems limited 'organizational learning to immediate project outputs (e.g. progress, results, efficiency, etc. as defined by the indicators) rather than extending it to issues of power and power relationships within the project community'. Northern-designed evaluations typically concentrate on the goals of donor agencies, ignoring the perspectives of intended beneficiaries (Crossley *et al.*, 2005, p. 37). Little of the knowledge extracted for donor-evaluation purposes is ever returned to partner staff (no less community members) in a locally-usable form.

Measurement Strategies

Donors have strongly favoured the use of pre-defined quantitative indicators in education, research and development evaluations. While project designs tend to emphasize impacts, evaluations often concentrate instead on immediate and readily-quantifiable inputs and outputs, such as the number of partnerships entered, courses developed, professionals trained, reports issued, workshops conducted and staff exchanged (Stone, 2004, p. 157). Impacts are long-term and frequently beyond the duration and full control of the partnership project (Schuller *et al.*,

2004, p. 188). Qualitative evaluations emphasize context and culture over global metrics (see Coxon and Munce, 2008, pp. 147, 162). This provides an important advantage because 'context is central to understanding' the relationship of transnational education to development (McGrath and Badroodien, 2006, p. 492; also Colclough 2012, pp. 160, 167). Qualitative process evaluations are particularly useful for uncovering (i) interpersonal and inter-institutional relationships, (ii) indigenous and marginalized voices, (iii) pathways to desired social and economic change and (iv) political and institutional realities which shape prospects for change (Mosse, 2004, pp. 656, 664). To yield meaningful insights, therefore, partnership evaluations need to be co-created frameworks which go beyond externally-imposed quantitative measures and global metrics. The inclusion of qualitative methods and indicators facilitates the cross-checking of findings through triangulated perspectives, thereby enriching evaluation as a meaningful explanatory exercise (Stern, 2004, pp. 38–39).

A Symmetrical Framework for Evaluating Tertiary Agricultural Education Institutions

A symmetrical TAE evaluation examines the extent to which the partnership has abided by the shared values and goals agreed upon by the partners at the outset. A thorough evaluation plan encompasses the dimensions of design, management, capacity building and sustainability.

Design. How symmetrical is participation in establishing partnership goals and governance arrangements? Do key administrators at all partner universities endorse the initiative (Calder and Clugston, 2004, p. 256)? Does it involve all those in each institution's core and periphery whose cooperation is needed in planning and project implementation? Are project responsibilities clearly assigned? Are intended beneficiaries involved in planning and task identification? Are stakeholders involved in defining goals, monitoring progress and evaluating outcomes (Stern, 2004, pp. 31–32)? To what extent are the incentives for each partner symmetrically aligned during the design phase (Catley-Carlson, 2004, p. 24)? Are the rewards and risks 'equitably shared'? Does the 'distribution of risks take account of the different abilities of partners' to tackle them (Stern, 2004, p. 35)? Are the timeframes for research, project inputs, outputs, impact and outcomes feasible and flexible?

To what extent does each institution demonstrate commitment to the partnership? Do senior TAE managers allocate adequate staffing, resources and rewards? Do participants, including senior officers, across the relevant units of partner institutions and external entities take initiative and engage in project activities?

Are the partners' initial perceptions of the benefits of transnational collaboration changing as the partnership unfolds? If the design is modified to reflect changing priorities during the implementation phase, is it accomplished jointly by the partners (Walsh and Kahn, 2010, p. 66)? To what extent are project components, strategies, and symmetrical arrangements refined and improved (Wanni *et al.*, 2010, p. 58) based upon feedback from and reflection by all partners (Stern, 2004, p. 38)?

Management. Mutual trust is a defining aspect of partnership symmetry. Is management of the partnership characterized by trust among partners and by effective conflict management (Stern, 2004, p. 32)? Specifically, are disagreements discussed openly and resolved in ways that strengthened the partnership (Boydell and Rugkasa, 2007, p. 224)? Do senior managers interact with project staff and community constituents? Is there sufficient exchange of key information and ideas among the partners (King, 2009, p. 44)? Budgeting is an important dimension of partnerships. Are programme budgets transparent and shared? Are budgets distributed equitably according to agreed-upon responsibilities? Are they modified as necessary (Stern, 2004, p. 31)?

Do the partnership managers conduct agreed-upon monitoring and evaluation exercises at regular intervals (Wanni *et al.*, 2010, p. 58)? Do the managers engage in bottom-up participatory evaluation? To what extent do partners jointly analyze the data collected (Walsh and Kahn, 2010, p. 67)?

Capacity building. Symmetrical partnership evaluations regularly assess institutional and human-resource capacity building. To what extent is each partner developing new institutional capacity (Catley-Carlson, 2004, p. 24)? Are the partners developing improved capacity for leadership and innovation (Calleson, 2005, pp. 319–320; Yarime *et al.*, 2012, p. 108)? Are curricula, course content and teaching methods being upgraded as a result of project activity or research findings? Are interactive, experiential, service-learning (i.e. learning based on community problem solving) and distance-learning approaches being introduced or enhanced (Morfit and Gore, 2009, p. 16)? Are the TAE partners' developing their capacity for monitoring, evaluation, and organizational learning (Crossley and Bennett, 2004, pp. 222–223)?

Other indicators of institutional capacity building centre on the community and societal involvement of the partners. Are faculty and staff engaged with communities, government ministries and the private sector? Have innovative outreach approaches been successfully introduced? Are researchers participating in external advisory bodies which shape national and local development? Are partners called upon by external agencies for consultancies, research services and training programmes (King, 2009, p. 44)?

Sustainability. Are personnel available to continue new programmes and activities after the partnership ends, and are they acquiring the requisite knowledge and skills? Are 'train the trainer' programmes in place and working (Morfit and Gore, 2009, p. 16)? Is local funding available to continue these programmes and activities? If not, are funding arrangements being explored?

Strategic feedback, continuous learning and transformational change are central features of the sustainability of partnerships and institutions. Partnership sustainability necessitates that shortcomings are identified and addressed and that evaluation findings are disseminated internally, among stakeholders and all partners, to other educators and to the evaluation community. Contextual and generic attainments become part of the university's reputation and are included in applications for external awards, recognition or certification.

16.7 An Example of a Tertiary Agricultural Education Partnership

The Innovative Agricultural Research Initiative (iAGRI) was a six-year partnership in TAE and research capacity in Tanzania funded by USAID from 2011 to 2017 (Ohio State University, 2017). The African partners were Sokoine University of Agriculture (SUA) and the Ministry of Agriculture, Livestock, and Fisheries (MALF) - both located in Tanzania - and RUFORUM, headquartered in Kampala, Uganda. The US partners were six land-grant universities in a consortium led by Ohio State University (OSU).[1] The aims of iAGRI were to strengthen the capacity of SUA and MALF to respond to current and emerging needs of producers and consumers in Tanzania's rapidly-expanding agricultural and food systems and, ultimately, to help the country achieve food security.

Scope of Partnership Activities

The iAGRI partnership carried out four types of activities: (i) degree training, (ii) collaborative research, (iii) institutional capacity building and (iv) transnational capacity building. Based on an assessment of training needs conducted by the partnership at SUA and MALF, project leaders awarded scholarships for degree training in various natural and social science sub-disciplines of agriculture to 20 Tanzanian doctoral and 115 masters students. Many of the trainees were junior-level teachers and researchers at SUA and MALF; some were employees of other Tanzanian government ministries, NGOs, or private businesses, while others were recent university graduates who were not yet employed. Half of the masters students studied in the United States, while the other half studied in Africa and India. The PhD students studied in South Africa, Tanzania and the United States. Most of those studying in the United States returned to Tanzania to gather data and were co-advised by a pair of US and Tanzanian advisors.

Teams of Tanzanian and US scientists carried out collaborative research on agriculture and nutrition. Research proposals - elicited, selected and funded by the partnership in two rounds - addressed agricultural and food issues and problems, with the aim of improving the country's food security. Participants reported research findings in academic papers, but did not stop there. Rather, the partnership encouraged and supported the researchers in developing and testing practical innovations for communities, companies, agencies and NGOs, drawing on the knowledge generated by their research. The institutional capacity-building activities of the partnership were designed to increase SUA's capacity in the areas of teaching and learning, research, university–business linkages, revenue enhancement, leadership and administration. The partnership undertook and completed over 20 institutional capacity-building projects. The partnership also sponsored study tours to other African universities for SUA officials and faculty members for the purpose of observing and bringing back ideas, procedures and processes for institutional improvements which could be adapted and implemented at SUA.

The partnership implemented transnational capacity-building activities designed to strengthen cooperation between SUA and universities in the United States and the Global South. These activities included placements of Tanzanian graduate students at universities in India, East and Southern Africa and the United States, where they developed contacts with colleagues and professors. These relationships persisted after students graduated and returned to employment at SUA, MALF and other Tanzanian institutions. Some masters graduates were later able to draw on relationships with their former advisors in the partnership to secure funding for doctoral studies. The partnership also strengthened important African transnational linkages between SUA and RUFORUM, which the partnership had contracted to manage and coordinate the partnership-funded study programmes of students of African universities (except for those studying in Tanzania, whose programmes were handled by the partnership office). This led to a renewal of SUA's RUFORUM membership and increased the organization's activities. The partnership's other transnational activities included sponsoring student participation in regional academic conferences in Africa and hosting several international research conferences at SUA.

An Asymmetric Beginning

To a large extent, the US partners determined the initial plan for the partnership, even though both sides would have preferred a more equal approach to the planning stage. Due to a sense of urgency within USAID to roll out the Obama administration's new *Feed the Future* initiative immediately after it was launched at the end of 2010, US universities wishing to compete for the partnership contract were given only five weeks from the release of the call for proposals until the deadline for submission. This period included the Christmas and New Year season, when both Tanzanian and US universities are closed. Consequently, researchers from OSU and Michigan State University (MSU) developed the proposal based on knowledge of SUA and Tanzania gained through previous research and training projects, including several with SUA. Proposal writers at OSU had previously collaborated with faculty members at SUA on several research projects and on a new masters-degree programme. One of the MSU proposal writers was a Tanzanian who had graduated from SUA and had in-depth knowledge of SUA and Tanzanian agriculture. A memorandum of understanding already existed between the two universities from an earlier collaboration, although it merely encouraged joint activities without specifying details (which were to be covered by supplemental agreements). The hurried timeframe for preparing the proposal made it impossible to have the official, on-site institution-to-institution dialogue essential for symmetrical input and power-sharing in goal setting, governance, management, performance monitoring and budgetary decision-making. Although insufficient time for dialogue constituted a major limitation, this was not the only impediment. Rules and procedures of the funding agency limited the scope for symmetry, particularly on budgetary matters.

Through a competitive selection process, USAID awarded the partnership contract to OSU. At that point, the lack of symmetry in the planning stage created a communication barrier which impeded progress both in establishing a partnership office at SUA and in the launching of activities. The programmatic and financial management of the partnership proved particularly contentious. SUA's top officials would have preferred to manage partnership activities through the university's ordinary administrative and financial structures and procedures. They would have preferred to assign day-to-day management responsibility to one or more SUA employees, overseen by a joint SUA–OSU partnership steering committee. Precedent for such an arrangement already existed at SUA and had been used in partnerships funded by other foreign donors. However, because USAID had no direct agreement with SUA, the agency awarded the grant to OSU, which had a long history of contracting with USAID.

Immediately after the award of the contract, OSU set up an office in Tanzania and hired staff to manage partnership activities. The director was a US professor from OSU who relocated to Tanzania for the duration of the partnership. The deputy director was a Tanzanian professor already employed by MSU to work on research and capacity-building projects in Africa. At the suggestion of SUA officials, OSU hired several SUA faculty members who managed key activities in the partnership office on a part-time basis, while continuing to carry out their regular duties at SUA. The deleterious effect of the partnership's initial asymmetry was overcome to some extent by this staffing arrangement and by frequent meetings between the partnership's local management and SUA administration.

Building Trust and Momentum

Frequent communication between the partnership office and SUA administration resulted in the gradual building of trust, which facilitated planning for degree training - the partnership's first major activity. SUA officials were eager to upgrade the academic qualifications of faculty members who lacked terminal degrees (i.e. the highest possible degree, which in agriculture and related disciplines is a PhD). In the first and second years of the partnership, iAGRI provided MSc- and PhD-degree scholarships for a number of SUA faculty members. The partnership's training committee, consisting of both Tanzanian and US partners, selected scholarship candidates according to a well-documented and publicized selection process. Additional trust-building occurred in the second and third years as the partnership office brokered dialogue between USAID and SUA, which resulted in a multi-million-dollar direct grant to SUA to upgrade the university's information technology infrastructure. Without the efforts of the partnership office nudging USAID to overcome internal bureaucratic hurdles to finalize the agreement, it is unlikely that the direct grant to SUA would have materialized. Building on the emerging spirit of cooperation, the Tanzanian and US partners jointly planned and launched a collaborative research programme in the second year. A competitive request for proposals was issued for research on food security. Each research team was required to have both Tanzanian and US researchers and to be co-led by a researcher from SUA and a researcher from one of the six US universities. The partnership funded the following 11 collaborative research proposals:

1. Improving Agricultural Productivity and Crop Nutritive Quality through a Gender Sensitive Approach
2. Improvement of Tomato Productivity and Quality in Tanzania
3. Integrated Salt-Affected Soil Management Options for Sustainable Rice Productivity in Irrigation Schemes of Tanzania
4. Agricultural Innovation for Smallholder Farmers through Locally Adapted Conservation Agriculture for Improved Food Security in the Context of Climate Change
5. From Soil Elements to Food Nutrients: The Nutrient Content of Foods for Human Consumption
6. Improved Soil Health and Germplasm to Advance Tomato Production in Tanzania
7. Low-Cost Drip Irrigation Technology
8. Assessment of Extension Service Delivery for Maize in Tanzania Using an Agricultural Innovation Systems Approach
9. Impacts of Drought-Insured Loans on Farmer Group Sustainability and Technology Adoption
10. Improving Maize Productivity by Drought, Striga, and Maize Lethal Necrosis Disease Management in Tanzania
11. Land Use and Climate Change Impacts on Sustainable Intensification in the Upper Ruvu River Basin

Some of the research teams' workplans included mentoring Tanzanian graduate students. Participants reported on team progress during several partnership-sponsored research conferences held at SUA.

Building Sustainable Capacity

Productive collaboration in the partnership's training and research activities created fertile ground for institutional capacity building. Two years after the start of the partnership, officials at SUA requested assistance in advancing a stalled initiative to update the university's administrative structure, which had remained largely unchanged for several decades despite growth in the student body and faculty. Central administration had continued to manage various administrative and academic functions which, in many universities, are devolved to faculties or schools. The prospect of restructuring created fear of loss of power and position in some of the university's units. The planning process had broken down due to internal disagreements and, over the prior four years, two successive taskforces had failed to move it forward. Given this stalemate, SUA officials requested help in overcoming the internal negativity and pessimism. They requested that the partnership help them build an internal culture which embraced change.

Through dialogue over a period of several months, SUA administration and the US partners designed a process of organizational learning, which was experiential rather than theoretical. It was intended to develop skills for both informal and formal leadership among faculty, staff and administrators. The main aim was for university employees at all levels to acquire the knowledge and experience needed to adapt to the changing societal and economy conditions in Tanzania. The means of learning was a series of organizational experiments, carried out by SUA employees and targeting specific institutional deficiencies. The learning process helped identify 'champions of change', brought them together in teams focused on a problem or issue, mentored and empowered the teams to try out solutions in small-scale organizational experiments and, then, helped them expand the solutions to institutional scale through changes in the university's strategies, structures and systems.

Co-creation was key in these experiments. The champions of change shared their ideas and insights with partnership managers regarding the most appropriate direction and activities for the experiments. The participatory design and implementation of the organizational experiments provided important non-financial incentives. Because of USAID's operating procedures, the partnership did not provide sitting allowances and other financial incentives to the champions of change. Some donor-funded projects at SUA offered such incentives, and the fact that iAGRI did not created some initial tension. That tension gradually faded, however, because the champions appreciated the partnership's symmetrical approach and took pride in the success of their activities.

The partnership developed a process of organizational experimentation which was hands-on, problem-focused, adaptive and iterative. The protocol developed for the organizational experiments was inspired by SUA's local context and by literature on complex adaptive systems, double-loop learning, emergent behaviour, problem-driven iterative adaptation, learning organizations and knowledge management (Dodgson, 1993; Senge, 2006; Tagg, 2007; Andrews *et al.*, 2013). This approach stands in contrast to traditional management, which is top-down and based on sequential stages of planning and implementation which heavily rely on borrowed solutions, often not well suited to the local context. Through dialogue and trial-and-error, the partnership created and refined a 12-step procedure (Table 16.1). The purpose was to ensure that organizational learning emerged from the experimentation and was used to refine the solutions to ensure their sustainability beyond the life of the partnership.

Dialogue among SUA employees and staff in the partnership office identified the deficiencies the experiments targeted. These dialogues often began as informal conversations, later expanding to a larger group. With technical and financial support from the partnership, SUA employees planned and implemented a total of 16 organizational experiments in the areas of teaching, research, university-business linkages, revenue enhancement, leadership and administration (Table 16.2).

Many, though not all, of the organizational experiments proved successful. Thirteen of the experiments resulted in successful solutions which were scaled up and implemented at SUA. Ten of the scaled-up solutions met standards developed by the partnership for organizational sustainability and eight of them met the partnership's standards for financial sustainability. These scaled-up solutions resulted in lasting improvements in the quality of teaching, learning and knowledge

Table 16.1. Stages, steps and indicators of iAGRI organizational experiments

Stages	Steps	Indicators
1. Mobilization	1. Identifying a problem	• A problem has been identified by a group of persons willing to try to solve it
	2. Agreeing to act	• The group has agreed to conduct an organizational experiment to solve the problem
	3. Setting goals	• The group has identified objectives for the experiment
	4. Reviewing the learning process	• The group has agreed on a process of experimentation, evidence-based learning and revision of the solution
2. Experimentation	5. Designing solution(s)	• A tentative solution has been identified and documented
	6. Analyzing costs and benefits	• An analysis of benefits, costs, and challenges of implementing the solution has been conducted
	7. Assigning tasks	• Tasks have been identified and agreement has been reached on who will carry out each one
	8. Implementing solution(s)	• The experiment has been implemented
3. Sustainability	9. Revising and scaling up	• Lessons have been extracted from the experiment and a refined version of the solution has been translated into changes in policies and/or procedures in the formal system of the university
	10. Managing the process and preparing budgets	• The formal system has adopted the solution(s) and has specified reporting relationships, assigned responsibilities and prepared budgets (if funding is required)
	11. Ensuring organizational sustainability	• The formal system has set performance standards against which it regularly monitors progress and continually adapts the solution to correct shortfalls and weaknesses
	12. Ensuring financial sustainability	• The scaled-up solution is adequately resourced with staff and funds for long-term continuation

Source: Constructed by authors from partnership documentation.

generation at SUA. Both the successes and failures added to the university's organizational learning and capacity development. More importantly, however, the organizational experiments helped shape SUA's organizational culture in a positive direction. In the fourth year of the partnership, the new spirit of cooperation resulted in agreement on an institutional restructuring plan which shifted some functions from central administration to the university's colleges and a newly-created school. The organizational learning at SUA created a sense of institutional self-confidence, which motivated officials to abandon the existing 10-year strategic plan, which had been largely ignored in planning and budgeting, and for which a monitoring mechanism had never been created. The new, scaled-down plan served as a realistic guide for annual planning processes.

In 2020, three years after the end of the partnership, the human and institutional change it brought about remains evident. Employees of SUA and MALF who had acquired masters and doctoral degrees through partnership-sponsored scholarships had returned to their institutions as teachers, researchers and administrators. Many of them continue to maintain professional contact with their graduate school advisors and colleagues. The innovations brought about through the organizational experiments and human and institutional enhancements continued to bring dynamism to SUA, both through the specific changes they brought about, as well as the change they helped create in the organizational culture of the university.

iAGRI Viewed through the Lens of Symmetry

Was iAGRI a symmetrical partnership? In actuality, the degree of symmetry varied over the

Table 16.2. Status of organizational experiments undertaken at Sokoine University of Agriculture

		Stage 1				Stage 2				Stage 3			
						Steps							
Category	Experiment	1	2	3	4	5	6	7	8	9	10	11	12
Teaching	Strengthening capacity to manage classroom facilities	✓	✓	✓	✓	✓	✓	✓	✓	✓	✓	✓	✓
	Strengthening capacity of remedial English language programme	✓	✓	✓	✓	✓	✓	✓	✓	✓	✓	✓	✓
	Strengthening capacity to utilize teaching assistants	✓	✓	✓	✓	✓	✓	✓	✓				
Research	Strengthening capacity for digital librarianship	✓	✓	✓	✓	✓	✓	✓	✓	✓	✓	✓	✓
	Strengthening capacity for providing statistical advisory services	✓	✓	✓	✓	✓	✓	✓	✓	✓	✓	✓	
External linkages	Strengthening capacity for entrepreneurship programming	✓	✓	✓	✓	✓	✓	✓	✓	✓	✓	✓	✓
	Strengthening capacity for horticultural teaching and outreach	✓	✓	✓	✓	✓	✓	✓	✓	✓	✓	✓	✓
	Strengthening capacity to provide commercial soil laboratory services	✓	✓	✓	✓	✓	✓	✓	✓	✓	✓		
	Strengthening capacity for poultry-industry outreach	✓	✓	✓	✓	✓	✓	✓	✓	✓	✓	✓	
	Strengthening capacity for farm-mechanization outreach	✓	✓	✓	✓	✓	✓	✓	✓				

Continued

Table 16.2. Continued.

		Stage 1				Stage 2				Stage 3			
		Steps											
	Enhancing capacity to strengthen national agricultural education and training system with focus on Ministry of Agriculture Training Institutes	✓	✓	✓	✓	✓	✓	✓	✓	✓	✓	✓	✓
Revenue enhancement	Strengthening capacity to generate income	✓	✓	✓	✓	✓	✓	✓	✓	✓	✓	✓	✓
	Strengthening capacity for alumni outreach	✓	✓	✓	✓	✓	✓	✓	✓	✓	✓	✓	✓
Leadership	Strengthening capacity for mentorship	✓	✓	✓	✓	✓	✓	✓	✓	✓	✓	✓	
	Strengthening capacity for leadership development	✓	✓	✓	✓	✓	✓	✓	✓	✓			
Administration	Strengthening capacity for electronic document handling	✓	✓	✓	✓	✓	✓	✓	✓	✓	✓	✓	✓
Totals		16	16	16	16	16	16	16	16	14	13	12	8

Source: Constructed by authors from partnership documentation.

life of the partnership, beginning at a low level and gradually increasing over time. Asymmetry in the preparation of the partnership proposal constituted an enormous impediment to trust-building, deterring the local partner (SUA) from fully embracing early activities of the partnership and slowing capacity-building progress. The decision by the outside partner (OSU) to create a parallel administrative structure and to handle budgetary planning and financial management independently also lessened the local partner's incentive to commit time and effort to building the partnership in the early stage and delayed critically-needed capacity building.

Although financial management remained asymmetric throughout the life of the partnership, enough cooperation and goodwill prevailed by the end of the first year to build trust and a sense of shared values with respect to partnership activities. A key to developing trust was building relationships both within the partnership office and among staff from the local and external partner institutions. In addition, frequent communication between the partnership office and the local partner's senior administration created goodwill and led to increased cooperation. Building trust created an environment in which the partners were able to engage in activities to build research and institutional capacity which were truly collaborative. A process of organizational experimentation and learning made it possible for the growing symmetry of the partnership to extend beyond the top levels of the partner institutions and to reach faculty and staff at the middle and lower levels as well. The penetration of symmetry within the institutions proved especially important for achieving results in collaborative research and in institutional capacity building.

An important lesson from iAGRI is that failure to engage in symmetric planning in the earliest phases of a partnership slows the implementation of activities and hinders capacity building. Another important lesson is that, even in partnerships with asymmetric elements, it is possible to build trust and increase symmetry over time by investing in relationships. Partners can learn from mistakes, institutionalize symmetry and improve outcomes through internal mechanisms which promote feedback, organizational learning and ongoing improvement. Ideally, these mechanisms should be built into the initial design of the partnership and activated throughout its life.

16.8 Conclusion

Transnational partnerships intended to facilitate TAE institutions in building their institutional and human-resource capacity through joint research, instruction or community outreach function best when founded on symmetrical planning and power-sharing. Symmetry in defining the scope and nature of collaborative activities and their management, budgeting and evaluation increases the likelihood that all partners will benefit. Partnerships, however, are not static arrangements; they are constantly evolving. The iAGRI case study shows that even when a partnership begins without the ideal degree of symmetry, it can still evolve towards shared goals, shared agenda-setting and shared management of activities. The extent of symmetry will be a function of the partners' commitment to the principles of mutual trust, mutual respect, co-creativity, transparency and mutual accountability.

Endnotes

* Gratitude is expressed to Dr. Amon Mattee for valuable comments and suggestions on an earlier draft of this chapter.

[1] The others are Michigan State University, Virginia Tech, Tuskegee University, the University of Florida and Iowa State University.

References

Accra Agenda for Action (2008) Third High-Level Forum on AID Effectiveness. 2–4 September, Accra, Ghana. Available at: https://www.afdb.org/fileadmin/uploads/afdb/Documents/AccraAgendaAaction-4sept2008-FINAL-ENG_16h00.pdf (accessed 3 June 2021).

African Union (2004) *Strategic Plan of the Commission of the African Union Vol. 2: 2004–2007 strategic framework of the commission of the African Union*. African Union, Addis Ababa, Ethiopia.

Andrews, M. Pritchett, L. and Woolcock, M. (2013) Escaping capability traps through problem driven iterative adaptation (PDIA). *World Development* 51, 234–244.

Ashcroft, K. and Raynor, P. (2011) *Higher Education in Development: Lessons from Sub-Saharan Africa*. Information Age Publishing, Charlotte, North Carolina, USA.

AUC (African Union Commission) (2015) *Agenda 2063: The Africa We Want.* Available at: https://au.int/sites/default/files/documents/36204-doc-agenda2063_popular_version_en.pdf (accessed 28 May 2021).

Austin, A.E. and Foxcroft, C. (2011) Fostering organizational change and individual learning through 'ground-up' inter-institutional cross-border collaboration. In Sakamoto, R. and Chapman, D.W. (eds.) *Cross-Border Partnerships in Higher Education: Strategies and Issues*. Routledge, New York.

Bailey, F. and Dolan, A.M. (2011) The meaning of partnership and development: Lessons for development education. *Policy and Practice: A Development Education Review* 13, 30–48.

Berkes, F., Colding, J. and Folke, C. (2003) Introduction. In Berkes, F., Colding, J. and Folke, C. (eds.) *Navigating Social-Ecological Systems: Building Resilience for Complexity and Change*. Cambridge University Press, Cambridge, UK.

Boydell, L.R. and Rugkasa, J. (2007) Benefits of working in partnership: A model. *Critical Public Health* 17, 217–228.

Brinkerhoff, J.M. (2002) *Partnerships for International Development: Rhetoric or Results?* Boulder, Colorado, USA.

Calder, W. and Clugston, R. (2004) Lighting many fires: South Carolina's sustainable universities initiative. In Corcoran, P.B. and Wals, A.E.J. (eds.) *Higher Education and the Challenge of Sustainability: Problematics, Promise, and Practice*. Kluwer Academic Publishers, Dordrecht, The Netherlands.

Calleson, D.C. (2005) Community-engaged scholarship: Is faculty work in communities a true academic enterprise? *Academic Medicine* 80, 317–321.

Catley-Carlson, M. (2004) Foundations of partnerships: A practitioner's perspective. In Liebenthal, A., Feinstein, O.N. and Ingram, G.K. (eds.) *Evaluation & Development: The Partnership Dimension*. Transaction Publishers, New Brunswick, New Jersey, USA.

Colclough, C. (2012). Education outcomes reassessed. In Colclough, C. (ed.) *Education Outcomes and Poverty: A Reassessment*. Routledge, London.

Coxon, E. and Munce, K. (2008) The global education agenda and the delivery of aid to Pacific education. *Comparative Education* 44, 147–165.

Crossley, M. and Bennett, J.A. (2004) Planning for case study evaluation in Belize, Central America. In Crossley, M. and Vulliamy, G. (eds.) *Qualitative Educational Research in Developing Countries: Current Perspectives*. Garland Publishing, New York.

Crossley, M., Herriot, A., Waudo, J., Mwirotsi, M., Holmes, K., *et al.* (2005) *Research and Evaluation for Educational Development: Learning from the PRISM Experience in Kenya*. Symposium Books, Oxford, UK.

Crossley, M. and Holmes, K. (2001) Challenges for educational research: International development, partnerships and capacity building in small states. *Oxford Review of Education* 27, 395–409.

DAC (Development Assistance Committee) (1996) *Shaping the 21st Century: The Contribution of Development Cooperation*. Organisation for Economic Co-operation and Development, Paris.

Damtew Teferra and Altbach, P.G. (2003) Trends and perspectives in African higher education. In Damtew Teferra and Altbach, P.G. (eds.) *African Higher Education: An International Reference Handbook*. Indiana University Press, Bloomington, Indiana, USA.

Damtew Teferra (2008) The international dimension of higher education in Africa: Status, challenges, and prospects. In Teferra, D. and Knight, J. (eds.) *Higher Education in Africa: The International Dimension*. Center for International Higher Education, Boston College, Chestnut Hill, Massachusetts, USA.

Damtew Teferra (2016) Shifting grounds in higher education partnerships. *The World View Inside Higher Education*, Center for International Higher Education, June 19. Available at: https://www.insidehighered.com/blogs/world-view/shifting-grounds-higher-education-partnerships (accessed 28 May 2021).

Defazio, D. Lockett, A. and Wright, M. (2009) Funding incentives, collaborative dynamics and scientific productivity: Evidence from the EU framework program. *Research Policy* 38(2), 293–305. Available at: https://doi.org/10.1016/j.respol.2008.11.008 (accessed 28 May 2021).

Dodgson, M. (1993) Organizational learning: A review of some literatures. *Organization Studies* 14(3), 375–394.

Eckel, P.D. and Hartley, M. (2011) Developing academic strategic alliances: Reconciling multiple institutional cultures, policies, and practices. In Tight, M. (ed.) *Higher Education: Major Themes in Education*, Vol. IV. Routledge, London.

Eddy, P.L. (2010) *Partnerships and Collaborations in Higher Education*. Jossey-Bass, San Francisco.

Esther Mebrahtu (2002) Perceptions and practices of monitoring and evaluation: International NGO experiences in Ethiopia. *Development in Practice* 12(August), 501–517.

Folke, C., Colding, J. and Berkes, F. (2003) Synthesis: Building resilience and adaptive capacity in socio-ecological systems. In Folke, C., Colding, J. and Berkes, F. (eds.) *Navigating Social-Ecological Systems: Building Resilience for Complexity and Change*. Cambridge University Press, Cambridge, UK.

Francisconi, C., Grunder, A. and Mulloy, M. (2011) *Building Sustainable U.S.-Ethiopian University Partnerships: Findings from a Higher Education Conference*. Institute for International Education White Paper, New York.

Gedde, M. (2009) *The International Health Links Manual: A Guide to Starting Up and Maintaining Long-term International Health Partnerships*. Tropical Health and Education Trust, London.

Gutierrez, D. (2008) Beyond disappointment: Transforming ideology and practice in North-South research partnerships. *NORRAG News* 41, 19–21. Available at: https://www.norrag.org/fileadmin/Full%20Versions/NN41.pdf (accessed 28 May 2021).

Hervy, A.-C., Kellogg, E., Pillay, P. and Demment, M. (2014) *African Higher Education: Opportunities for Transformative Change for Sustainable Development*. United States Agency for International Development, Washington, DC.

Holm, J.D. and Malete, L. (2010) *The Asymmetries of University Partnerships between Africa and the Developed World: Our Experience in Botswana*. Paper delivered at the 2010 Going Global 4: The British Council's International Education Conference, London, 24–26 March.

Ishengoma, J. (2016) Strengthening higher education space in Africa through North-South partnerships and links: Myths and realities from Tanzania public universities. *Comparative and International Education* 45(1), 1–17. Available at: https://doi.org/10.5206/cie-eci.v45i1.9282 (accessed 28 May 2021).

Johanson, R. and Saint, W. (2007) *Cultivating Knowledge and Skills to Grow Africa Agriculture: A Synthesis of an Institutional, Regional, and International Review*. Agriculture and Rural Development Department, World Bank, Washington, DC.

Jones, G.A. and Oleksiyenko, A. (2011) The internationalization of Canadian university research: A global higher education matrix analysis of multi-level governance. *Higher Education* 61(1), 41–57.

Kaddu, S. and Haumba, E.N. (2016) Documenting and disseminating agricultural indigenous knowledge for sustainable food security in Uganda. *Fifth RUFORUM Biennial Regional Conference* 14(14), 221–230.

Kellogg, E.D. and Hervy, A.-C. (2009) *Contributions of Higher Education Investments to Development and Implications for African Higher Education*. Paper presented at the Conference on Reshaping Human and Institutional Capacity Building through Higher Education Partnerships, Accra, Ghana, 26 August.

KFPE (Swiss Commission for Research Partnerships with Developing Countries) (2012). *A Guide for Transboundary Research Partnerships: 11 Principles*. Swiss Commission for Research Partnerships with Developing Countries, Bern.

King, K. (2008) The promise and peril of partnership. *NORRAG News* 41, 5–6.

King, K. (2009) Higher education and international cooperation: The role of academic collaboration in the developing world. In Stephens, D. (ed.) *Higher Education and International Capacity Building: Twenty-five Years of Higher Education Links*. Symposium Books, Oxford, UK.

Kinser, K. and Green, M. (2009) *The Power of Partnerships: A Transatlantic Dialogue*. American Council on Education, Washington, DC.

Koehn, P.H. and Obamba, M. (2014) *The Transnationally Partnered University: Insights from Research and Sustainable Development Collaborations in Africa*. New York, Palgrave Macmillan.

Koehn, P.H. and Rosenau, J.R. (2010) *Transnational Competence: Empowering Professional Curricula for Horizon-Rising Challenges*. Paradigm Publishers, Boulder, Colorado, USA.

Koehn, P.H. and Uitto, J.I. (2017) *Universities and the Sustainable Development Future: Evaluating Higher-Education Contributions to the 2030 Agenda*. Sustainability and Development Series, Routledge, London and New York.

Lindow, M. (2009) Innovation flowers in a crowded, crumbling Africa university. *Chronicle of Higher Education* 11(September), A23–A24.

Maassen, P., Pinheiro, R. and Cloete, N. (2007) *Bilateral Country Investments and Foundations Partnership Projects to Support Higher Education across Africa*. Center for Higher Education Transformation, Wynberg, UK.

McGrath, S. and Badroodien, A. (2006) International influences on the evolution of skills development in South Africa. *International Journal of Educational Development* 26(5), 483–494.

Mihyo, P.B. (2008) Strengths and weaknesses in institutional partnerships between northern and southern academic institutions. *NORRAG News (December)*, 112–114.

Minde, I.J. and Jayne, T. (2017) Endeavoring to nurture partnerships. In *Rethinking African Partnerships for Global Solutions.* Alliance for African Partnership, Michigan State University. Available at: https://aap.isp.msu.edu/files/7115/0911/5508/AAP_Thought_Piece_2017August.pdf (accessed 28 May 2021).

Minde, I. and Monson, J. (2019) Challenges of creating alliances across borders: Midterm reflections from the Alliance for African Partnership. *Journal of Global Ethics* 15(2), 155–167. Available at: https://doi.org/10.1080/17449626.2019.1641543 (accessed 28 May 2021).

Morfit, C. and Gore, J. (2009) *HED/USAID Higher Education Partnerships in Africa 1997–2007*. Higher Education for Development, Washington, DC.

Mosse, D. (2004) Is good policy unimplementable? Reflections on the ethnography of aid policy and practice. *Development and Change* 35 (4), 639–671.

Naidoo, I.A. supported by Soares, A.R. Fox, A. and Wang, X.G. (2013) *Evaluation and Its Role in Advancing a Measurement Discourse Experience from the Evaluation Office of the UNDP*. UNDP Evaluation Office. Available at: http://web.undp.org/evaluation/documents/directors-desk/papers-presentations/EO-ppt-IDEAS_2013.pdf (accessed 28 May 2021).

Neave, G. (2012) *The Evaluative State, Institutional Autonomy and Re-engineering Higher Education in Western Europe: The Prince and His Pleasure*. Palgrave Macmillan, London.

Nebeker, C. (2011) *Reactive and Proactive Approaches to Facilitating International Collaborations. In International Research Collaborations: Much to be Gained, Many Ways to Get in Trouble,* edited by Melissa S. Anderson, M.S. and Steneck, N.H. Routledge, New York.

Oanda Ogachi, I. (2016) Engaging the African academic diaspora: How ready are universities in Africa? *International Journal of African Higher Education* 2(1). Available at: https://doi.org/10.6017/ijahe.v2i1.9263 (accessed 29 May 2021).

Obamba, M.O. and Mwema, J.K. (2009) Symmetry and asymmetry: New contours, paradigms, and politics in African academic partnerships. *Higher Education Policy* 22(3), 349–371.

Obamba, M.O., Mwema, J.K. and Riechi, A.R. (2011) *University Partnerships Can Reduce Poverty*. ANIE Policy Brief, Association of African Universities, Accra, Ghana.

OECD (Organisation for Economic Co-operation and Development) (2005) *Paris Declaration on Aid Effectiveness*. OECD Publishing, Paris. Available at: https://doi.org/10.1787/9789264098084-en (accessed 3 June 2021).

OECD (2012) *Aid Effectiveness 2011: Progress in Implementing the Paris Declaration*. OECD Publishing. Available at: http://dx.doi.org/10.1787/9789264125780-en (accessed 28 May 2021).

Ohio State University (2017) *Innovative Agricultural Research Initiative (iAGRI): Final Report.* Available at: https://ipa.osu.edu/sites/international/files/imce/iAGRI%20final%20report%2011.2017.pdf (accessed 29 May 2021).

Oleksiyenko, A. and Sa, C.M. (2010) Resource asymmetries and cumulative advantages: Canadian and US research universities and the field of global health. *Higher Education* 59 (3), 367–385.

Pryor, J. Kuupole, A. Kutor, N. Dunne M. and Adu-Yeboah, C. (2009) Exploring the fault lines of cross-cultural collaborative research. *Compare* 39 (6), 769–782.

Samoff, J. and Bidemi, C. (2004) The promise of partnership and continuities of dependence: External support to higher education in Africa. *African Studies Review* 47 (1), 67–199.

Schuller, T., Hammond, C. and Preston, J. (2004) Reappraising benefits. In Schuller, T., Preston, J., Hammond, C., Brassett-Grundy, A. and Bynner, J. (eds.) *The Benefits of Learning: The Impact of Education on Health, Family Life and Social Capital*. Routledge Falmer, London.

Senge, P.M. (2006) *The Fifth Discipline: The Art and Practice of the Learning Organization* Doubleday, New York.

Singh, M. (2007) Universities and society: Whose terms of engagement? In Sorlin, S. and Vessuri, H. (eds.) *Knowledge Society vs. Knowledge Economy: Knowledge, Power, and Politics*. Palgrave Macmillan, Hampshire, UK.

Smith, H. (2000) Transforming education through donor-funded projects: How do we measure success? In Mebrahtu, T., Crossley, M. and Johnson, D. (eds.) *Globalisation, Educational Transformation and Societies in Transition*. Symposium Books, Oxford, UK.

Stern, E. (2004) Evaluating partnerships. In Liebenthal, A., Feinstein, O.N. and Ingram, G.K. (eds.) *Evaluation & Development: The Partnership Dimension*. Transaction Publishers, New Brunswick, New Jersey, USA.

Stone, D. (2004) Research partnerships and their evaluation. In Liebenthal, A., Feinstein, O.N. and Ingram, G.K. (eds.) *Evaluation & Development: The Partnership Dimension*. Transaction Publishers, New Brunswick, New Jersey, USA.

Tagg, J. (2007) Double loop learning in higher education. *Change* 39(4), 36–41.

Tikly, L. (2005) The new partnership for African development: Implications for skills development. In Zajda, J. (ed.) *International Handbook on Globalisation, Education and Policy Research*. Springer, Dordrecht, The Netherlands.

Toure, K. (2008) Developing Quality Partnerships for Quality Research. *NORRAG News* (December), 55–58.

Trani, E.P. and Holsworth, R.D. (2010) *The Indispensable University: Higher Education, Economic Development, and the Knowledge Economy*. Rowman & Littlefield, Lantham, Maryland, USA.

UNCTAD (United Nations Conference on Trade and Development) (2008) *The Least Developed Countries Report 2008: Growth, Poverty and the Terms of Development Partnership*. United Nations, New York.

United Nations (2015) *Transforming Our World: The 2030 Agenda for Sustainable Development*. United Nations, New York.

Utting, P. (2006) Introduction. In Utting, P. (ed.) *Reclaiming Development Agendas: Knowledge, Power and International Policy Making*. Palgrave Macmillan, Basingstoke, UK.

van de Water, J. Green, M.F. and Koch, K (2008) *International Partnerships: Guidelines for Colleges and Universities*. American Council on Education, Washington, DC.

Walsh, L. and Kahn, P. (2010) *Collaborative Working in Higher Education: The Social Academy*. Routledge, New York.

Wanni, N. Hinz, S. and Day, R. (2010) *Good Practices in Educational Partnerships Guide: UK-Africa Higher & Further Education Partnerships*. The Africa Unit, London. Available at: https://assets.publishing.service.gov.uk/government/uploads/system/uploads/attachment_data/file/31917/10-1031-africa-unit-good-practices-guide-final.pdf (accessed 28 May 2021).

Warden, R. (2015) UN development goals: A bigger role for universities? *University World News* 17 October. Available at: https://www.universityworldnews.com/post.php?story=20151014211927167 (accessed 28 May 2021).

Williams, P. (2002) The competent boundary spanner. *Public Administration* 80(1), 103–124.

Yarime, M. Trencher, G. Mino, T. Scholz, R.W. Olson, L. *et al.* (2012) Establishing sustainability science in higher education institutions: Towards an integration of academic development, institutionalization, and stakeholder collaborations. *Sustainability Science* 7(Supplement 1), 101–113.

Yusuf, S., Saint, W. and Nabeshima, K. (2009) *Accelerating Catch-up: Tertiary Education for Growth in Sub-Saharan Africa*. World Bank, Washington, DC.

Part IV

Implementing the Transformation

17 A Transformation Agenda for Tertiary Agricultural Education in Africa

David Kraybill[1]*, John Lynam[2] and Adipala Ekwamu[2]
[1]Ohio State University, United States; [2]Independent Scholar, based in Kenya; [3]Regional Universities Forum for Capacity Building in Agriculture, Uganda

17.1 Introduction

The central argument echoed throughout this volume is the need to transform Africa's tertiary agricultural education (TAE) entities in response to fundamental shifts in the external environment, including the demand for graduates, the challenges facing the agricultural sector and the expansion of Africa's food systems. Institutional transformation at both the university and the faculty level requires an ability, firstly, to adapt to changes in the external environment and, secondly, to design and implement an internal process of institutional change. Such a process, in turn, requires an ability to diagnose rapid changes in African food systems - where the demand for graduates with new skills originates - and to understand the internal factors which constrain TAE entities from responding optimally to these changes. At the same time, this external environment forces TAE faculties to be more outward-focused and to engage with other actors in the food system more effectively. This chapter synthesizes the change process defined and explored throughout this volume, providing a review of the external environment, the components of internal transformation processes and the ways in which TAE entities can take advantage of different types of networks and partnerships.

17.2 Tertiary Agricultural Education in a Rapidly Changing Environment

Institutional change evolves through the synergy of organizations' adaptive responses to changes in the external environment. Rapid external change presents opportunities for institutions to transform, especially in competitive environments where agents and processes of change are in place to facilitate fast and dynamic responses. Adapting to change allows institutions to become more effective in achieving their mandates, and this, in turn, leads to specialization and differentiation. TAE entities operate within a complex external economic and institutional environment (Ch. 3, this volume). TAE's principal mission is to produce the human capital needed to generate the knowledge base for agricultural development, which has expanded over time to encompass the complete food system. At the same time, the African food system is undergoing rapid structural change, which has major implications for the demand for graduates and the skills required by the sector (Ch. 2, this volume). Moreover, such changes also have implications for the research needed to improve productivity across the food system, which requires recasting the methods and organization of delivering advisory services. As a

*Corresponding author: kraybill.1@gmail.com

 Transforming Tertiary Agricultural Education in Africa (eds D. Kraybill, J. Lynam and Adipala Ekwamu)
DOI: 10.1079/9781789246544.0017

food system implies, TAE entities operate within a rapidly-changing institutional ecosystem, where institutional linkages are critical to the successful adaptation of TAE programmes to the changing needs of these different actors.

An Evolving Food System

In the post-independence era, African economies have shifted away from their former dependence on export crops and have moved towards the development of domestic food markets, driven by the rapid rate of urbanization. Since the turn of the millennium an urban middle class has also been created, which has diversified food demand from a principal dependence on food staples to a diet comprising more meat, dairy products, vegetables and fruit - i.e. 'higher-value food products' - which also incorporate advances in convenience foods. The demand for convenience, however, often comes at the expense of a healthy diet (Ch. 2, this volume). This rapidly-changing demand creates backward linkages through changes in marketing systems, which then trigger changes in production systems. Growth and structural change in the food system is being led by private-sector development, particularly investment in market infrastructure for both inputs and products. Dairy, horticulture and meat products require specific storage and processing infrastructure in order to maintain quality in supply chains serving urban centres. For grain staples, bulking and storage are improving the efficiency of African supply chains. The growing middle class has also sparked a significant change in retail market infrastructure, particularly the growth in supermarkets.

The challenge created by rapid, market-led structural change is, firstly, to what degree the growth process will be equitable and, secondly, to what degree the production response will be inclusive and sustainable (Ch. 2, this volume). Continued growth of the rural population, even with high rates of rural–urban migration, has put pressure on farm size in many parts of Africa. This, together with persistent poverty concentrated in rural areas, produces a policy challenge in terms of how to integrate smallholders into growing markets and, where this is not possible, to create employment in the non-farm rural economy. At the same time, diversification into higher-value crops offers the potential for agro-ecological intensification and an adaptive response to climatic fluctuations. The basis for this will be improved soil management prompted by higher returns from commercialization. Ensuring more equitable growth will require public-sector programmes targeting smallholders and their integration into markets. Food systems will also need to be resilient in the face of climate change and incorporate environmentally-friendly approaches. Such a focus on inclusive development in the food system will drive: (i) shifts in research priorities, (ii) the development of new extension methodologies to link smallholders with higher-value supply chains and (iii) changes in TAE to provide the human capital needed to lead inclusive and sustainable development of the food system.

A Changing Labour Market for Graduates

Structural change in the African food system produces rapid change in the labour market and the demand for TAE graduates. One means of charting such transformation is through value chains. For higher-value products - such as dairy, horticulture, vegetable oils and meats - each value chain is organized to address raw-material supply, processing, transportation, storage and quality standards as the product progresses from the farm to consumer. This creates demand for new skills and knowledge, but it also raises the question of how TAE can best be organized to respond to these changing needs. Although faculties are organized by disciplines, cross-pollination is needed to promote a more holistic, multidisciplinary understanding of how different disciplines can contribute their expertise within a value-chain context. This understanding can be attained through cross-disciplinary courses or curricula, but is best achieved within individual faculties. Disciplines within agricultural faculties traditionally focused narrowly on crop science but have broadened over time (Ch. 5, this volume). Further expanding curricula to incorporate a food-system framework may lead to dispersion and loss of reputation in key disciplines. How national TAE systems can organize themselves most effectively to meet the changing demand and expanding scope in disciplines will be a central agenda for the future, particularly because this implies increased staffing levels and, hence, costs. An effective food-system approach may require (i) new faculties, (ii) inter-faculty mechanisms, (iii) faculty differentiation across universities and (iv) national coordination frameworks.

The Institutional Landscape

The development of an efficient, sustainable and equitable food system led by market forces and private-sector investment will require effective public-sector agencies working interactively within a coherent policy framework (Ch. 3, this volume). A functioning national agricultural innovation system (AIS) is required, and it must also adapt to the rapidly-changing needs of the food system and environment. These demands come at a time when African nations continue to underinvest in agriculture, and existing funds are often directed into input subsidies. Underinvestment is a common challenge for national agricultural research systems, and human capital constraints are especially binding (Ch. 5, this volume). An evolving food system requires particular innovation in the scope, delivery and methodologies of its agricultural advisory services (Ch. 10, this volume). The combination of underfunding and decentralization has led to the devolution of Africa's national extension systems into a varied set of public, private and civil-society service providers.

This institutional landscape, combined with the needs of a changing food system, creates particular opportunities for TAE institutions. In order to prepare their graduates for the workplace, TAE institutions must develop their own capacity in research, extension and community development. Curricula should incorporate training in 'soft skills' and team dynamics, as well as much-needed problem-solving skills. TAE institutions require both research and extension programmes which support the creation of knowledge and methods to direct changes in the food system and enable graduates to fulfil their roles as the next generation of leaders, both within and beyond their institutions (Chs. 9 and 10, this volume). Closer linkages across institutions will be necessary to facilitate the testing of the new methods and knowledge developed, all of which will contribute to the effectiveness of national AIS. Ultimately, increased innovation should facilitate TAE institutions in becoming more directly involved in the development of national food system policies.

The University Landscape

TAE is principally organized around faculties of agriculture, which are embedded in a university system founded on specific funding and governance structures. The university landscape in Africa has changed significantly over the past couple of decades (Ch. 4, this volume). The number and size of public universities have expanded rapidly, and private universities have emerged from insignificant levels to constitute a major component of the tertiary education sector. This has created a competitive market for undergraduate students, not least because tuition fees now provide a significant source of funding. TAE institutions have also grown, but not nearly at the rates of universities overall (Ch. 5, this volume). TAE is less competitive in attracting students, despite the expansion and introduction of satellite campuses in more rural areas. The issue, therefore, is to determine the best ways for TAE institutions to respond to the changing demand for their graduates across the food system, to organize themselves effectively to meet this demand and, in particular, to compete for the number and type of students needed to make these changes possible.

Responding to rapid change in the food system requires a certain level of autonomy in developing programmes, allocating resources and setting staffing priorities. Moreover, these issues must be framed within the context of the university's and other faculties' needs. Many countries have universities which have traditionally focused solely on agriculture but more recently have expanded into other disciplines to attract more students. Examples include Sokoine University of Agriculture (Tanzania), Egerton University (Kenya) and Jomo Kenyatta University of Agriculture and Technology (Kenya). Yet with the increase in the number of universities, diversification strategies are giving way to specialization strategies, particularly in the area of graduate training. Many universities have responded to the changes in the food system by creating new faculties focusing on food science and nutrition or natural resources and the environment. Such faculties complement traditional faculties of agriculture and veterinary science, and in the newer universities they often exist without agricultural and veterinary science faculties. A food-system approach would suggest potential synergies in offering joint programmes across faculties. The possibilities need to be identified and tested to establish whether there may be grounds for more effective coordination across universities at a national level.

Critical Needs for Human Capital

Human capital underpins both the growth and structural change occurring across Africa's food systems in terms of technical change, institutional innovation, strategic investment and new knowledge production. The challenge arises from the intense pressure the needs of an expanding and diversifying food system exert on TAE institutions. Four principal needs can be identified.

1. ***Filling the supply gap for public-sector staff.*** The demand for well-qualified staff outstrips supply in the public sector because of the development of new faculties and high rates of staff retirement and attrition in national agricultural research institutes (NARIs), which have lost staff to the more prestigious and lucrative TAE and private sectors (Ch. 5, this volume).
2. ***Expanding graduate programmes.*** The expansion in the size and scope of graduate programmes drives the need for more MSc- and PhD-qualified faculty members. It will also lead to greater differentiation across the TAE sector, with some universities specializing in high-level graduate programmes and research (Ch. 8, this volume).
3. ***Maintaining the quality of programmes.*** The rapid expansion of universities, including TAE institutions and programmes, has strained institutions' ability to maintain the quality of their programmes. Increasing class sizes, teaching loads, undergraduate enrolments and demands on university infrastructure have eroded the quality of educational programmes. Yet the quality of teaching staff, programmes and research determines the quality of graduates, and ultimately the effectiveness of TAE faculties and NARIs (Ch 10, this volume).
4. ***Meeting the demand for new skills and knowledge in an evolving food system.*** New knowledge across a range of disciplines is required. At the same time, market-driven structural change calls for very different skillsets, including entrepreneurial and problem-solving skills, which are especially in demand in the private sector, and 'soft skills' - such as leadership, team building, negotiation and communication skills - which are increasingly called for in the public and private sectors (Chs. 6 and 7, this volume).

These basic needs for human capital in a growing and diversifying food system make the case for transformational change in the TAE sector. Such change can be framed in two parts. The first occurs within universities and faculties of agriculture themselves and spans leadership, teaching capacity, curricula development, financing, quality assurance and future planning. For faculties of agriculture, particularly those developing graduate programmes, improved capacities in research and community outreach are also critical to this change process. The second component reinforces these internal change processes and involves enhanced external linkages to promote a more functional national AIS through participation in regional and global networks, partnerships and programmes. These two aspects of the change process are explored in the next two sections.

17.3 The Transformation Agenda

This volume offers an analysis of trends in agricultural and food systems and higher education, describing the critical inputs TAE can provide in the form of human capital, knowledge, technology and outreach. The trends are markedly different from those in place when most of Africa's TAE institutions were created. As a result, institutions must adapt. To address the daunting trends exerting pressure on human populations and the natural environment, TAE institutions must transform themselves in order to guide the transformation of the society and economy in which they operate, and ensure the quality and relevance of the services they provide. Relevance is enhanced by aligning training and delivery modalities with the needs and expectations of stakeholders, especially employers and students. Quality is improved by creating strategies, structures, systems and organizational cultures which facilitate active learning by students, lifelong learning by graduates and - ultimately - the creation and dissemination of innovative, effective and impactful knowledge and technologies.

Leadership, Management and Quality Assurance

Transformative TAE institutions are managed by visionary and competent leaders. The supply-driven approach of the past - where higher-education institutions developed programmes without substantial input and feedback from external stakeholders - has led to loss of effectiveness, status, influence and funding. In contrast, the demand-driven approach to academic programme planning adopted by the most innovative institutions in Africa today is evidence-based and entrepreneurial. Academic leaders in these institutions don't wait for change to be forced upon them but, instead, proactively lead their institutions in new directions. Such leaders possess the critically-important dynamic capabilities identified by Teece (2018) as sensing, seizing and transforming, as described below.

Sensing. The first dynamic-leadership capability is based on data and analysis regarding the training, knowledge and technology needs of the country and region where the TAE institution is located. From this information, dynamic leaders create a vision and narrative which drives internal institutional change. The sensing capability manifests itself through (i) the alignment of institutional strategies with national, regional and continental priorities, (ii) the development of linkages with industry, (iii) interactions with communities and (iv) knowledge of best practices, such as learner-centred education and theme-driven research (Ch. 7, this volume). Quality-assurance standards and procedures are also important components of the sensing capability (Ch. 6, this volume).

Seizing. The second dynamic-leadership capability is the ability to move an academic institution or a team towards action. Once an evidence-based and compelling vision and narrative have been developed, dynamic leaders communicate, motivate, assemble and coordinate the required human, material and financial inputs. The seizing capability is the capacity and willingness to engage internal and external stakeholders, maintain open communication even in times of difficulty and conflict, build partnerships and networks and create sound policies and internal systems (Ch. 7, this volume).

Transformation. The third dynamic-leadership capability is successfully implementing fundamental change. Leaders possessing this capability reshape organizational culture, create shared values, empower colleagues and subordinates, legitimize experimentation, embrace innovation and scale up the plans developed through sensing and seizing. TAE institutions can be transformed by rallying champions of change, supporting young talent, adopting a collaborative style of leadership and ensuring that internal feedback, evaluation and auditing are effective (Ch. 7, this volume). Most importantly, they transform their institutions into organizations capable of learning from their own successes and failures, thereby constantly improving over time.

Curriculum, Teaching and Learning

The competitiveness of higher-education institutions in attracting talented students is determined in large part by their graduates' success in the job market. Employability, including self-employment, is a fundamental societal goal of higher education; other important goals include preparing students for responsible citizenship, training them for leadership in thought and action and providing them with a foundation for lifelong learning. Education in transformative TAE institutions is based on scientifically-informed understanding of the processes and stages of learning. These institutions use pedagogical methods which promote active learning. Rather than relying on the traditional lecture method - pouring knowledge into passive minds - educators facilitate active inquiry by their students towards mastery of knowledge and skills, including soft skills, such as critical thinking and oral and written communication. The student, as an engaged learner, is at the centre of the education process, and the teacher assumes the role of 'guide on the side', not 'sage on the stage' (King, 1993).

TAE becomes transformative when teachers plan their instruction focusing on learning outcomes - i.e. what students should know or be able to do upon completion of the course or module which they did not know or could not do before it began. Rather than focusing purely on knowledge outcomes as in the past, educators committed to transformative TAE build their instructional planning on a broader set of learning objectives referred to in the educational literature as 'competencies'. In addition to knowledge, these include skills and attitudes. Where best practices are observed, the competencies to be taught and learned in an academic programme, course or module are based on research into stakeholders' expectations and graduates' labour-market successes and challenges.

Defining the competencies to be taught is not a 'desk exercise' which can be performed independently by individual teachers. Information on the competencies required for labour-market success is a scarce and costly public good. Therefore, higher-education systems and institutions should invest in research on the skills and knowledge required in the labour market and make the findings available to teachers and curriculum planners. Competencies to be targeted can be identified through dialogue with employers and entrepreneurs, graduate tracer studies and studies of major occupations. This research should aim to identify gaps in competencies not being adequately addressed by current academic programmes, from which learning objectives can be determined. Thereafter, transformative TAE institutions focus on identifying the most effective curricula, teaching methods and learning environments to support students in achieving those learning objectives. To gain the desired level of mastery, students need active and experiential learning activities, including laboratory exercises, fieldwork, industry and community exposure, internships, individual and group project- and problem-based learning and service learning.

The transformative TAE institution also provides training in active-learning methods for its academic staff. Such training is essential for shifting the focus from *teacher*-centred to *learner*-centred education. Many TAE teachers have no formal training in pedagogy and are uninformed about current best practices. A critical pedagogical knowledge and skills gap therefore exists among educators. This gap should be filled through on-site training workshops focused on learning processes and best practices in teaching. High-quality teaching is recognized and valued in transformative TAE institutions through rewards and accountability measures. Rewards may take the form of public recognition at annual ceremonies or include monetary awards, such as travel grants, book and computer allowances, personnel promotions and salary increases. Accountability measures include curricular review and peer and student review of teaching.

Research and Innovation

The transformative TAE institution produces a steady stream of research-based knowledge and innovation on agricultural productivity, food security, household livelihoods, community vitality and related topics. The development of research skills is central to graduate training and essential for preparing the next generation of researchers. Innovation is the process of translating research findings into solutions to economic, social and environmental problems. Traditionally, researchers viewed the publication of research findings as an end-product. In contrast, the transformative TAE institution views research as an interim output. Research publications are essential for communicating findings to scientific colleagues, but they are not sufficient to provide society with a reasonable return on its investment in the research process. Whenever possible, transformative TAE institutions extend the research process by producing innovations and sharing ideas to improve human well-being and environmental sustainability based on the knowledge they have generated.

Transformative TAE institutions actively nurture and catalyze research, innovation and dissemination. The leaders of these institutions understand that the volume and quality of innovations can be increased by establishing an internal value chain for research, beginning with the initial inquiry and ending with dissemination - and, where appropriate, commercialization - of information and technologies. Transformative TAE institutions assist researchers in convening community and private stakeholders to stimulate research-based innovation. Such engagement informs researchers of problems to be investigated and provides insights into how the new knowledge can be translated into solutions. The internal research value chain in transformative TAE institutions includes provision of research services, such as (i) notifying researchers of grant opportunities, (ii) convening research teams to prepare proposals, (iii) promoting and supporting research partnerships, (iv) providing efficient financial management and timely dispersal of research funds, (v) supporting the mentorship of young researchers by senior researchers, (vi) securing funding for research infrastructure and equipment, (vii) managing intellectual property, (viii) supporting prototyping of innovations, (ix) providing editorial and communication support for the dissemination of findings and (x) convening research and innovation conferences.

Many of the pressing issues in agriculture and food systems are multi-faceted and require investigation by researchers from multiple disciplines. Conceptually, the AIS is one framework for organizing multidisciplinary research, but it's not the only one. In transformative TAE institutions, researchers work in multidisciplinary and multi-agency teams, and leaders support and reward the resulting teamwork. Transformative TAE institutions are perpetual research collaborators. Their researchers collaborate intensively with NARIs to develop research proposals and carry out research projects. Many TAE graduate students conduct their research in NARI labs and test sites, where they engage in research collaborations with local, regional and international institutions and networks (discussed in the next section). Consequently, the research publications of transformative TAE institutions are frequently co-authored by researchers from multiple countries. They also collaborate with the private sector, communities and NGOs to develop particular innovations and build innovation ecosystems.

Outreach and Community Development

Engaging with organizations, groups and individuals outside the university is a high priority of transformative TAE institutions. Traditionally, the outreach function of TAE institutions was largely a one-way activity in which research findings were published in academic journals and then transmitted to farmers and other end-users. In transformative TAE, the outreach function is a two-way information-sharing activity in which institutions actively engage with a variety of external groups or communities, including farmers, businesses, citizens' groups, NGOs and local and national governments (Ch. 10, this volume). In the transformative TAE institutions, outreach is a prioritized activity, equal to teaching and research. It is promoted in universities through both the formal system (e.g. mission and vision statements, strategy documents, policies and incentive and support systems) and the informal system (e.g. the institutional culture). Outstanding outreach is recognized and rewarded through staff promotion, tenure, annual performance evaluation, salary increases and public recognition (e.g. through annual award ceremonies). Whenever possible, faculty and students are encouraged to link outreach to teaching and research rather than to view them as separate activities. The transformative TAE institution is entrepreneurial in its pursuit of funding for outreach. Funding comes from multiple sources including research grants requiring outreach, the national government, local governments, NGOs, businesses, international organizations and foreign aid donors.

Community development is a highly-effective way of discovering and understanding local economic, social and cultural needs. The information and understanding gained by faculty and students through community engagement aids in (i) identifying competencies needing to be taught in academic programmes, (ii) providing experiential learning opportunities for students and (iii) identifying relevant research topics, sites and populations to be studied by both students and faculty. The transformative TAE institution integrates community engagement as the third pillar of its mandate, alongside teaching and research (Ch. 10, this volume). Engaging with communities has a catalyzing influence on the mindsets of faculty and students. It changes the internal culture of an institution, motivating it to be outward looking, innovative, entrepreneurial and socially responsive. University interactions with the community build knowledge-driven and mutually-beneficial external partnerships. Curricula, course content and teaching methods evolve as faculty and students gain a deeper understanding of the knowledge and skills needed to solve pressing societal problems. Research agendas become more relevant as faculty, students and communities work together to conduct research which produces information, procedures and technologies to solve real-world problems. Inspired by their community engagement experiences, graduates of transformative TAE institutions leave their universities committed to lifelong learning and inspired to continue to make a difference in the world.

Entrepreneurship

An important way in which transformative TAE builds the capabilities of women and youth is through entrepreneurship training and business incubation. Entrepreneurship programmes have expanded greatly in universities and agricultural technical and vocation education and training (ATVET) institutions in recent years (Ch. 12, this volume). While both types of institutions lie within the TAE spectrum, they have tended to operate independently in the past. That is beginning to change, and transformative ATVET institutions and universities now collaborate to draw on each other's strengths to create a wider range of training and incubation programmes. TAE makes its largest contribution to entrepreneurship in the early stages of business startups. This period is fraught with challenges and high failure rates. TAE institutions train and support early-stage entrepreneurs through courses and workshops on the principles and practices of entrepreneurship, and by establishing and operating business incubators, hubs and accelerators. They also connect and partner with the private sector and other actors in the innovation ecosystem, and they help reform national and local policies affecting entrepreneurs.

Transformative TAE institutions maintain contact with their entrepreneurship graduates, viewing them as a valuable source of information for identifying and understanding the competencies required for successfully initiating and growing business startups. This information, gathered through tracer studies and occupational studies of graduates and other entrepreneurs, is used by curriculum planners and teachers to develop learning objectives, plan course content and select teaching methods for entrepreneurship courses. Problem-based, practical and experiential learning is a primary pillar of entrepreneurship programmes in transformative TAE. After completing core courses in developing a business plan, students and participating community members are given the opportunity to launch an enterprise in a business incubator operated by the university or other organization. Mentoring of student entrepreneurs is provided by established entrepreneurs. In particular, transformative TAE institutions look to their successful entrepreneurship graduates to 'pay forward' their experience by serving as mentors to subsequent cadres of students.

The transformative TAE institution is an active member of Pan-African networks promoting entrepreneurship. These networks, such as the Regional Universities Forum for Capacity Building in Agriculture (RUFORUM) and Comprehensive Africa Agriculture Development Programme (CAADP), are sources of activities and programmes to (i) train the trainers, (ii) train entrepreneurs and (iii) model entrepreneurship and incubation programmes, all of which can be adapted by TAE institutions to suit their local needs and contexts. RUFORUM aims to stimulate the growth of entrepreneurship programmes through its Agribusiness Innovation Model, an entrepreneurship programme designed to prepare students to develop their own business enterprises. A pilot version is currently operating at Gulu University in Uganda and Egerton University in Kenya (Ch. 12, this volume). RUFORUM also sponsors the Entrepreneurship Challenge Programme to assist TAE institutions in establishing mini-incubation hubs. In addition, RUFORUM helps managers of nascent business incubators connect with mentors to build their capacity to manage and nurture their businesses. CAADP is similarly piloting a model agricultural entrepreneurship programme, currently operating in six African countries (Ch. 12, this volume).

RUFORUM is also catalyzing innovative entrepreneurship programmes linking universities and ATVET institutions. Egerton University in Kenya, for example, has developed a partnership with three ATVET providers. Egerton is assisting with curriculum development and teacher training, and is collaborating on research and other activities. One of the mutual benefits of these partnerships is preparing ATVET graduates to advance to universities to undertake further studies.

Inclusivity

In transformative TAE institutions, a commitment to inclusivity permeates the processes underlying student recruitment, student admissions and services, staff hiring, academic programmes, course content, teaching methods and research and outreach. The aim of inclusivity in higher education is to increase the representation of students and staff from under-represented groups and to transform organizational culture and mindsets - and hence, academic and institutional services - to reflect greater diversity of ideas, interests and realities. For TAE in Africa, inclusivity means facilitating the inclusion of women, individuals with disabilities and economically-disadvantaged students.

Transformative TAE institutions embrace equality of women as essential to human rights, strengthening households and communities and improving agricultural productivity. Gender equality is embedded in the institutional policy and organizational culture. Formal and informal leaders shape the organizational culture through their attitudes and actions, influencing the mindsets of staff and students towards mainstreaming gender equality across their institutions. Many African TAE institutions have large gender imbalances in rates of enrolment and graduation, as well as faculty and senior administration. To address these imbalances, the transformative institution mainstreams gender through policies, structures and processes, including targeted scholarships, policy and practices addressing sexual and other violence toward women, increased hiring of female educators and leaders, gender-informed pedagogy, gender sensitization training, mentoring and relevant infrastructure improvements (Ch. 11, this volume).

Researchers in transformative TAE institutions examine a wide range of research topics through the lens of gender. They are aware that research on virtually all agricultural and food system issues will have greater impact if the gender dimensions are taken into account. Consequently, most research teams in transformative TAE institutions include a gender specialist. This has become achievable because, through training and sensitization, many of institutions' agronomists, plant breeders, soil scientists, economists, engineers and other scientists have attained secondary specializations in gender. Recognizing that talk about gender mainstreaming is insufficient to bring about change, transformative TAE institutions have adopted concrete measures to make it a reality and light the path for other institutions to follow.

17.4 Networking and Partnerships in Facilitating Transformation

The internal processes leading to transformation of TAE institutions are enhanced and facilitated by strategic external institutional linkages. These can involve individual teaching staff, which has been the predominant type of external involvement, but linkages built among organizations themselves - whether formal or informal - are more enduring. Such linkages (i) expand the knowledge of teaching staff, (ii) enhance best practice in both teaching and administration and (iii) establish the brand and reputation of institutions. Establishing such linkages requires a certain level of operational autonomy within universities and is more easily established at the faculty level. Network and partnership arrangements which facilitate the transformation process take three principal forms: (i) participation in the national AIS, (ii) membership in regional networks and (iii) partnership arrangements with northern and southern universities. These are discussed in turn below.

National Agricultural Innovation System

TAE institutions are fundamental to national AIS, especially in producing the critical human capital required for the innovation process. As a core component of producing trained MSc and PhD graduates, TAE must develop capacity in both research and outreach (i.e. extension services and community development). The research and outreach conducted by TAE institutions, which tends to focus on more basic, disciplinary research and methodology development, complements the more applied and adaptive research undertaken by national agricultural research and extension agencies. Students conducting research for their MSc or PhD degrees are an asset to both research and extension activities. Interactions between TAE and research and extension offer potential synergies in knowledge creation, particularly by combining disciplinary skills with field work in different agroecological settings. This practical grounding in farming and larger food systems builds students' capacity for identifying relevant problems and research questions, ultimately contributing to an ongoing research agenda around which faculties can organize their funding and activities.

In recent decades, the private sector has become the engine of rapid change in the food system. This includes the entry of supermarkets into retail trade, the establishment of value chains for high-value commodities (such as dairy and horticulture), the diversification of rapidly-growing urban markets and the commercialization of smallholder agriculture. This evolving food system requires trained personnel with business and problem-solving skills, in addition to new disciplinary skills. Closer and more established links with the private sector are a critical part of the transformation process (Chs. 12 and 13, this volume). These links take a different form in the off-farm portion of the food system and can include innovation parks, student attachments, joint research ventures and policy studies. At the same time, such links give faculties insights into changing labour-market needs and skill requirements.

Membership in Regional Networks

With its links to natural resource management and food systems, agriculture encompasses a broad range of quite distinct disciplines, usually organized as departments. Plant pathology, agricultural economics or soil science encompass theory and techniques from closely-related disciplines outside agriculture, but they are all necessary to an understanding of farming systems. The scope of agricultural and food sciences is broad, and individual faculties have difficulty accumulating strong capacity across the range of disciplines. This is especially the case at the graduate level, where class sizes are usually small, and staffing, funding and infrastructure are often constrained. Such constraints across national or regional TAE institutions drive the development of networks, along with the trend toward specialized disciplinary centres of excellence.

In the age of the Internet, the connectivity and information flows on which networks rely are now pervasive. Beyond information flows, networks differ in terms of the other functions they provide to members. Networks thus range from disciplinary, research networks involving individual faculty members, to university networks providing advocacy and conference functions, to networks of TAE faculties providing a range of capacity strengthening functions (Ch. 14, this volume). By operating at a regional and sub-regional level, RUFORUM, for example, is able to generate economies of scope and foster specialization in areas with significant capacity constraints, especially in the development of high-quality, graduate programmes. RUFORUM has developed several critical programmes which aid individual faculties in filling capacity gaps by drawing on high-quality capacity in other parts of the region. These programmes include the collaborative PhD programme based on both coursework and a thesis, a competitive research grant programme focusing on developing master's-level research skills, a collaborative programme in research methods (which are vital for research degrees) and a programme in innovative community development. Interaction at the faculty level creates a positive competitive environment, which promotes quality in key areas. Faculties are able to draw on high-quality programmes from other faculties and evolve models of best practice in teaching, administration and research. The challenge is that networks such as RUFORUM require resources to operate their programmes. While members do provide partial funding from their limited budgets, funding from development assistance agencies is also required. Ensuring a stable funding level during this period of increasing and changing demand on TAE faculties remains a critical bottleneck.

University Partnerships

Partnerships with universities in the larger countries in the Global South - particularly China and India - and universities in the Global North are a direct means for African TAE institutions to become integrated into global networks. Such partnerships are also a key means of filling capacity gaps. In attempting to meet global standards and practice in management, research and curricula, transnational partnerships are an important element in the transformation process for TAE institutions. However, to be most effective in contributing to such institutional change, an equal balance of power is needed to achieve what is termed 'symmetrical partnerships' (Ch. 16, this volume). The principal reason for lack of symmetry in such partnerships is that funding is derived from bilateral aid donors in the North, which fund their own universities and tend to give the Northern partner control over the allocation and management of the finances. Such control tends to reinforce the structural and material asymmetries between universities in the North and the South, particularly the large differences in salaries and overheads. Nevertheless, with sufficient time, trust and understanding, a symmetrical partnership can be achieved.

These partnerships take two basic forms. The first is organized around joint research projects, and the second is specifically devoted to strengthening the capacity of the Southern partner. By its nature, the first category of partnership leads to greater asymmetries, but these can be managed for mutual benefit. For TAE institutions, these benefits include developing human resources, strengthening research methods, becoming integrated into global scientific networks and generating research funding. Nevertheless, the key to building a strong research capacity is a strategic and relevant research agenda derived from the evolving African food system. Research grant programmes tend to be organized around themes or problem areas. Relevance to local circumstances can be improved when research is designed around these problem areas but is tailored to the needs of local agricultural and food systems. It is beneficial for partnerships to focus on the process of designing research agendas and projects.

Partnerships based directly on capacity strengthening tend to be longer-term arrangements. They can focus on single departments, often introducing new areas of study, or they can be directed to the faculty as a whole. Such partnerships tend to focus on a combination of human-resource development and organizational strengthening, including administration, finance and curricula development. Joint research, often through competitive grant schemes, is usually a fixture of these type of programmes. The organizational strengthening component can be a key factor in the transformation of TAE institutions. Nevertheless, adapting the change process to institutional circumstances is critical (Ch. 16, this volume). Buy-in from administration and staff is also vital in effecting change, and developing trust between the partners over time is mandatory. Change processes must be appropriately defined and sequenced over time as trust is established, ensuring that activities are prioritized and appropriately monitored. It should also be noted that effective change is a dynamic process, and as progress is made in one area, it will inevitably generate demand for change in others.

17.5 Conclusion

The current and emerging challenges of agriculture and food systems in Africa are enormous and include, among other factors, (i) rapid growth in demand for food, (ii) rapid commercialization of agriculture, (iii) lagging productivity in production, (iv) changing consumption patterns, (v) climate change, (vi) inadequate transportation infrastructure, (vii) limited electrification and (viii) poorly-developed agro-industries. The stakes are high in addressing these challenges, in terms of both human well-being and environmental sustainability. These challenges present TAE with an opportunity to produce human capital, knowledge and technology to fill resource gaps and provide sustainable solutions. To seize this opportunity, TAE institutions must transform themselves internally. The transformation agenda described in this volume is demand-driven, strategic, collaborative and adaptive. The goal of the agenda is to unlock the talents of faculty, students and graduates, and facilitate collaboration with farmers, consumers, communities, businesses and governments in order to solve the practical problems of agricultural and food systems. Human talents and ingenuity are unlocked and fostered in an institutional environment which embraces innovation and change, rewards achievement and provides the necessary environment, infrastructure and equipment. The task of TAE leaders is to create strategies, structures and systems in support of such an environment.

Demand continues to grow for undergraduate and graduate programmes in agricultural and food systems in Africa. To meet this demand, the number of PhD-qualified teachers and researchers must increase. Many TAE institutions are only beginning to develop staff and curricula at the PhD level, and are in need of financial resources and technical support for further development. The current level of funding for TAE is inadequate to meet the demand for degree-level training. Investment by governments, the private sector and donors is essential to fill the funding gap. In a world of competing claims on the financial resources of governments, companies and donors, TAE is assessed as a potential recipient of investment based on the extent to which it produces graduates, knowledge and technology which shape agricultural and food systems in sustainable ways. The transformation agenda outlined in this volume is intended to support TAE institutions in achieving greater relevance, quality and success, ultimately allowing them to attract higher levels of investment.

References

King, A. (1993) From sage on the stage to guide on the side. *College Teaching* 41(1), 30–35.

Teece, D.J. (2018) Managing the university: Why "organized anarchy" is unacceptable in the age of massive open online courses. *Strategic Organization* 16(1), 92–102. Available at: https://doi.org/10.1177/1476127017732760 (accessed 13 June 2021).

Index

Note: Page numbers in *italics* denote tables and figures.
Page numbers followed by 'n' denote notes.